Lecture Notes in Statistics **120**

Edited by P. Bickel, P. Diggle, S. Fienberg, K. Krickeberg,
I. Olkin, N. Wermuth, S. Zeger

Springer

New York
Berlin
Heidelberg
Barcelona
Budapest
Hong Kong
London
Milan
Paris
Santa Clara
Singapore
Tokyo

Maia Berkane
Editor

Latent Variable Modeling and Applications to Causality

Springer

Maia Berkane
Department of Mathematics
Harvard University
Cambridge, MA 02138

CIP data available.
Printed on acid-free paper.

Camera ready copy provided by the editor.
Printed and bound by Braun-Brumfield, Ann Arbor, MI.
Printed in the United States of America.

9 8 7 6 5 4 3 2 1

ISBN 0-387-94917-8 Springer-Verlag New York Berlin Heidelberg SPIN 10490443

Preface

This volume gathers refereed papers presented at the 1994 UCLA conference on "Latent Variable Modeling and Application to Causality." The meeting was organized by the UCLA Interdivisional Program in Statistics with the purpose of bringing together a group of people who have done recent advanced work in this field. The papers in this volume are representative of a wide variety of disciplines in which the use of latent variable models is rapidly growing. The volume is divided into two broad sections. The first section covers Path Models and Causal Reasoning and the papers are innovations from contributors in disciplines not traditionally associated with behavioural sciences, (e.g. computer science with Judea Pearl and public health with James Robins). Also in this section are contributions by Rod McDonald and Michael Sobel who have a more traditional approach to causal inference, generating from problems in behavioural sciences.

The second section encompasses new approaches to questions of model selection with emphasis on factor analysis and time varying systems. Amemiya uses nonlinear factor analysis which has a higher order of complexity associated with the identifiability conditions. Muthen studies longitudinal hierarchichal models with latent variables and treats the time vector as a variable rather than a level of hierarchy. Deleeuw extends exploratory factor analysis models by including time as a variable and allowing for discrete and ordinal latent variables. Arminger looks at autoregressive structures and Bock treats factor analysis models for categorical data. Berkane brings differential geometric approaches to covariance structure models and shows how estimation can be improved by use of the different measures of curvatures of of the model. Bentler considers the problem of estimating factor scores and proposes equivariant estimators as an alternative to conventional methods such as maximum likelihood. Jamshidian and Browne offer new algorithms and computational techniques for parameter estimation.

All these contributions present new results not published elsewhere. I believe this volume will provide the readers with opportunities to capture the various connections between techniques, viewpoints and insights that would not be possible otherwise.

I would like to thank the following sponsors who made the conference happen through their support and help: The Drug Abuse Research Center, the Cognitive System Laboratory, the National Institute on Alcohol Abuse and Alcoholism, the National Research on Evaluation, Standards and Student Testing, the National Research Center for Asian American Mental Health, and the National Science Foundation.

Maia Berkane

Lexington, MA
1996

Contents

Embedding common factors in a path model

Roderick P. McDonald
University of Illinois

1. Introduction

In models containing unobservable variables with multiple indicators--common factors, true scores--it seems to be common practice not to allow, except in special cases, nonzero covariances of residuals of indicators of the common factors and other variables in the model. Indeed, those models, such as LISREL and LISCOMP, that separate a "measurement" model from a "structural" model, exclude these by their defining matrix structure. Yet it is not at all obvious that such covariances should not be allowed, and it seems desirable to look for some explicit principles to apply to such cases, rather than settle the matter by default. This question will be examined in the context of the reticular action model. The following section sets out some necessary preliminaries from the case of path models without common factors, then section 3 treats the problem of embedding a block of common factors in a path model. Section 4 gives a numerical example.

2. Path models without common factors

Let **v** be a qx1 random vector of observable variables, that is, of variables whose values can be realized in a sampling experiment. We suppose **v** is so scaled that its expected value is null. A special case of the Reticular Action Model (RAM--see McArdle and McDonald, 1984) may be written as

$$\mathbf{v} = \mathbf{B}\mathbf{v}+\mathbf{e} \tag{1}$$

with

$$\mathrm{Cov}\{\mathbf{v}\} = \Sigma, \tag{2}$$

$$\mathrm{Cov}\{\mathbf{e}\} = \Psi, \tag{3}$$

where the elements β_{jk}, ψ_{jk}, of $\mathbf{B},\Psi$ are functions of a tx1 vector of parameters. We will refer to the components of **e** neutrally as <u>error terms</u>. From (1) without further assumptions we have

$$\mathbf{e} = (\mathbf{I}-\mathbf{B})\mathbf{v}, \tag{4}$$

and

$$\mathbf{v} = (\mathbf{I}-\mathbf{B})^{-1}\mathbf{e}. \tag{5}$$

We define

$$\mathbf{A} = [\alpha_{jk}] = (\mathbf{I} - \mathbf{B})^{-1},$$

and immediately have, further,

$$\Sigma = (\mathbf{I} - \mathbf{B})^{-1}\Psi(\mathbf{I} - \mathbf{B})'^{-1}, \tag{6}$$

or

$$\Sigma = \mathbf{A}\Psi\mathbf{A}', \tag{7}$$

and

$$\Psi = \mathbf{A}^{-1}\Sigma\mathbf{A}'^{-1}. \tag{8}$$

We write

$$\Phi = \mathrm{Cov}\{\mathbf{v},\mathbf{e}\},$$

and by (3) and (5) we have

$$\Phi = \mathbf{A}\Psi \tag{9}$$

and

$$\Phi = \Sigma(\mathbf{I}-\mathbf{B})'. \tag{10}$$

If the jth row of **B** is null, then $e_j=v_j$. With the variables suitably ordered, we may write (1) as

$$\begin{bmatrix} \mathbf{x} \\ \mathbf{y} \end{bmatrix} = \begin{bmatrix} 0 & 0 \\ \Gamma & \mathbf{B}^* \end{bmatrix} \begin{bmatrix} \mathbf{x} \\ \mathbf{y} \end{bmatrix} + \begin{bmatrix} \mathbf{x} \\ \mathbf{d} \end{bmatrix}, \tag{11}$$

or

$$\mathbf{y} = \mathbf{B}^*\mathbf{y} + \Gamma\mathbf{x} + \mathbf{d}, \tag{12}$$

following a widely accepted though unnecessary convention that separates variables corresponding to null rows of **B** in (1) into an mx1 vector **x** of <u>exogenous</u> variables and the remainder as a p x 1 vector **y** of <u>endogenous</u> variables.

It is easily seen that there is a correspondence between a specification of the structural model (1) and the specification of a path diagram, such that (a) the nodes (vertices) of the diagram correspond to the variables $v_1, \ldots, v_q$, (b) a possibly nonzero β_{jk} corresponds to a directed arc (an arrow) from the node v_k to the node v_j (c) a possibly nonzero α_{jk} corresponds to a directed path (a chain of unidirectional arcs) from the node v_k to the node v_j, (d) a possibly nonzero ψ_{jk} corresponds to a nondirected arc (a line) between the nodes v_j and v_k, including a nondirected arc from v_j to itself corresponding to a variance. We will always suppose that Diag{**B**} is null; thus, there cannot be a directed arc from a variable to itself.

We consider only recursive models here. A <u>recursive</u> model is one in which (possibly after permuting components of **v**) **B** is a (lower) triangular matrix. (In the corresponding path diagram, there are no closed loops formed by unidirectional arcs.) It is <u>nonrecursive</u> if and only if it is not recursive as just defined. (Some writers--e.g. Fox, 1984; Bollen, 1989--require also that in a recursive model Ψ is diagonal. Others--e.g. Wright, 1934; Duncan, 1975; Kang and Seneta, 1980--do not.) It is well-known that if **B** is lower triangular then **A** is lower triangular with unities on the diagonal (and conversely). We rewrite (1) as

$$v_j = \boldsymbol{\beta}_{(j)}'\mathbf{v}_{(j)} + e_j, \tag{13}$$

and (5) as

$$v_j = \boldsymbol{\alpha}_{(j)}'\mathbf{e}_{(j)} + e_j, \quad j=1, \ldots, q, \tag{14}$$

where $\boldsymbol{\beta}_{(j)}'$, $\boldsymbol{\alpha}_{(j)}'$ contain the j-1 elements of the jth row of **B** and of **A** except the diagonal unity, and $\mathbf{v}_{(j)}$, $\mathbf{e}_{(j)}$ contain the corresponding components of **v**, **e**.

We note that while certainly at most j-1 components of $\boldsymbol{\beta}_{(j)}'$, $\boldsymbol{\alpha}_{(j)}'$ are nonzero, in applications some number less than j-1 may be nonzero because of the pattern of **B** and the patterning of **A** that is in general induced by the pattern of **B**. We therefore rewrite (13) as

$$v_j = \boldsymbol{\beta}_{r(j)}\mathbf{v}_{r(j)} + e_j \tag{15}$$

and (14) as

$$v_j = \alpha'_{s(j)} e_{s(j)} + e_j, \qquad (16)$$

in which $\beta_{r(j)}$ contains only the $r_j \leq j-1$ nonzero elements of $\beta_{(j)}$ and $\alpha_{s(j)}$ contains only the $s_j \leq j-1$ nonzero elements of $\alpha_{(j)}$ and $v_{r(j)}$, $e_{s(j)}$ the corresponding components of $v_{(j)}$ and $e_{(j)}$.

We will say that v_k directly precedes v_j if in the path diagram there is a directed arc from v_k to v_j. Correspondingly, β_{jk} is then possibly nonzero. We will say that v_k precedes v_j if there exists at least one directed path from v_k to v_j. Correspondingly, α_{jk} is then possibly nonzero. We will say that two variables are ordered/unordered if one precedes/does not precede the other.

We are now in a position to list a number of possible "rules" suggested in the literature governing the specification of nondirected paths. The word "rule" is used here deliberately to avoid pre-judging the status of these as assumptions or definitional statements. They are at least commonly offered as rules for the guidance of a user of path models. We consider the following:

1. The Direct Precedence Rule (DPR) states that if $\beta_{jk} \neq 0$, then $\phi_{kj}=0$.
That is, the error-term of variable v_j is orthogonal to each variable in its own equation (Land, 1969; Kang and Seneta, 1980; Wright, 1960; Schmidt, 1982; Duncan, 1975; James, Mulaik and Brett, 1982).
2. The General Precedence Rule for Variables (GPRV) states that

$$\text{if } \alpha_{jk} \neq 0, \text{ then } \phi_{kj}=0$$

That is, if v_k precedes v_j, directly or indirectly, v_k and e_j are orthogonal. (See McDonald, 1985. Duncan, 1975, pp. 27-28 states that in a recursive model "each dependent variable is predetermined with respect to any other dependent variable that occurs later in the causal ordering.")
3. The General Dependence Rule for Errors (GPRE) states that

$$\text{if } \alpha_{jk} \neq 0, \text{ then } \psi_{jk} = 0$$

(Kang and Seneta, 1980; McDonald, 1985). That is, the errors e_k, e_j of ordered variables are orthogonal.
4. The Exogenous Variable Rule (EVR) states that if v_k is exogenous and v_j is endogenous, then $\phi_{kj} = 0$ or, equivalently, $\psi_{jk} = 0$. (See, for example, Fox, 1984; Bollen, 1989. Note that we do not here assume that the variables are ordered. Note also that the components of **e** are not interpreted as exogenous variables in the application of this rule.)
5. The Orthogonal Errors Rule (OER) states that

$$\psi_{jk} = 0$$

unless both v_j and v_k are exogenous (Fox, 1984; Land, 1969; Bollen, 1989).

This list of six rules appears to cover the main recommendations offered, authoritatively, in the literature. In principle one or more of these "rules" may be used by an investigator to determine the location of the nondirected arcs when a substantive hypothesis has been represented by a specification of directed arcs. The problem is that it does not seem obvious how to choose between them. McDonald (in prep.) proves the following mathematical propositions, intended to assist in an evaluation of these rules. The proofs follow fairly readily from the simple algebra of model (1).

Proposition 1: The direct precedence rule is a sufficient condition that a (recursive or nonrecursive) path model (without unobservable variables with multiple indicators) is just identified, if all nonzero β_{jk}, ψ_{jk} are unconstrained.

Corollary: Under any stronger rule the model is overidentified.

Proposition 2: In a recursive model, the general precedence rules for variables and for errors imply each other.

Proposition 3: In a recursive model the orthogonal errors rule implies the general precedence rules which imply the direct precedence rule. The orthogonal errors rule implies the exogenous variables rule. (Here "implies" can be read as "includes" and "is at least as strong as".)

Note that we do not have a relationship of implication between the general precedence rules and the exogenous variables rule, or, more importantly, between the latter and the direct precedence rule. By Proposition 1 the direct precedence rule gives a weak sufficient condition for identifiability. In contrast, it is well-known that the exogenous variables rule does not.

3. Path models containing common factors

In the more recent literature the term "latent variable" has been used, commonly without explicit definition, in a number of distinct senses, including but not confined to its original sense as a synonym for a latent trait or common factor. To avoid confusion, we will avoid the term "latent variable" and confine attention to cases where the random variables in the model are (i) observable variables, (ii) error-terms, and (iii) latent traits or common factors (including in particular "true scores") defined by the weak principle of local independence (McDonald, 1981) and hence necessarily possessing more than one indicator. We define a selection matrix $\mathbf{J}$ as a p x q matrix of zeros and unities that selects a proper subset $\mathbf{v}$ from the components of $\mathbf{v}^*$, q x 1, by

$$\mathbf{v} = \mathbf{J}\mathbf{v}^*. \tag{17}$$

We then have the general form of the Reticular Action Model as

$$\mathbf{v} = \mathbf{J}(\mathbf{I} - \mathbf{B})^{-1}\mathbf{e}, \tag{18}$$

yielding

$$\mathrm{Cov}\{\mathbf{v}\} = \Sigma_v = \mathbf{J}(\mathbf{I}-\mathbf{B})^{-1}\Psi(\mathbf{I}-\mathbf{B})'^{-1}\mathbf{J}', \tag{19}$$

where now $\mathbf{B}$, Ψ are path coefficient matrices and error covariance matrices of $\mathbf{v}^*$.

The mathematical statement that

$$\mathbf{y} = \Lambda\eta + \varepsilon \tag{20}$$

with

$$\mathrm{Cov}\{\varepsilon\} = \Delta_y, \text{ diagonal}, \tag{21}$$

so that

$$\mathrm{Cov}\{\mathbf{y}\} = \Lambda\Sigma_{\eta\eta}\Lambda' + \Delta_y, \tag{22}$$

where

$$\Sigma_{\eta\eta} = \mathrm{Cov}\{\eta\},$$

defines what is meant by common factors -- the vector η -- the indicators of those common factors-- the vector $\mathbf{y}$--and the residuals or unique components of the indicators--the vector ε. In special cases of this model the common factors are sometimes interpreted as "true scores," and ε as a vector of "errors of measurement," though the justification for such an interpretation is hardly ever clear, and it does not rest on mathematical grounds. The present discussion is limited to recursive models. We consider the consequences of embedding a common factor model within a path model, with particular reference, again, to the specification of nondirected paths. It is common practice not to include nondirected arcs between indicators of a common factor and other variables in the model. Indeed, models such as LISREL are designed to exclude such paths by separating the model into a "structural" model and "measurement models". A proof will now be outlined of the possibly counterintuitive

Proposition 4: The general precedence rule is a sufficient condition that the covariance structure of a set of indicators $\mathbf{y}$ of a set of factors η in a path model be given by (22), independent of the covariances of their residuals with other variables in the model.

Proof: In the model (18), with $\mathbf{B}$ lower triangular, and at least a partial order of $v_1^*, \ldots, v_q^*$, let η be an mx1 vector of unordered common factors and $\mathbf{y}$ be an nx1 vector of their indicators. Then without loss of generality we may partition $\mathbf{v}'$ as $[\mathbf{z}', \eta', \mathbf{y}', \mathbf{f}']$, (where possibly preceding variables in $\mathbf{z}$ and possibly following variables in $\mathbf{f}$ may include common factors and their indicators). By definition of the factor model, we have, with obvious notation for submatrices,

$$\mathbf{B}_{\eta\eta} = 0, \quad \mathbf{B}_{yy} = 0, \quad \mathbf{B}_{yz} = 0, \quad \mathbf{B}_{fy} = 0.$$

By repeated applications of the well-known identity for the inverse of a partitioned triangular matrix, it is easy though tedious to show that (5) yields corresponding submatrices of $\mathbf{A}$ such that

$$\mathbf{A}_{y\eta} = \Lambda_{y\eta}, \quad \mathbf{A}_{fy} = 0,$$

and

$$\mathbf{A}_{yz} = \Lambda_{y\eta}\mathbf{A}_{\eta z}, \tag{23}$$

and in the corresponding error covariance matrix we have

$$\Psi_{y\eta} = 0, \quad \Psi_{yy} = \Delta_y,$$

again by definition of the common factor model. On substituting in (7), with a little further algebra, we find that

$$\Sigma_{yy} = \mathrm{Cov}\{\mathbf{y}\} = \Lambda\Sigma_{\eta\eta}\Lambda' + \Delta_y + \Psi_{yz}\mathbf{A}'_{yz} + \mathbf{A}_{yz}\Psi_{zy}, \tag{24}$$

$$\Sigma_{y\eta} = \mathrm{Cov}\{\mathbf{y},\eta\} = \Lambda\Sigma_{\eta\eta} + \Psi_{yz}\mathbf{A}'_{\eta z}, \tag{25}$$

where

$$\Sigma_{\eta\eta} = \mathbf{A}_{\eta z}\Psi_{zz}\mathbf{A}_{\eta z}' + \Psi_{\eta z}\mathbf{A}'_{\eta z} + \mathbf{A}_{\eta z}\Psi_{zy} + \Psi_{\eta\eta}. \tag{26}$$

At this point we might suppose that we need the additional assumption that $\Psi_{zy} = 0$ if we are to embed the common factor model in a path model without altering its structure to the more complex form of (24), with (25). To complete the proof of the proposition we need to show that under the general precedence rule this assumption is not necessary.

If the structural model (22) is identified, then $\Sigma_{\eta\eta}$, Λ_y, and Λ are determined by Σ_{yy}. Included in these conditions will be the condition that m<n and Λ is of full column rank almost everywhere, i.e., except on sets of points of the parameter space of measure zero (where, for example, elements in a column of Λ vanish or two or more columns are linearly dependent). Hence, by (23),

$$\mathbf{A}_{\eta z} = (\Lambda'\Lambda)^{-1}\Lambda'\mathbf{A}_{yz}. \tag{27}$$

On substituting (27) in (24), (25), (26), and noting that by the general precedence rule we have

$$\mathbf{A}_{yz}\Psi_{zy} = \mathbf{O}, \tag{28}$$

it follows that (24) reduces to (22), (25) reduces to

$$\Sigma_{y\eta} = \Lambda\Sigma_{\eta\eta}, \tag{29}$$

and (26) reduces to

$$\Sigma_{\eta\eta} = \mathbf{A}_{\eta z}\Psi_{zz}\mathbf{A}'_{\eta z} + \Psi_{\eta\eta}. \tag{30}$$

This completes the proof.

We note that further substitution in (7) yields

$$\mathrm{Cov}\{\mathbf{y},\mathbf{z}\} = \Lambda\Sigma_{\eta z} + \Psi_{yz}\mathbf{A}'_{zz}, \tag{31}$$

and

$$\mathrm{Cov}\{\mathbf{f},\mathbf{y}\} = \Sigma_{f\eta}\Lambda' + \mathbf{A}_{fz}\Psi_{zy} + \mathbf{A}_{ff}\Psi_{fy}. \tag{32}$$

Since $\mathbf{A}_{zz}$ is nonsingular (being triangular with unities in the diagonal) the second term in (31) vanishes if and only if

$$\Psi_{yz} = \mathbf{0}. \tag{33}$$

Similarly, since $\mathbf{A}_{ff}$ is nonsingular, the third term in (32) vanishes if and only if

$$\Psi_{fy} = \mathbf{0}. \tag{34}$$

If both these matrices were assumed to vanish, we would have

$$\Sigma_{\eta z} = (\Lambda'\Lambda)^{-1}\Lambda'\Sigma_{yz}, \tag{35}$$

and

$$\Sigma_{\eta f} = (\Lambda'\Lambda)^{-1}\Lambda'\Sigma_{yf}, \tag{36}$$

and hence sufficient conditions that the entire model is identified. These are in fact known identifiability results in the LISREL model, which makes the assumptions (33) and (34) as part of its specification. However, the conditions (33) and (34) do not follow from the general precedence rule or from the definition of common factors. It is not obvious that we would wish to introduce these further assumptions on any grounds except as strong sufficient conditions for identifiability. It is easy, however, to find cases of this model that do not assume (33) and are, like the common factor model itself, identified except on sets of points of measure zero. An example is given shortly. Since it is reasonable by the general precedence rule not to expect to account for the relations between indicators of a factor and a nonpreceding variable by the directed paths of the model, it is possible to fit such cases without assumption (33) and investigate identifiability, rather than adopt a strong assumption solely to secure it. Relations of the type $\Psi_{fy} \neq \mathbf{O}$ require further investigation for similar reasons. We note that corresponding results cannot be obtained from the exogenous variable rule.

To conclude this section we note that attention has been confined to the case of latent variables that fit the definition of common factors. It may be that well-motivated cases can be found that do not fit this definition. Inter-battery factors would be one type, perhaps, and principles of conditional structure other than local independence could be used to define further classes of "latent variable" (see McDonald, 1981). Care is needed when we seek to recognize such cases. Bollen (1989, pp. 171-2) gives an example of a causal chain of three variables in which the intermediate variable Y_1 is subject to an "error of measurement." He describes this as an instance of cases with "only one indicator per latent variable." The model is identified except on sets of points of measure zero. However, inspection of the path diagram, reveals, not surprisingly, that the "true score" η_1 is the common factor of Y_1 and Y_2, that is, a latent trait in the classical sense, and the interpretation of the model as a causal chain from X_1 to Y_2 through Y_1 rests on extra-mathematical grounds.

4. A numerical example

Consider a model with an exogenous observed variable and an exogenous common factor defined by three indicators. In Appendix 1 it is shown that this model is identified except on points where $[\lambda_{32}, \lambda_{42}, \lambda_{52}]$ is proportional to $[\psi_{31}, \psi_{41}, \psi_{51}]$. Using a data-generating program, the covariance matrix in Table 1 was generated, with a sample size of 1000. Model (a), the generating model, gives the ML estimates and corresponding SEs shown in Table 2. An attempt to fit the model by ML without the nondirected paths ψ_{31}, ψ_{41}, ψ_{51} failed to converge, so GLS results were sought. (The analyses used COSAN, Fraser & McDonald, 1988). These are given as Model (b) in Table 2. Model (a) gives a chi-square of .326 with 1 df, p=.568. Model (b) gives a chi-square of 218.47 with 4 df, p=.00, which is clearly unacceptable. Since data can be constructed conforming to model (a), it seems that we might reasonably allow the possibility of fitting such a model to empirical data. It should be remarked at this point that the results of Sections 2 and 3 are all on matters logically prior to sampling and parameter estimation, which, however, have clear implications with respect to samples concerning consistency and goodness of fit.

Perhaps the point to be stressed is that global measures of fit are essentially tests of the assumptions about nondirected paths. Until there is a clear basis for the specification of such paths, neither global tests of significance nor goodness-of-fit indices can be taken seriously.

References

Bollen, K.A. (1989). Structural equations with latent variables. N.Y.:Wiley.

Duncan, O.D. (1975). Introduction to structural equation models. N.Y.: Academic Press.

Fox, J. (1984). Linear statistical models and related methods. N.Y.:Wiley.

Fraser, C. & McDonald, R.P. (1988). COSAN: Covariance structure analysis. Multivariate Behavioral Research, 23, 263-265.

James, L.R., Mulaik, S.A., & Brett, J. M. (1982). Causal analysis: assumptions, models, and data. Beverley Hills: Sage.

Kang, K.M. & Seneta, E. (1980). Path analysis; an exposition. In Krishnaiah, P.R., ed., Developments in statistics, vol. 3, N.Y.: Academic Press, pp. 19-49.

Land, K.C. (1969). Priciples of path analysis. In Borgatta E.F. (Ed.), Sociological Methodology, pp.3-37.

McArdle, J.J. & McDonald, R.P. (1984). Some algebraic properties of the reticular action model for moment structures. British Journal of mathematical and statistical Psychology, 37, 234-251.

McDonald, R.P. (1981). The dimensionality of tests and items. British Journal of mathematical and statistical Psychology, 34, 100-117

McDonald, R.P. (1985). Factor analysis and related methods. Hillsdale: Lawrence Erlbaum Associates, Ch. 4.

McDonald, R.P. (in prep.) Haldane's lungs: a case study in path analysis.

Schmidt, P. (1982). Econometrics, in Encyclopedia of statistical Sciences, vol. 2, 441-451, N.Y.; Wiley.

Wright, S. (1934). The method of path coefficients. Annals of mathematical Statistics, 5, 161-215.

Wright, S. (1960). Path coefficients and path regressions: alternative or complementary concepts? Biometrics, $$, 189-202.

Table 1

1	.983				
3	-.098	.979			
4	-.534	.448	.971		
5	.018	.257	.217	.967	
6	.511	.309	-.098	.184	.715

Sample covariance matrix--model with paths from indicators to an exogenous variable.

Table 2

	True	Model (a) (ML)	Model (b) (GLS)
β_{61}	.7	.754(.062)	.530(.026)
β_{62}	.5	.509(.035)	.264(.028)
λ_{32}	.8	.751(.042)	.563(.033)
λ_{42}	.6	.612(.042)	.900(.034)
λ_{52}	.4	.343(.035)	.263(.033)
ψ_{21}	-.4	-.450(.098)	-.536(.033)
ψ_{31}	.2	.241(.058)	NA
ψ_{41}	-.3	-.264(.077)	NA
ψ_{51}	.2	.178(.032)	NA
ψ_{11}	1	.984(.044)	.843(.043)
ψ_{22}	1	fixed	fixed
ψ_{33}	.36	.410(.052)	.435(.029)
ψ_{44}	.64	.612(.040)	.147(.046)
ψ_{55}	.84	.849(.040)	.816(.026)
ψ_{66}	.25	.244(.018)	.251(.014)

Estimates -- model with paths from indicators to an exogenous variable

Appendix 1: Proof of identifiability of model with nondirected paths from indicators to exogenous variable.

We have that

$$\mathbf{B} = \begin{array}{cccccc} & & & & & \\ & & & & & \\ & \lambda_{32} & & & & \\ & \lambda_{42} & & & & \\ & \lambda_{52} & & & & \\ \beta_{61} & \beta_{62} & & & & \end{array}$$

$$\mathbf{\Psi} = \begin{array}{cccccc} \psi_{11} & \psi_{21} & \psi_{31} & \psi_{41} & \psi_{51} & \psi_{61} \\ \psi_{21} & 1 & & & & \\ \psi_{31} & & \psi_{33} & & & \\ \psi_{41} & & & \psi_{44} & & \\ \psi_{51} & & & & \psi_{55} & \\ \psi_{61} & & & & & \psi_{66} \end{array}$$

which by (6) yields:

$\sigma_{11} = \psi_{11},$ (A1)

$\sigma_{34} = \lambda_{32}\lambda_{42},\ \sigma_{35} = \lambda_{32}\lambda_{53},\ \sigma_{45} = \lambda_{42}\lambda_{52}$ (A2)

yielding $\lambda_{32},\ \lambda_{42},\ \lambda_{52},$

$\sigma_{33} = \lambda^2_{32} + \psi_{33},\ \sigma_{44} = \lambda^2_{42} + \psi_{44},\ \sigma_{55} = \lambda^2_{52} + \psi_{55},$ (A3)

yielding $\psi_{33},\ \psi_{44},\ \psi_{55},$

$\sigma_{31} = \lambda_{32}\ \psi_{21} + \psi_{31}$

$\sigma_{41} = \lambda_{42}\ \psi_{21} + \psi_{41}$ (A4)

$\sigma_{51} = \lambda_{52}\ \psi_{21} + \psi_{51},$

$\sigma_{63} = \lambda_{32}\ \beta_{61}\psi_{21} + \lambda_{32}\beta_{62} + \beta_{61}\psi_{31}$

$\sigma_{64} = \lambda_{42}\ \beta_{61}\psi_{21} + \lambda_{42}\beta_{62} + \beta_{61}\psi_{41}$ (A5)

$\sigma_{65} = \lambda_{52}\ \beta_{61}\psi_{21} + \lambda_{52}\beta_{62} + \beta_{61}\psi_{51}.$

From (A4) we have

$\psi_{31} = \sigma_{31} - \lambda_{32}\psi_{21}$

$\psi_{41} = \sigma_{41} - \lambda_{42}\psi_{21}$ (A6)

$\psi_{51} = \sigma_{51} - \lambda_{52}\psi_{21},$

whence (A5) yields

$\lambda_{32}\beta_{62} + \sigma_{31}\ \beta_{61} = \sigma_{63}$

$\lambda_{42}\beta_{62} + \sigma_{41}\ \beta_{61} = \sigma_{64}$ (A7)

$\lambda_{52}\beta_{62} + \sigma_{51}\ \beta_{61} = \sigma_{65},$

any two of which may be solved for $\beta_{62},\ \beta_{61}$. Further

$\sigma_{61} = \beta_{61}\psi_{11} + \beta_{62}\psi_{21},$ (A8)

whence we obtain ψ_{21}, and thence

$\psi_{31},\ \psi_{41},\ \psi_{51}$ by (A6).

Finally,

$\sigma_{66} = \beta^2_{61}\ \psi_{11} + \psi_{21}\ (\beta_{62}\beta_{61} + \beta^2_{61}) + \psi_{66}$ (A9)

yields ψ_{66}. The solution of (A7) fails on points where

$[\lambda_{32}\ \lambda_{42}\ \lambda_{52}] \propto [\psi_{31}\ \psi_{41}\ \psi_{51}].$

Measurement, Causation and Local Independence in Latent Variable Models

Michael E. Sobel
University of Arizona
Tucson, AZ 85721

Abstract

Latent variable models are used extensively in the social and behavioral sciences for a variety of purposes, including measurement, description, and explanation. The latent variables may be continuous or discrete and the indicators of these may be continuous and/or discrete as well. Crossing the levels of measurement of the indicators with the assumptions on the level of measurement and distribution of the latent variables yields a variety of distinct latent variable models. A theme that unifies these different models is the assumption (or axiom) of conditional independence, which states that the indicators are independent, given the latent variable(s). The use of this axiom has been justified on various grounds, ranging from convenience to considerations of causality. This paper examines several of these justifications. First, the use of this axiom in the context of measurement and prediction is examined. Examples where the use of the axiom is scientifically plausible and implausible are considered, and the implications of the use of this assumption in both situations is discussed. I also show that the usual practice of viewing factor loadings as scaling factors that translate between units of measurement is incorrect. Second, the principle of the common cause is sometimes given as a justification for the use of the conditional independence assumption. The argument here is that the latent variables are the causes of the observed variables (indicators), and these indicators are not causes of one another. Hence (according to the argument), the association between the indicators is supposed to vanish when conditioning on the values of the latent variables. I show that this principle is not sound, and therefore cannot be used to justify the axiom of conditional independence. I also show that a modified version of this principle can be used to justify the conditional independence assumption in some instances.

1. INTRODUCTION

Independence and conditional independence are key concepts in probability theory and statistical practice (Dawid, 1979; Smith, 1988; Whittaker, 1990), and various interpretations of these concepts have been offered. In statistical modeling, it is often assumed

*For helpful comments on a previous version of this manuscript, I am grateful to Gerhard Arminger, James Robins, and Michael Trosset.

regressors. In some models for metric panel data, the stochastic error is decomposed into three independent parts (Hsiao, 1995). In latent variable models, the subject of this paper, one observes several observed variables that are regarded as indicators of one or more unobserved variables. For example, in Sobel (1981), the observed variables are expenditures, and these are taken to indicate several latent dimensions of lifestyle. In these types of models, the indicators of the latent variables are usually assumed to be independent, given the value of the latent variables.

From a mathematical standpoint, independence assumptions in statistical modeling may be viewed as part of the model definition. However, from a scientific standpoint, one wants to know the implications of such assumptions for the problem under investigation and/or whether such assumptions are reasonable in this context. Thus, statisticians and scientists often offer extramathematical justifications for the use of such assumptions.

This paper examines several common extramathematical justifications for the use of the conditional independence assumption in latent variable models. Section 2 considers the implications of the conditional independence assumptions for measurement. I show how using this assumption can lead to erroneous conclusions about that which is being measured and to predictions of the wrong quantities of interest. For the sake of concreteness, a particular latent variable model, factor analysis, is used in the examples. However, the conclusions of this section do not hinge upon the use of this particular latent variable model; that is, the same conclusions would be reached in the context of the latent class model, the Rasch model, or the latent profile model, for example. Section 3 examines the principle of the common cause (Reichenbach, 1956), which some authors (for example, Anderson, 1984; Bartholomew, 1987; Bollen, 1989) have invoked to justify use of the axiom of conditional independence. In the latent variable context, the argument is that the latent variables are causes of the indicators, the association between the indicators being spurious (from a causal point of view). Hence (according to the argument), the indicators should be conditionally independent. The argument itself is vague, but once it is suitably operationalized, it is evident that this principle is not sound, and therefore cannot be used to justify the axiom of conditional independence. With modification, it is possible to construct an argument for the use of the axiom; however, substantive considerations limit the applicability of this argument. Section 4 concludes.

2. Measurement, Prediction and Conditional Independence in Latent Variable Models

Latent variable models are sometimes used to estimate values of the latent variables, when direct measures of these are not obtained (or unobtainable) or contain measurement error. For example, in Fuller (1987, pp. 364-369) the observations are a sample of paper materials. For each observation (unit of paper) the smoothness is measured, at a number of different laboratories. This is the observed data, and it is measured with error, that is, the measurements vary across laboratories. The latent variable is the actual smoothness, and Fuller uses factor analysis to estimate the values of the latent variable. Second, latent variables are also used as predictors of other variables of interest. Covariance structure models with latent independent variables (though the latent variables are not actually assigned values) illustrate this case. While this section focuses on the first case, the relevance of the material to the second case should be obvious.

To better understand the use of latent variables in measurement, it is useful to begin with the following rather different setup: a researcher observes realizations of the quantities $(\underline{Y_i}', z_i) = (Y_{i1}, ..., Y_{ij}, ..., Y_{iJ}, z_i)$ for a sample of size n, $i = 1, ..., n$, from a population $\mathcal{P}$; z_i is the measurement of interest and $(Y_{i1}, ..., Y_{iJ})$ are predictors. For fixed i, z_i may be regarded as a fixed constant; the predictors are drawn from a joint distribution. Under random sampling from $\mathcal{P}$, the joint distribution of $(Y_1, ..., Y_J, Z)$ is of interest. Specifically, the investigator wants to ascertain

$$F_{Z|\underline{Y}}(z \mid \underline{y}), \tag{1}$$

the conditional distribution function of Z, given the indicators, or in the absolutely continuous case, the conditional density, $f_{Z|\underline{Y}}(z \mid \underline{y})$.

The foregoing setup is of interest when the z_i are expensive to obtain, but the Y_{ij} are cheap to obtain (Sobel & Arminger, 1986). Here one may want to know whether or not the Y_{ij} can be used to obtain "adequate" predictions of the z_i. (Obviously, notions of adequacy are subjective, and assessments of adequacy depend upon the purposes of the investigator.) If the answer is affirmative, in future studies, the investigator will obtain the noisy measurements Y_{ij} and use these to predict the measurement of interest. Another situation of interest occurs when complete data are available for a subsample of cases, but z_i is missing for the other cases (Sobel & Arminger, 1986, Arminger & Sobel, 1990). Still another situation is the case where, for ethical reasons, it is not possible to measure z_i directly; in this case, if a previous investigator had obtained such measures on both the noisy and true measurements, or, for example, a government agency "merged" data sources to obtain this information, and reported the joint distribution of the noisy and true measures, future Y_{ij} can be used to predict future z_i.

For predictive purposes, the investigator might use the conditional mean

$$E(Z \mid \underline{Y}), \tag{2}$$

or, if the distribution is unimodal, the mode. The investigator will also want to know if one or more of the indicators is unecessary, that is whether or not $Z \| \underline{Y}_A \mid \underline{Y}_B$, where $\underline{Y}_A$ is a subset of $Y_1, ..., Y_J$ and $\|$ denotes "independence"; if so, there is no reason to obtain the measurements $\underline{Y}_A$, as these have no additional predictive value.

Before proceeding to a discussion of the case where z_i is not observed, several comments about the setup above are in order. First, technically one does not need to ask what the indicators indicate, as the z_i are known quantities, and the indicators play only the role of predictors of these. Thus, for example, if weight is measured on J different scales, and the measurements from the scales (the indicators) are used to predict height, this may be of no concern, if "adequate" predictions are obtained and the sole aim of the analysis is to predict height. A more likely scenario, however, is that the measures of weight do not provide "adequate" predictions of height, and the researcher is also interested in "measuring" (as opposed to only predicting) height. Suppose for the moment that the researcher does not understand the difference between height and weight. In that case, if the researcher believed the predictors were accurately measured (in some sense), the conclusion that these predictors do not actually measure height, a valuable scientific lesson in and of itself, might be reached. Presumably, this could generate greater theoretical understanding, as well as lead to a search for a better set of indicators, that is, a better measuring instrument. Note, however, there is no guarantee, even if the highest standards of adequacy are used, that the indicators directly measure the quantities of interest. For example, measures of weight will (under suitable conditions) predict mass well, even though weight and mass are distinct. Finally (although this would be rather unusual in the context above) if there are two measurements of interest, say z_{i1} and z_{i2}, the indicators can be used to estimate the joint conditional distribution of the true measurements, and this can be used to assign predicted values.

In latent variable models, realizations from the vector $\underline{Y_i}$ are observed, but z_i (or the vector $\underline{z_i}$) is not observed. As before, the investigator wants to know (1) (for some measurements of interest as yet to be defined). Further, there are typically no previous studies from which to estimate (1) directly.

The latent variable approach is as follows: for fixed i, it is hypothesized that

$$Y_{ij} = g_j(\underline{z_i}, \varepsilon_{ij}), \tag{3}$$

where the ε_{ij} are mutually independent errors. Assuming also that $\underline{\varepsilon}_i' = (\varepsilon_{i1}, ..., \varepsilon_{iJ})$ are independent and identically distributed in $\mathcal{P}$, where the distribution is of known form (up to parameters), under random sampling from $\mathcal{P}$,

$$F_{\underline{Y}|\underline{Z}}(\underline{y} \mid \underline{z}) = \prod_{j=1}^{J} F_{Y_j|\underline{Z}}(y_j \mid \underline{z}), \tag{4}$$

that is, the indicators are conditionally independent, given the latent variables (Lord & Novick, 1968). The distribution (1) or the corresponding density (when this exists, as below) is then obtained by hypothesizing a distribution (which can, if desired be thought of as a prior distribution) of $\underline{z}$ and applying Bayes theorem

$$f_{\underline{Z}|\underline{Y}}(\underline{z} \mid \underline{y}) = \prod f_{Y_j|\underline{Z}}(y_j \mid \underline{z}) f(\underline{z}) / \int \cdots \int \prod f_{Y_j|\underline{Z}}(y_j \mid \underline{z}) f_{\underline{Z}}(\underline{z}) d\underline{z} \tag{5}$$

to obtain the posterior density (5)(Bartholomew, 1981); all the information in the indicators about the value of the latent variable(s) is contained in (5) (or more generally (1)), and therefore (5 (or (1)) is the basis for predicting the values of the latent variable(s).

It is important to understand the role of the assumptions on the prior and the conditional distribution in the foregoing development. Without some assumptions, it is of course not possible to know the joint distribution of the latent and observed variables, when only the latter are observed. From a mathematical standpoint, the assumptions above suffice to define this joint distribution. Using these assumptions and data on the observables only, (5) can be estimated. But these assumptions also define the latent variables as variables that render the observed variables conditionally independent. From a scientific standpoint, it will often be unacceptable to allow the mathematical assumptions to define the quantitities of interest (the unobserved variables), except when it can be argued that these assumptions are empirically reasonable. For example, given a latent variable of interest, the data collection methods may render the conditional independence assumption plausible. When this assumption is not plausible, latent variables that satisfy the model postulates can typically be constructed, but there should generally be no scientific reason for holding interest in these latent variables.

To illustrate the utility of the conditional independence assumption when it is reasonable, and some of the implications of using this assumption when it is not reasonable, as in a number of common applications in the social and behavioral sciences, a simple example is offered where two plausible, but different theories of error, lead to latent variable models from which very different conclusions are obtained. Even in the best case where the use of this assumption is justified, latent variable models do not (contrary to common belief) generally allow calibration between units of measurement, nor do such models yield unambiguous predictions of the quantities of interest. Note also that in practice, matters are often more complicated than the examples suggest, the implications of which are obvious.

To fix ideas, consider the use of the factor model for measuring the weight of a sample of small steel balls. Here the true measurement of interest is clearly Platonic (Sutcliffe 1965). That is, the weight of an object is a well defined (but not intrinsic) property of the object; given the weight, the conditional independence assumption may or may not be true. Similarly, for certain diseases, the disease state is well defined; given the disease, the indicators on which diagnosis is based may or may not be conditionally independent. For other "diseases" that are not well defined, for example, lupus and various forms of mental disorders, the disease state itself is typically defined (operationally) by the diagnosis itself. In such cases, when the researcher is willing to commit to the idea that there is an underlying physiological entity (even if it cannot be currently identified), the disease conception is Platonic. But when the investigator is unwilling to assert the existence of such an entity, the underlying conception is not Platonic.

Now suppose that weight is measured using J scales, and let Y_{ij} denote the measurement of object i on scale j. Some of the scales may be metric, others British. Then

$$Y_{ij} = \lambda_j z_i + \epsilon_{ij}, \tag{6}$$

where $\lambda_j z_i$ is the actual weight of object i on scale j in the appropriate units, λ_j is a fixed constant for scale j that scales z_i to the relevant metric, that is, pounds or kilograms, and ϵ_{ij} is the true error of measurement for unit i on scale j. It is important to note that (6) merely states an identity that defines the error ϵ_{ij} with respect to the actual weight; in particular, (6) is not a regression model. (The actual weight depends on the acceleration of gravity, and as this is not a constant, weight is not an intrinsic property of an object; in latent variable terminology, weight is not a latent trait. For the purposes at hand, it is assumed that each of the J scales is subjected to the same gravitational attraction.)

Now consider the conditional expectations $E(Y_j \mid Z = z)$, $j = 1, ..., J$. From (6),

$$E(Y_j \mid Z = z) = \lambda_j z + E(\epsilon_j \mid Z = z). \tag{7}$$

Suppose that

$$\epsilon_{ij} = d_j + \tau_j z_i + \varepsilon_{ij}, \tag{8}$$

where the ε_{ij} are independent and identically distributed with mean 0 and $E(\varepsilon_j \mid Z) = 0$. Then

$$Y_{ij} = d_j + \gamma_j z_i + \varepsilon_{ij}. \tag{9}$$

In (9), d_j is a constant bias unique to the $(j)th$ instrument, $\gamma_j = \lambda_j + \tau_j$ is the population regression coefficient, and ε_{ij} is the usual regression error.

No assumptions about the joint distribution of the errors (over scales) in (9) have been made. Suppose now that these errors are not independent, specifically:

$$\varepsilon_{ij} = \alpha_j \delta_i + \delta_{ij}, \tag{10}$$

where $\delta_{ij} \| \delta_{ij'}$ for all j and j' with $j \neq j'$. (For the subsequent development, it is also assumed that the random variable δ_i is iid δ, and the random vectors $(\delta_{i1}, ..., \delta_{iJ})$ are iid $\underline{\delta}_j$, with $E(\delta \mid Z = z) = 0, E(\underline{\delta}_j \mid Z = z) = \underline{0}$.

Substitution of (10) into (9) yields:

$$Y_{ij} = d_j + \gamma_j z_i + \alpha_j \delta_i + \delta_{ij}. \tag{11}$$

Comparison of (6) with (9) is instructive. When, for example, z_i is observed, and $\tau_j = 0$, $\gamma_j = \lambda_j$. In this case, regression analysis (or multivariate regression) can be used to estimate the conversion factors λ_j, even though $\varepsilon_{ij} \neq \epsilon_{ij}$. If $\tau_j \neq 0$, the conditional expectation of the true error of interest (ϵ_{ij}) depends on z_i. In regression analysis, however, the error is chosen so that it's conditional expectation does not depend on the values of explanatory variables; choosing the error in this fashion identifies the population regression coefficients. (The assumption that the error is "mean independent" of the explanatory variables could just as well be replaced with the weaker assumption that errors and stochastic regressors are uncorrelated or the stronger assumption that errors and regressors are independent.) If the focus is solely on predicting future z_i (or on using z_i as a predictor), only the conditional distribution (or conditional mean) of the dependent variable, given the predictor(s), is of concern; how this is parameterized is unimportant. When interest centers on λ_j, however, the population regression coefficient γ_j is simply not the parameter of interest.

Now consider the case where z_i is not observed. When $\delta_i = 0$, the errors in (9) are independent over scales, and (9) is a one factor model for weight. The difference between this case and the case where z_i is observed is that to identify the model, an additional constraint must now be imposed. If, for example, $\gamma_{j'}$ is set to 1, the identified parameters are the ratios $\gamma_j / \gamma_{j'}$. Even in the case where $\lambda_j = \gamma_j$ for all j, the implication now is that only ratios of the λ_j parameters can be ascertained. (For the example here, however, insofar as weight is measured in arbitrary units, one learns that one pound = .454 kilograms, and this is the key issue.) But when $\gamma_j \neq \lambda_j$, the identifiable ratios $\gamma_j / \gamma_{j'} \neq \lambda_j \lambda_{j'}$. That is, the use of the errors ε_{ij} (as opposed to the true error ϵ_{ij}) implicitly identifies the parameters of the regression as those of interest, when in fact this is not so (as in the case where z_i is observed). The implication is that the common interpretation of factor loadings as scaling factors between different units of measurement (for example, Henry (1986), Williams & Thomson (1986), and Bollen (1989, p. 184) is not correct. As in the previous case, if interest focuses solely on predicting z_i, (5) can be used to predict $\gamma_{j'} z_i$; if a prediction of z_i is required, the factor model is simply too weak as a measurement instrument.

Now consider the case where z_i is unobserved, $\delta_i \neq 0$ and the axiom of conditional independence is invoked. Then the one factor model (9) no longer holds, but the two factor model (11) does hold. Under this model, the naive researcher would conclude that weight is "multidimensional". To identify the two factor model, suppose that $\gamma_{j'}$ and $\alpha_{j'}$ are set to 1; the latent variables predicted using (5) are then $z_i^* = \gamma_{j'}^{-1} z_i$ and $\delta_i^* = \alpha_{j'}^{-1} \delta_i$.

Suppose instead an alternative theory of error. The jth instrument "measures"

$$\theta_{ij} = z_i + \delta_i + d_j + \delta_{ij}, \tag{12}$$

where the quantities on the right are as previously defined, and computes:

$$Y_{ij} = \beta_j \theta_{ij}. \tag{13}$$

In this case, equating (13) and (6) gives the true error:

$$\epsilon_{ij} = \beta_j d_j + (\beta_j - \lambda_j) z_i + \beta_j \delta_i + \beta_j \delta_{ij}. \tag{14}$$

Note that it is not necessary, as in the previous case, to require $E(\delta_i) = 0$.

The corresponding latent variable model is obtained by substitution of (12) into (13):

$$Y_{ij} = \beta_j d_j + \beta_j (z_i + \delta_i) + \beta_j \delta_{ij}. \tag{15}$$

In this case, because $\delta_{ij} \| \delta_{ij'}$ for $j \neq j'$, a one factor model is obtained, with latent variable (assuming $\beta_{j'}$ is set to 1) $\beta_{j'}^{-1}(z_i + \delta_i)$.

Comparison of (11) and (15) is useful. When $\delta_i = 0$, a one factor model is obtained under either (11) or (15); however, it will not be possible to ascertain which of the two error processes generated the model. When $\delta_i \neq 0$ and the conditional independence assumption is invoked, under the first error process, the two factor model (11) is obtained, leading to the incorrect conclusion that the weight is multidimensional. Under the second error process, this erroneous conclusion is not reached, as a one factor model is still obtained. However, the wrong latent variable is obtained. In this case, use of (5) would yield predictions of $(\beta_{j'})^{-1}(z_i + \delta_i)$.

Given our knowledge of weight as a characteristic of the steel balls, and our knowledge of the conditions under which the J measurements are obtained (for example, by different operators in different labs), the assumption $\delta_i = 0$ may be quite plausible. In this case, the one factor model can be used to predict (up to a multiple) weight, and several (not all) of the complications previously discussed vanish. In the social and behavioral sciences, however, latent variable models are often used in cases where substantive knowledge and/or the data collection methods render the assumption $\delta_i = 0$ implausible. For example, Mare and Mason (1980) use latent variable methods to reconcile children's reports of parental socioeconomic status; here (assuming parental socioeconomic status is Platonic), it would seem reasonable to expect, for example, consistent overstatement, leading to correlated errors). Bielby (1986) considers the case of two measures of schooling; although Bielby's example is hypothetical, and he therefore does not state the context in which the two measurements are obtained, suppose that the measures have been obtained by reinterviewing respondents, or by means of a wife's report and a husband's report. In such cases, one would expect (treating true schooling Platonically) $\delta_i \neq 0$. Suppose now that respondents report a multiple of θ_{ij}. For the case with two indicators considered by Bielby, both loadings can be set to an arbitrary constant, for example, 1. Let us suppose that in fact, both loadings are really equal (one can imagine they are not, but with two indicators, this cannot be ascertained), with actual common value β. A one factor model is thus obtained, with latent variable $\beta^{-1}(z_i + \delta_i)$, where $E(\delta_i) = \mu_i$. If $\delta_i = \mu_i$, (5) can be used to predict the constant $\beta^{-1}(z_i + \mu_i)$; otherwise, the random variable $\beta^{-1}(z_i + \delta_i)$ is predicted.

With two indicators, two factor models cannot be meaningfully considered. But consider the case where there are a sufficient number of indicators, and a two factor model is obtained (as in 11) because the errors are correlated. In this case, if one knew the first factor was z_i and chose the correct rotation, a multiple of z_i could be predicted. But if an incorrect rotation is chosen, this is not the case. This illustrates some further ways that the rules used to identify a model might commit us to measurements of the wrong quantities; of course, this was also implicit in the case previously considered.

Another case where the assumption $\delta_i \neq 0$ does not seem plausible occurs when the J measurements are different items on the same questionnaire, as in attitudinal studies. Evidently this case is more complicated than those previously considered, for whereas variables such as the number of years of schooling and weight are Platonic, attitude is not usually conceptualized in this fashion. Thus, it is necessary to first examine various notions of attitude.

To that end, one might argue simply that there seems to be something (one or more factors) in common uniting the distinct items. For both mental (it is easier to think about a few variables and their relationships with other variables than to think about many such variables) and statistical reasons (for example, "multicollinearity" among the observed variables), it is useful to reduce the many items to a few variables. In this case, the attitude(s) towards some phenomenon of interest is simply a convenient name for the latent variable(s), and there seems little point in worrying about the response process that generates errors (as above). One simply obtains a model with as many factors as needed (hopefully few) and uses these in other analyses, for example, in a mean and covariance structure model, to predict other variables of interest. Note that this argument is compatible with either the idea that attitudes are Platonic or non-Platonic, though this issue is simply not of interest in this context.

An intermediate and commonly invoked argument is that attitude is a conceptual entity (construct). Such concepts are purported to be useful in basic science, which is held by some writers to be concerned with studying the relationships among constructs. Constructs may be unidimensional, but constructs may also be multidimensional. Moreover, no committment to the idea that constructs are Platonic is made, that is, a construct need not correspond to any physiological or physical entity. In this case it does not seem reasonable to argue that there is an error ϵ_{ij}. Nevertheless, the response process that generates the error ε_{ij} should be of interest to the investigator. For if the construct is unidimensional, but the first error process previously discussed is considered, a one factor model will be rejected, provided there are a sufficient number of items. The researcher may go on to obtain a two factor model and conclude that the hypothesis of unidimensionality is untenable, that is, in fact, the construct is two-dimensional. This may be the case, but it may also be the case that δ_i represents nothing more than response uncertainty. Insofar as it is not clear to begin with what, if anything, is being measured, and there is no theory of the cognitive process generating the responses, choosing between the two interpretations (an important issue) is not easy. Note that in the case of weight (or number of years of schooling), this should not be as problematic; that is, a non-naive researcher would not conclude, for example, that years of schooling is multidimensional. In the case of the second error process previously discussed, a one factor model is obtained; the researcher is likely to conclude that the concept is unidimensional, and that a latent trait is measured, that is a "consistent and stable human" characteristic (Lord & Novick

1968, p. 537). (Note that nothing precludes this interpretation in the two-dimensional case). When $\delta_i = d_i$, one can hold to this argument, though it might not be clear what this trait is, but when δ_i is not a degenerate random variable, such an argument cannot be sustained. However, it is important to note that the model cannot distinguish between these two cases.

Finally, one could adopt the view that attitude is Platonic; for example, one might argue that a portion of the brain is activated to produce the response to the attitude items. If this view is taken, interest should ultimately focus on measuring the true physiological quantity (quantities) of interest (that is, z_i or $\underline{z_i}$). When z_i ($\underline{z}_i$) is not directly measured, understanding the response process itself is critical, for this determines (in part, as shown earlier) how many factors are found and what latent variable(s) is (are) predicted using (5). The implication is that when the true quantities cannot be measured (or are not measured) and the response process is not well understood, latent variable models may produce very misleading results.

As above, the interpretation of δ_i (a different issue than whether or not δ_i can be separated from the other latent variable(s)) is more problematic than in the case where the weight of the steel balls is measured, for the balls do not change their weight (or when they do, the relevant conditions can be described and checked against the conditions under which the measurements are taken). For the case of a Platonic attitude, δ_i could be an error (in the sense above). For example, there might be a physiological trait that is invariant over occasions, but the response process itself may actually be intrinsically probabilistic, (as opposed to the case where the process is deterministic but not all determinants are known and probability theory is therefore used). Lazarsfeld (1959) makes an argument of this type, albeit in an attempt to justify the use of the conditional independence assumption (over occasions). See also Suppes (1985) and Salmon (1984) on this subject. A second possibility is that the response process depends upon a number of variables specific to occasions (it is important under this view that there be a physiological aspect that is invariant across occasions and that this is what is called attitude). Under either of these interpretations, it is important to understand the process by which the error is generated (for the reasons given previously). In addition, the magnitude of error should be of interest; for example, under (15), if $E(\delta_i) = 0$ and the variability is small, it would not be unreasonable to use (5) to obtain predictions of (a multiple of) z_i. If this is not so, it is more important to separate δ_i from z_i; in general, however, this is not possible, although under some conditions, the use of repeated measures would allow this.

Two other possibilities lead to the view that δ_i should not be interpreted as error. First, there may be a physiological trait (as above), but attitude is not regarded as this trait, but rather as that which this trait, in conjunction with the other variables, produces. This view has not proven to be popular in psychological measurement. Another possibility is simply that there is no trait involved at all. Here, it is more reasonable to argue that the characteristic being measured may have a central tendency, but there is fluctuation, purely random and/or systematic. Blood pressure would appear to be an example; here a trait conception does not seem useful. Under either of these views, several of the problems above do not arise. Of course, the model itself cannot be used to adjudicate among any of these views.

In short, when the latent variable is well defined, the properties of the measuring instruments well understood, and the data collected in a "suitable" manner, the assumption of conditional independence is often reasonable and latent variable models can be used to obtain predictions of (multiples of) the quantities of interest. However, as demonstrated above, the models cannot be used (unless certain other assumptions are also true) to translate units on one scale of measurement into units on another scale. When the data are collected in a manner that renders the conditional independence assumption unreasonable, as in much sociological and psychological research, and the quantities of interest and/or the process generating the indicators less than well understood, various other problems emerge, as illustrated above. Nor are these problems resolved by adopting the usual view that latent variables are non-Platonic. The idea that latent variable models are measuring traits of individuals is also problematic. First, the trait conception may be erroneous, that is, there is no intrinsic property to be measured. Second, even if such intrinsic properties exist, the latent variables defined by the model are (generally) not these intrinsic properties.

3. Causation and Conditional Independence

The principle of the common cause (Reichenbach, 1956) is sometimes used to justify invoking the axiom of conditional independence. Roughly, the principle states that if X and Y are two associated random variables, then either (say X is prior to Y) X causes Y or there is a prior variable Z that causes both X and Y. In the case where X does not cause Y, the $X - Y$ association is supposed to vanish, conditioning on Z. (For variants on this idea, see also Simon (1954), Suppes (1970), and Geweke's (1984) essay on Granger-Sims causation). In this vein, a number of authors (for example, Anderson (1984) and Bartholomew (1987)) have applied this principle to latent variable models, arguing that latent variables (Z or a vector $\underline{Z}$) cause the indicators and the association between the indicators is due to these common causes. Thus, if one wants to find the latent variable(s) causing the indicators, the indicators should be independent, given the latent variable(s), that is, the assumption of conditional independence follows from the commitment to latent variables that are causes.

Before examining this principle further, several points are in order. First, latent variable models are used in many contexts where it is clear that one would not adopt this type of argument. An example is the latent variable termed "socioeconomic status", as indicated by respondent's level of education and subsequent occupational status (Wheaton, Muthén, Alwin & Summers 1977); here, most social scientists would argue that education causes subsequent occupational status. Thus, the use of the conditional independence axiom in applications of this type cannot be justified by reference to the principle of the common cause. Second, assuming the principle is sound, if one finds a variable Z such that $X \underline{\|} Y \mid Z$, it does not necessarily imply that this particular Z is a common cause of X and Y; the principle as stated above asserts then when X does not cause Y, such a Z exists; it does not assert that a unique Z exists, nor does it assert that any Z is a common cause. In practice, however, one uses the principle to ascertain whether or not a particular Z satisfies the conditional independence relation, and if so, Z is taken to be the common cause. When Z is observed, there may be stronger grounds for believing that one has found a common cause than in the case where Z is generated by a latent variable model. This is so for two reasons. First, an observed Z is likely to be chosen using substantive considerations. Second, it is possible, under very mild conditions, to obtain latent variables that satisfy the conditional independence axiom. For example, in factor analysis, with J indicators, it is always possible to find a $J-1$ dimensional latent variable that satisfies the postulates of the model (Guttman, 1957). For binary indicators, Suppes & Zanotti (1981) show that a latent variable exists if and only if the indicators have a joint distribution; Holland & Rosenbaum (1986) have generalized this result.

The principle of the common cause itself is often justified by arguing that the conditional independence condition substitutes for the inability to conduct a proper experiment. The argument is as follows: first, it is impossible to manipulate X (or the data were not collected from an experimental study), but were it possible to do so (or the data were obtained from an experimental study), one would find the outcome (Y) is unaffected (in some sense that is usually not specified). The $X-Y$ association is thus spurious. Second, were one to manipulate Z, both Y and X would be affected. It is then concluded that the $X-Y$ association is due to Z, that is, X and Y are conditionally independent, given Z. A stock example is a set of barometer readings associated with the presence or absence of rain (Y). Here the argument is that if the investigator were to tamper with the barometers, the outcome would be unaffected. However, alterations of the underlying weather conditions cause changes in the barometer readings.

In this section, the justification offered in the preceeding paragraph is examined. Note that there the causal relation is viewed counterfactually. Therefore, consideration of this justification requires a framework in which the causal relation is treated counterfactually. (For material on this notion of causation, as well as others, see Sobel (1994a, 1994b, 1995). I draw heavily on Rubin (1974, 1977, 1978, 1980), who has proposed methods for causal inference based on this view of the causal relation, as well as Holland (1986, 1988). To keep matters simple, only the case where there are two indicators X and Y of a latent variable Z is considered. However, the results generalize in an obvious fashion to the case with many indicators and more than one latent variable. The setup used is essentially that in Pratt and Schlaifer (1988). The basic components are: a) a population $\mathcal{P}$, with elements i, b) variables X and Y, with values $x \in \Omega_1$ and $y \in \Omega_2$, respectively, c), for each $x \in \Omega_1$, and for every element of $\mathcal{P}$, a random variable Y_{xi}; that is, for each i, there are as many random variables as elements in Ω_1. To keep matters simple, assume for the moment the Y_{xi} are independently and identically distributed as Y_x. (In some setups, the random variable Y_{xi} is replaced with y_{xi}, which need not be viewed as random. However, to tie the material to latent variable models, the setup herein is more straightforward.) In this paper, a notion of counterfactual causation will be examined, causation in distribution (Sobel, 1994a; 1994b):

Definition 1. X causes Y in distribution in $\mathcal{P}$ if for some pair x and x' in Ω_1, the random variables Y_x and $Y_{x'}$ have different distributions. Otherwise, X does not cause Y in distribution in $\mathcal{P}$.

In the statistical literature, attention typically focuses on the average effect of a treatment. Definition 1 is easily modified for this case by defining causation in mean, and replacing the word "distribution" with "expected value". Note that causation in mean implies causation in distribution, but not conversely. In this paper, given the interest in conditional independence, it is more natural to focus attention on causation in distribution.

Definition 1 can now be used to (partially) examine the previously offered justification for the principle of the common cause. If the principle is sound, it gives support for the use of the conditional independence assumption, but if the principle is not sound, it does not support the use of this assumption. To begin, suppose that X does not cause Y in distribution in $\mathcal{P}$. The observed data is a random sample from the joint distribution (Y, X), with distribution function $F(y, x)$; the conditional distribution function is then $F_{Y|X}(y \mid x) = \Pr(Y_x \leq y \mid X = x) = F_{Y_x|X}(y \mid x)$. In general, when the data come from an observational study, Y and X will be associated, but this does not bear upon the issue of whether X causes Y in distribution, for the distribution above only describes the distribution of Y_x in the subpopulation with $X = x$. However, consider the special case where observations are randomly assigned to levels of the cause; this is the case $X \| Y_x$. Under random assignment, $F_{Y_x|X}(y \mid x) = \Pr(Y_x \leq y) = F_{Y_x}(y)$, that is the distribution of Y_x is identical to the distribution of Y_x in the subpopulation with $X = x$. Under the further assumption that X does not cause Y in distribution in $\mathcal{P}$, $F_{Y_x}(y) = F_Y(y)$, that is $X \| Y$. This conclusion lends support to the first part of the argument for the principle of the common cause; subsequently, I return to this matter.

The second part of the argument asserts that Z causes both X and Y. When both parts of the argument are true, it is concluded that X and Y are independent, given Z.

There are several ways in which the two arguments can be operationalized. An obvious choice is to apply definition 1 to the three statements a) Z causes X in distribution in $\mathcal{P}$, b) Z causes Y in distribution in $\mathcal{P}$, c)X does not cause Y in distribution in $\mathcal{P}$. Then it is ascertained whether or not these three statements imply $X \| Y \mid Z$; the answer is no, and counterexamples are readily made up. Note also that the conditional indpendence assumption fails to imply any of these three statements.

Notwithstanding this negative answer, perhaps the principle can be usefully modified to provide a related justification for the conditional independence assumption in latent variable models. To that end, recall that the initial problem was to decide whether X "causes" Y when it was not possible to conduct a randomized experiment. If attention is focused on this issue, and not on the causal role of Z, statements a) and b) can be ignored. However, since a), b), and c) do not jointly imply conditional independence, c) alone will not imply conditional independence. Thus, it is evident that if progress is to be made, statement c) will need to be strengthened.

To that end, suppose the latent variable Z is a covariate that is sufficient for treatment assignment, that is, $Y_x \| X \mid Z$ for all x, whence the conditional distribution

$$F_{Y|X,Z}(y \mid x, z) = F_{Y_x|(X,Z)}(y \mid x, z) = F_{Y_x|Z}(y \mid z). \tag{16}$$

That is, while assignment to the cause X is not random, assignment is random within subpopulations of the latent variable(s). Suppose also that X does not cause Y in distribution in subpopulations of $\mathcal{P}$ where the latent variable Z takes value z:

$$F_{Y_x|Z}(y \mid z) = F_{Y_{x'}|Z}(y \mid z) \tag{17}$$

for all pairs (x, x') in $\Omega_1 \times \Omega_1$. Note that this implies X does not cause Y in distribution in $\mathcal{P}$. It is evident that (16) and (17) imply $X \| Y \mid Z$; conversely, (16) and $X \| Y \mid Z$ imply (17). Thus, if the assumption that the latent variable Z is sufficient for treatment assignment is made, conditional independence is equivalent to the condition that X does not cause Y in distribution in subpopulations of $\mathcal{P}$ defined by Z.

The idea that the indicator Y does not cause the indicator X has yet to be discussed. If X is prior to Y, no discussion is necessary, but if this is not the case, it is also reasonable to require that Y not cause X.

Under the assumptions in the preceding paragraph, conditional independence holds: therefore

$$F_{X|Y,Z}(x \mid y,z) = F_{X_y|(Y,Z)}(x \mid y,z) = F_{X_{y'}|(Y',Z)}(x \mid y',z) = F_{X|Z}(x \mid z). \quad (18)$$

If it is also assumed that Z is sufficient for treatment assignment to Y, that is, $X_y \| Y \mid Z$, it is evident from (18) that Y does not cause X in distribution in subpopulations of $\mathcal{P}$ defined by Z, hence Y does not cause X in distribution in $\mathcal{P}$. That is, under the additional assumption that Z is sufficient for treatment assignment to Y, the statement Y does not cause X in distribution in $\mathcal{P}$ is a deduction from previous assumptions, not an assumption itself.

The foregoing solution replaces the notion that the latent variable Z is a common cause with the idea that the latent variable is a covariate sufficient for treatment assignment (to either of the two indicators). With this replacement, conditional independence is equivalent to the absence of causal relations amongst the indicators (in the sense described above). That is, although the principle of the common cause cannot be used to justify the conditional independence assumption in latent variable models, if Z is treated as above, and the assumption that the indicators do not cause one another is maintained, as in the principle of the common cause, the indicators are conditionally independent, given the latent variable.

The mathematical treatment above shows how the principle of the common cause fails to justify the use of the conditional independence assumption in latent variable models and demonstrates how the principle may be modified so that the conditional independence assumption is justified; for future reference, I refer to this modification as the principle of the common covariate.

While the idea that X does not cause Y (and Y does not cause X) has been examined formally, closer inspection raises several concerns. I illustrate by way of example.

Consider the case where the latent variable "ability" is sufficient for treatment assignment, X is the first test score, Y the second test score, and X does not cause Y in distribution in $\mathcal{P}$. By the preceeding results, conditional independence holds. While it is reasonable to believe that the first test score does not cause (in some sense) the second test score, the statement that X does not cause Y in distribution in $\mathcal{P}$ means that if all persons in $\mathcal{P}$ had score 150 on the first test, the distribution on the second test is identical to that in the case where all persons in $\mathcal{P}$ had score 80 on the first test. Intuitively, this does not seem plausible, meaning that the first score causes the second.

To reconcile this apparent contradiction, recall the example of the barometer and the rain. In the hypothetical experiment used to justify the argument that the barometric reading does not actually cause the rain, the experimenter tampers with the normal functioning of the barometer, and then discovers that X does not cause Y in distribution in $\mathcal{P}$. In some sense, this experiment does not seem to bear on the initial issue. One wants to find out if the barometer causes the rain when the barometer is functioning normally, not when the way things work are fundamentally changed; for example, if the experimenter breaks the barometer, how are we to really find out if the barometer (when it is not broken) causes the rain? However, when the barometer functions normally, it is not possible to induce all values $x \in \Omega_1$, that is, one might argue that Y_x is not well defined over $\mathcal{P}$; this is because under normal conditions, X is caused by Z. Similarly, if persons score 150 (or 80) on the first test because the connection between the test and what the test measures is changed or broken, this is different than the case where we consider the hypothetical response to the second test had the person scored 150 on the first test (Sobel, 1990). For related material, albeit in a different context, see Angrist, Imbens & Rubin (1993).

The preceding paragraph illustrates that the first argument for the principle of the common cause may also be faulty (at least for some examples of interest), insofar as reference to a possibly irrelevant experiment is made. In such instances, one may not wish to attempt to justify the conditional independence assumption using the principle of the common covariate.

In short, the principle of the common cause cannot be used to justify application of the conditional independence assumption. The principle of the common covariate justifies the use of this assumption; however, the use of this principle rests on untestable assumptions, and in some examples, application of this principle seems unwarranted.

4. DISCUSSION

This paper examines the use of the conditional independence assumption in latent variable models. Scientifically, the use of this assumption is difficult to justify in a number of contexts of interest, for example, when survey respondents report on a number of items (which are usually placed together in the survey) on the same occasion. The implications of using this assumption when it is not warranted are also discussed. The discussion does not imply the conclusion that latent variable models are necessarily useless in scientific endeavors; such models have proven useful in many domains of inquiry, especially educational testing. See also the case where factor analysis is used to reconcile measurements of a well defined object taken over different laboratories, as in the example from Fuller (1987). What is less evident is the utility of these models in attitudinal studies and for measurement of Platonic latent variables, of the variety discussed in the text. By implication, this is also the case when latent variables derived from these types of studies are used in structural equation models (Sobel, 1994b). A second rationale for the axiom of conditional independence, hinging on the principle of the common cause, was also examined. The principle itself is not sound, and hence cannot be used to justify the conditional independence assumption, but with modification, a variant of this principle, (called the principle of the common covariate) can be applied, under conditions described in section 3, to justify the use of conditional independence. To apply this principle, it is necessary to invoke (typically) untestable assumptions. Further, even when such assumptions hold, the principle is not universally applicable.

REFERENCES

Angrist, J.D., Imbens, G.W., & Rubin,D.B. (1993). Identification of causal effects using instrumental variables. unpublished manuscript, Harvard University.

Anderson, T. W. (1984). *An introduction to multivariate statistical analysis* (2nd ed.). New York: Wiley.

Arminger, G., & Sobel, M. E. (1990). Pseudo maximum likelihood estimation of mean and covariance structures with missing data. *Journal of the American Statistical Association*, 85, 195-203.

Bartholomew, D. J. (1981). Posterior analysis of the factor model. *British Journal of Mathematical and Statistical Psychology*, 34, 93-99.

Bartholomew, D. J. (1987). *Latent variable models and factor analysis*. London: Griffin.

Bielby, W. T. (1986). Arbitrary metrics in multiple-indicator models of latent variables. *Sociological Methods and Research*, 15, 3-23.

Bollen, K. A. (1989). *Structural equation models with latent variables*. New York: Wiley.

Dawid, A. P. (1979). Conditional independence in statistical theory (with discussion). *Journal of the Royal Statistical Society* Ser. B, 41, 1-31.

Fuller, W. A. (1987). *Measurement error models*. New York: Wiley.

Geweke, J. (1984). Inference and causality in economic time series models. In Z. Griliches & M. D. Intrilligator (Eds.), *Handbook of Econometrics* (Vol. 2), (pp. 1101-1144). Amsterdam: North Holland.

Guttman, L. (1957). Simple proofs of relations between the communality problem and multiple correlation. *Psychometrika*, 22, 147-157.

Henry, N. W. (1986). On "Arbitrary metrics" and "Normalization issues". *Sociological Methods and Research*, 15, 59-61.

Holland, P. W. (1986). Statistics and causal inference (with discussion). *Journal of the American Statistical Association*, 81, 945-70.

Holland, P. W. (1988). Causal inference, path analysis, and recursive structural equation models (with discussion). In C. C. Clogg (Ed.), *Sociological methodology, 1988.* (pp. 449-493). Washington, D. C.: American Sociological Association.

Holland, P. W., & Rosenbaum, P. (1986). Conditional association and unidimensionality in monotone latent variable models. *The Annals of Statistics*, 14, 1523-1543.

Hsiao, C. (1995). Panel analysis for metric data. In G. Arminger, C. C. Clogg & M. E. Sobel (Eds.) (1995). *Handbook of statistical modeling for the social and behavioral sciences*, (pp. 361-400). New York: Plenum.

Lazarsfeld, P. F. (1959). In S. Koch (Ed.), *Psychology: a study of a science*, (Vol. 3), (pp. 476-543). New York: McGraw-Hill.

Lord, F. M., & Novick, M. R. (1968). *Statistical theories of mental test scores.* Reading, Mass: Addison-Wesley.

Mare, R. D., & Mason, W. M. (1980). Children's reports of parental socioeconomic status. *Sociological Methods and Research*, 9, 178-198.

Pratt, J. W., & Schlaifer, R. (1988). On the interpretation and observation of laws. *Journal of Econometrics*, 39, 23-52.

Reichenbach, H. (1956). *The direction of time.* Berkeley: University of California Press.

Rubin, D. B. (1974). Estimating causal effects of treatments in randomized and non-randomized studies. *Journal of Educational Psychology*, 66, 688-701.

Rubin, D. B. (1977). Assignment to treatment groups on the basis of a covariate. *Journal of Educational Statistics*, 2, 1-26.

Rubin, D. B. (1978). Bayesian inference for causal effects: The role of randomization. *The Annals of Statistics*, 6, 34-58.

Rubin, D. B. (1980). Comment on "Randomization analysis of experimental data: The Fisher randomization test" by D. Basu. *Journal of the American Statistical*

Association, 75, 591-93.

Salmon, W. C. (1984). *Scientific explanation and the causal structure of the world.* Princeton, N. J: Princeton University Press.

Simon, H. A. (1954). Spurious correlation: A causal interpretation. *Journal of the American Statistical Association*, 49, 467-492.

Smith, J. Q. (1988). Models, optimal decisions and influence diagrams. In J. M. Bernardo, M. H. DeGroot, D. V. Lindley & A. F. M. Smith (Eds.), *Bayesian statistics 3*, (pp. 765-776). Oxford: Oxford University Press.

Sobel, M. E. (1981). *Lifestyle and social structure: Concepts, definitions, analyses.* New York: Academic Press.

Sobel, M. E. (1990). Effect analysis and causation in linear structural equation models. *Psychometrika*, 55, 495-515.

Sobel, M. E. (1994a). Causal inference in artificial intelligence. In P. Cheeseman & R. W. Oldford (Eds.), *Selecting models from data*, (pp. 183-196). New York: Springer-Verlag.

Sobel, M. E. (1994b). Causal inference in latent variable models. In A. von Eye & C. C. Clogg (Eds.), *Latent Variables Analysis*, (pp. 3-35). Thousand Oaks, CA: Sage.

Sobel, M. E. (1995). Causal inference in the social and behavioral sciences. In G. Arminger, C. C. Clogg & M. E. Sobel (Eds.), (1995). *Handbook of statistical modeling for the social and behavioral sciences.* (pp. 1-38). New York: Plenum.

Sobel, M. E., & Arminger, G. (1986). Platonic and operational true scores in covariance structure analysis: An invited comment on Bielby's "Arbitrary metrics in multiple indicator models of latent variables". *Sociological Methods and Research*, 15, 44-58.

Suppes, P. (1970). *A probabilistic theory of causality.* Amsterdam: North-Holland.

Suppes, P. (1984). *Probabilistic metaphysics.* Oxford: Basil Blackwell.

Suppes, P., & Zanotti, M. (1981). When are probabilistic explanations possible? *Synthese*, 48, 191-99.

Sutcliffe, J. P. (1965). A probability model for errors of classification. I: General considerations. *Psychometrika*, 30, 73-96.

Wheaton, B., Muthén, B., Alwin, D. F., & G. F. Summers. (1977). Assessing reliability and stability in panel models. In D. R. Heise (Ed.), *Sociological methodology, 1977.* (pp. 84-135). San Francisco: Jossey-Bass.

Whittaker, J. (1990). *Graphical models in applied multivariate statistics.* Chichester: Wiley.

ON THE IDENTIFICATION OF NONPARAMETRIC STRUCTURAL MODELS

Judea Pearl
Cognitive Systems Laboratory
Computer Science Department
University of California, Los Angeles, CA 90024

Abstract

In this paper we study the identification of nonparametric models, that is, models in which both the functional forms of the equations and the probability distributions of the disturbances remain unspecified. Identifiability in such models does not mean uniqueness of structural parameters but rather uniqueness of policy-related predictions that such parameters would normally support.

We provide sufficient and necessary conditions for identifying predictions of the type "Find the distribution of Y, assuming that X is controlled by external intervention," where Y and X are arbitrary variables of interest. Whenever identifiable, such predictions can be expressed in closed algebraic form, in terms of observed distributions. We also show how the identifying criteria can be tested qualitatively using the graphical representation of the structural model, thus simplifying and generalizing the standard identifiability tests of linear models (e.g., rank and order). Finally, we provide meaningful and precise definitions of effect decomposition for both parametric and nonparametric models.

1 Introduction

In the literature on structural equation models, one usually asks whether or not certain or all of the model parameters are identified, that is, does $P(y;\theta_1) = P(y;\theta_2)$ imply $\theta_1 = \theta_2$?[1] Implicit in such questions is the premise that a model is more useful when its parameters

[1]The term "simultaneous equations" is often used in the literature, interchangeably with "structural equations". We prefer the latter and will define precisely what makes a set of equations "structural" (see Subsections 1.2 and 2.1). We will also take the liberty of using the symbol $P(\cdot)$ to denote the probability functions for both continuous and discrete variables.

are identified. This premise, coupled with the fact that parameter identification plays such a central role in modeling, indicates that analysts attribute to parameters an important metaprobabilistic meaning, one that cannot be expressed in distribution functions.

Indeed, if we adopt an orthodox statistical attitude[2] and pretend that the sole purpose of models is to provide a compact representation for distribution functions, then we should pay no attention to questions of identification. After all, if two distinct and equally parsimonious models are observationally equivalent (i.e., $P(y;\theta_1) = P(y;\theta_2)$ while $\theta_1 \neq \theta_2$) then they should yield the same statistical predictions and any one of the models can be taken as a working hypothesis; whether the model chosen is unique need not concern us. Most modelers, however, are driven by the understanding that the choice of parameters is not arbitrary but has empirical implications that lie outside the distribution functions. To such modelers, the issue of identifiability becomes one of crucial importance.

What are the empirical implications that give the parameters their distinct metaprobabilistic meaning and under what circumstances are those implications uncovered? The standard literature on simultaneous equation models is remarkably vague on this issue. The meaning of the parameters is sometimes described informally as "telling us how data were generated" and sometimes as "measuring the average change in the response variable per unit change in the explanatory variable." Lacking formal definitions for the notions of "generate" and "change" leaves the meaning of the parameters ambiguous and is largely responsible for the long-standing confusion between structural parameters (e.g., path coefficients) and statistical parameters (e.g., regression coefficients).

The purpose of this paper is threefold. First, we will provide a formal definition for the empirical content of structural parameters in terms of control queries, that is, queries about the outcomes of hypothetical controlled experiments. Second, we will extend the notion of identifiability from parametric to nonparametric models by requiring that answers to such control queries, rather than the parameters per se, be identified uniquely from the observed data. In this way, we capture the intent and the ultimate purpose of parameter identification without ever dealing with parameters. Finally, we will devise mathematical procedures for testing whether identifiability holds in a given nonparametric model and show that, in many cases, identifiability can be tested by inspection of the topological features of the diagram associated with the model.

Before moving on to the formal part of the paper (Section 2), it seems appropriate to illustrate the agenda using a simple example.

1.1 Parametric vs. nonparametric models, an example

Consider the following set of structural equations:

$$X = f_1(U) \tag{1}$$
$$Z = f_2(X, V) \tag{2}$$
$$Y = f_3(Z, U, W) \tag{3}$$

where X, Z, and Y are observed variables, f_1, f_2, and f_3 are unknown arbitrary functions, and U, V, and W are unobservables that we can regard either as latent variables or as

[2]Unfortunately, most statisticians and some social scientists still adhere to this attitude.

disturbances. For the sake of this discussion, we will assume that U, V, and W are mutually independent and arbitrarily distributed. Such a set of equations would obtain, for example, when the modeler is not willing to commit to any particular functional form but feels strongly about the qualitative nature of the data-generating process – for example, that X is a factor determining Z, that X and Y are influenced by some common factor U, and so on. Graphically, these influences can be represented by the path diagram of Figure 1. Note that the arcs in Figure 1 should be labeled with the functions themselves, not with coefficients as in traditional path analysis where all relationships are assumed linear.

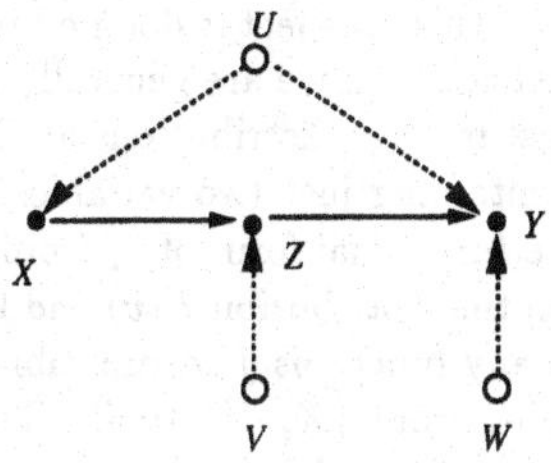

Figure 1:
Path diagram corresponding to Eqs. (1)-(3), where $\{X, Z, Y\}$ are observed and $\{U, V, W\}$ are unobserved.

The problem we pose is as follows:
We have drawn a long stream of independent samples of the process defined by Eqs. (1)-(3) and have recorded the values of the observed quantities X, Z, and Y, and we now wish to estimate the unspecified quantities of the model to the greatest extent possible.

To sharpen the scope of our problem, let us consider its solution in the case of a linear version of the model, which is given by[3]

$$X = U \tag{4}$$
$$Z = aX + V \tag{5}$$
$$Y = bZ + cU + W \tag{6}$$

with $\{U, V, W\}$ uncorrelated, zero-mean disturbances. It is not hard to show that all three parameters, a, b and c, can be determined uniquely from the correlations among the observed quantities X, Z and Y. In particular, multiplying Eq. (5) by X and taking expectations gives

$$a = \frac{E(XZ)}{E(X^2)} \tag{7}$$

Further multiplying Eq. (6) by X, then by Y, taking expectations and solving, gives

[3]An equivalent version of this model would obtain by eliminating cU from the equation of Y and allowing U and W to be correlated.

$$b = \frac{E^2(XY) - E(Y^2)\,E(X^2)}{E(XZ)\,E(XY) - E(ZY)\,E(X^2)} \tag{8}$$

$$c = \frac{E(XY)\,E(ZY) - E(XY)E(Y^2)}{E(X^2)\,E(ZY) - E(XY)\,E(XZ)} \tag{9}$$

Thus we see that, given the right set of assumptions on the disturbances, together with the parametric form of the equations, it is possible to estimate all model parameters from the observed distribution; such models are called "identifiable."

Returning to our nonparametric model of Eqs. (1)-(3), a natural generalization would be to require that, for the model to be identifiable, the functions $\{f_1, f_2, f_3\}$ be determined uniquely from the data. However, this prospect is doomed to failure from the start. When the equations are in nonparametric form, we are generally unable to identify the form of the functions, even when the disturbance distributions are known precisely. Consider the simplest case of an equation containing just two variables, for example, $Z = f_2(X,\ V)$. We are unable to determine the functional form of f_2 from the joint distribution $P(x, z)$ of X and Z, even if we are given the distribution $P(v)$ and know that V is independent of X. In other words, there are many functions f_2 compatible with the given distributions, each defining a different mapping from $\{X,\ V\}$ to Z. Thus, it might appear that the problem of identifying nonparametric models is hopeless and that nothing useful can be inferred from such loosely specified model as the one given in Eqs. (1)-(3).

However, parametric and functional identification is not an end in itself, even in linear models, but rather serves to answer practical questions of prediction and control. Therefore, the right question to ask is not whether the data permit us to identify the form of the equations but rather whether the data can constrain the equations to the extent of providing unambiguous answers to questions of interest, of the kind answered by traditional parametric models.

If the model is used merely for probabilistic prediction (i.e., to determine the probabilities of some variables given a set of observations on other variables), then such predictions can be estimated directly from the observed distributions. Moreover, if dimensionality reduction is needed (e.g., to improve estimation accuracy) these distributions can be encoded in a variety of simultaneous equation models, all of the same dimensionality. For example, the correlations among X, Y and Z in the linear model, M, of Eqs. (4)-(6) might as well be represented by the model M':

$$X = U \tag{10}$$

$$Z = a'X + V \tag{11}$$

$$Y = b'Z + dX + W \tag{12}$$

which is as compact as Eqs. (4)-(6). Although obviously the choice is not unique, it is nevertheless compatible with the observations and, upon setting $a' = a$, $b' = b$ and $d = c$, will yield the same probabilistic predictions as would the model of Eqs. (4)-(6) . Still, when viewed as data-generating mechanisms, the two models are clearly not equivalent; each tells a different story of the processes generating X, Y and Z and, naturally, each predicts different changes that would result from subjecting these processes to external interventions.

1.2 Causal effects: The structural interpretation of simultaneous equation models

The difference between the two models above illustrates precisely where the structural reading of simultaneous equation models comes into play.[4] Model M', defined by Eqs. (10)-(12), proclaims X to be a direct participant in the process which determines the value of Y, while model M, defined by Eqs. (4)-(6), views X as an indirect factor; its effect on Y is mediated by Z. This difference is not manifested in the data but only in the way the data would change in response to outside interventions. For example, suppose we wish to predict the expectation of Y after we intervene and fix the value of X to some constant x, denoted $E(Y|set(X = x))$. Substituting $X = x$ in Eq. (11) and (12), model M' yields

$$E[Y|set(X = x)] = E[b'a'x + b'V + dx + W] \quad (13)$$
$$= (b'a' + d)x \quad (14)$$

while model M yields

$$E[Y|set(X = x)] = E[bax + bV + cU + W] \quad (15)$$
$$= bax \quad (16)$$

Equating $a' = a$, $b' = b$ and $d = c$ (to match the data, as in Eqs. (7)-(9)), we see clearly that the two models assign different magnitudes to the (total) effect of X on Y; model M predicts that a unit change in x will change $E(Y)$ by an amount ba, while model M' puts this amount at $ba + c$.

At this point, it is natural to ask whether we should not substitute the constant x for U in Eq. (6) prior to taking expectations in Eq. (15). If we permit the substitution of Eq. (5) in Eq. (6), so the argument goes, why not substitute Eq. (4) as well? After all, there is no harm in upholding a mathematical equality, $U = X = x$, which the modeler deems valid. This argument is fallacious. Structural equations are not meant to be treated as immutable mathematical equalities. Rather, they are introduced into the model to describe an equilibrium condition, only to be violated when that equilibrium is perturbed by outside interventions. The power of structural models is that they also encode the information necessary for determining the new equilibrium. If the intervention consists merely of holding X constant, at x, then the equation $X = U$, which represents the pre-intervention process determining X, should be overruled and replaced with the equation $X = x$. The solution to the new set of equations then represents the new equilibrium. Thus, the essential characteristic of structural equations, which sets them apart from ordinary mathematical equations is that they do not stand for one, but for many sets of equations, each corresponding to a subset of equations taken from the original model. Every such subset represents some physical reality, one that is configured by overruling all but the processes corresponding to the selected equations.

[4] I ask the reader to bear with me as I review concepts that might seem obvious. The reviewer of this paper has commented: "Nor is it clear to me that structural equations do anything more than summarize distributions" and has proposed specifically that the model in Eqs. (4)-(6) be replaced with that of Eqs. (10)-(12). To me, these comments suggest that even renowned scholars and experienced modelers do not find the interpretation of structural equations to be as obvious as one would expect; neither have I found these issues addressed forthrightly in the standard literature. Subsection 3.4 discusses some of the prevailing confusions and the reasons that they have not been resolved thus far.

Taking the stand that the primary value of structural equations lies not in summarizing distribution functions but in encoding causal information for predicting the effect of interventions [Haavelmo 1943, Marschak 1953], it is natural to view such predictions as the proper generalization of structural coefficients when dealing with nonparametric model. For example, the proper generalization of the coefficient b in the linear model M would be the answer to the control query: "What would be the change in the expected value of Y if we were to intervene and change the value of Z from z to $z+1$," which is different, of course, from the observational query "What would be the difference in the expected value of Y if we were to find Z at level $z+1$, instead of z." Observational queries can be answered directly from the joint distribution $P(x, y, z)$, while control queries require causal information, such as the one encoded in structural equations, as well. To distinguish between the two types of queries, we use the "hat" symbol ($\hat{}$) to indicate externally controlled quantities. For example, we write

$$E(Y|\hat{x}) = E[Y|set(X = x)] \tag{17}$$

for the controlled expectation and

$$E(Y|x) = E(Y|X = x) \tag{18}$$

for the standard conditional expectation. The inequality $E(Y|\hat{x}) \neq E(Y|x)$ can easily be seen in the model of Eqs. (4)-(6), where $E(Y|\hat{x}) = abx$ while $E(Y|x) = (ab + c)x$. Indeed, the passive observation of $X = x$ should not violate any of the equations, and would justify substituting Eqs. (4) and (5) in (6) before taking the expectation.

In the case of linear models, the answers to questions of direct control are encoded in the so-called "path coefficients" or "structural coefficients," and these can be used to derive the total effect of any variable on another. For example, the value of $E(Y|\hat{x})$ in the model defined by Eqs. (4)-(6) is abx, i.e., x times the product of path coefficients along the path $X \longrightarrow Z \longrightarrow Y$. In the nonparametric case, the computation of $E(Y|\hat{x})$ would naturally be more complicated, even when we know the functions f_1, f_2, and f_3. It is nevertheless well defined, and requires the solution (for the expectation of Y) of a modified set of equations in which f_1 is "wiped out" and X is replaced by the constant x:

$$Z = f_2(x, V) \tag{19}$$

$$Y = f_3(Z, U, W) \tag{20}$$

Thus, the computation of $E(Y|\hat{x})$ requires the evaluation of

$$E(Y|\hat{x}) = E\{f_3[f_2(x, V), U, W]\}$$

where the expectation is taken over U, V, and W. This computation will be carried out in Section (2.3). Similar modifications of the model are required for the computation of $E(Z|\hat{x})$, $E(X|\hat{z})$, or $E(X|\hat{y})$, and can easily be shown to yield $E(Z|\hat{x}) = E(Z|x)$, $E(X|\hat{z}) = E(X)$, and $E(X|\hat{y}) = E(X)$, respectively.

What then would be an appropriate definition of "identifiability" for nonparametric models? Consistent with our focus on control queries, a reasonable definition of identifiability is that answers to such queries are *unique*. Accordingly, we will define a model to be *identifiable* if there exists a consistent estimate for every control query of the type "Find $P(r|\hat{s}) = P[R = r|set(S = s)]$," where R and S are subsets of observables and r and s are any realization of these variables. The set of probabilities $P(r|\hat{s})$ is called the "causal effect" of S on R, as it describes how the distribution of R varies when S is changed by external control.[5] Naturally, we should allow for some queries to be identifiable while the system as a whole is not. Hence, we say that $P(r|\hat{s})$ is identifiable in model M if every choice of model parameters (i.e., the functional forms and the distributions) that is compatible with the observed distribution P would yield the same value for $P(r|\hat{s})$.

For example, we might inquire whether the model defined by Eqs. (1)-(3) is identifiable. The answer is yes; we will see that this model permits the identification of all control queries. For example, the methods developed in Section 2 will enable one to conclude immediately that:

1. $P(x|\hat{y}, \hat{z}) = P(x)$,
   consistent with the intuition that consequences can have no effect on their causes; and

2. $P(z|\hat{x}) = P(z|x)$,
   because V is independent of X, hence Z is not confounded with X; and

3. $P(y|\hat{z}) = \sum_x P(y|z, x)P(x)$,
   because x is an appropriate covariate for adjustment; and

4. $P(y|\hat{x}) = \sum_z P(z|x) \sum_{x'} P(y|x', z)P(x')$,
   for reasons to be explained in Section 2.

These answers are unique because all terms on the right-hand sides are functions of the observable distribution $P(x, y, z)$. Hence, any choice of functions (f_1, f_2, and f_3) and distributions (of U, V, and W) compatible with the observed distribution P would necessarily yield the same answers to the control queries above.

Remarkably, many aspects of nonparametric identification, including tests for deciding whether a given control query is identifiable, as well as formulas for estimating such queries, can be determined graphically, almost by inspection, from the path diagram. These aspects will be developed and demonstrated in the body of the paper.

2 Computing Causal Effects

2.1 Definitions and Notation

2.1.1 Models, Graphs, and Theories

We consider models consisting of a set of n (recursive) equations

$$X_i = f_i(X_1, X_2, ..., X_{i-1}; U_1, ..., U_m), \; i = 1, 2, ..., n, \tag{21}$$

[5] Technically, the adjective "causal" is redundant. It merely serves to emphasize, however, that the changes in S are enforced by external control, and do not represent stochastic variations in the observed value of S. The phrase "the effect of S on R" has improperly been applied to $P(r|s)$, in which s stands for uncontrolled statistical observations.

where $X_1, ..., X_n$ are observed variables, and $U_1, ..., U_m$ are unobserved (or latent) disturbances.[6] The f_i are unspecified deterministic functions with restricted sets of arguments, and the distribution of the disturbances may be constrained by independence restrictions but is otherwise unspecified.

Restrictions on the arguments of the equations[7] can be represented by a directed graph G in which each node corresponds to an observed variable and an arrow from node X_i to node X_j indicates that X_i is an argument of f_j. The restrictions on the dependencies among the U variables will also be represented graphically, by adding a "confounding path" (a dashed curved arc with double arrows) between any two variables X_i and X_j whenever a dependency exists between the U variables in f_i and those in f_j. Thus, for example, the model described by Eqs. (1)-(3) is completely specified by the graph of Figure 2. Unlike the path diagram of Figure 1, G does not represent the disturbances

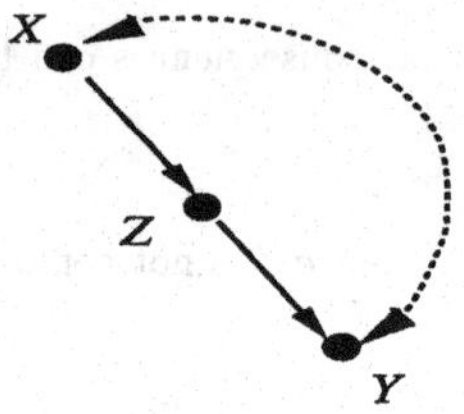

Figure 2:
A graph G, representing the restrictions specified in the nonparametric model of Eqs. (1)-(3).

explicitly; only the dependencies induced by these disturbances are represented. Each confounding path may represent several unobserved disturbances common to a given pair of equations. In most cases, modelers prefer to specify the induced dependencies directly without making the disturbances explicit. However, in order to read off the dependencies embodied in a given graph, it sometimes may be convenient to restore the U variables. In such cases, we will add a "dummy" root node (hollow circle) for each dashed arc, as shown by the node marked U in Figure 1. A theorem by [Pearl & Verma 1991] states that such "dummy" root nodes can faithfully represent any pattern of dependencies among any set of latent variables.

The structural model M defined by Eq. (21) (equivalently, by the corresponding graph $G(M)$) delineates a set of *grounded models* which we call *theories*.[8] Each theory $T =< \{f_i\}, P(u) >$ in M corresponds to a specific choice of function f_i and a specific choice of disturbance distribution $P(u) = P(u_1, ..., u_m)$, both satisfying the restrictions

[6]The recursive nature of Eq. (21) corresponds to a lower triangular matrix B in linear models [Bollen 1989] and therefore excludes feedback mechanisms. This restriction is not essential to the discussion of the basic concepts, though it simplifies the test for identifiability. The restriction that all unobserved variables be exogenous (i.e., do not appear on the left-hand side of any equation) can easily be relaxed: endogenous latent variables can always be eliminated by substitutions, thus restoring the form of Eq. (21).

[7]In linear models, these correspond to the "zero-coefficient" restrictions, while independencies among the U's are specified by zero entries in the covariance matrix of the disturbances.

[8]Koopmann and Reiersol (1950) used the term "structure" for our "theory."

imposed by M. For each theory T of M, there is a corresponding unique probability distribution $P_T(x) = P_T(x_1, ..., x_n)$, which we say to be "generated" by T.

Definition 1 *We say that* $P(x)$ *is* compatible *with* M *iff there exists a theory* T *of* M *that generates* $P(x)$, *i.e.,*

$$P(x) = P_T(x)$$

A model M *is said to be* universal *if it is compatible with every arbitrary* $P(x)$; *otherwise, it is said to be* falsifiable. □

Clearly, every model whose corresponding graph is complete (i.e., every pair is connected by an arrow) is universal, since such a model can generate any given $P(x)$, using mutually independent disturbances. Figure 2 is an example of a universal model which will become falsifiable upon removing any of the arcs.

2.1.2 Queries and Identifiability

A query q is any quantity that can be computed from a given theory; i.e., a functional of T. For example, the queries

$$\begin{aligned} q_1: &\quad f_2(X = 1, V = 3.06) = ? \\ q_2: &\quad P(U = 1|Y = 0.8) = ? \\ q_3: &\quad P(X = 1|Y = 3) = ? \\ q_4: &\quad E_u[P(Y = 1|X = 1, u) = ? \end{aligned} \tag{22}$$

can be computed from any theory of the structural model described in Eqs. (1)-(3), because, once we choose the functions $\{f_1,\ f_2,\ f_3\}$ and the distribution $P(u, v, w)$, the answers to each of these queries is well defined. Note, moreover, that the answers to queries $q1$ and $q2$ depend critically on the specific choice of theory, while $q3$ depends solely on the distribution $P(x, y, z)$. Query $q4$, which the reader may recognize as $P(Y = 1|set(X = 1))$ (Section 1), appears at first glance to depend on the choice of theory. We will see, however, that, by virtue of the structural restrictions communicated by Figure 2, query q_4 will have the same answer in all theories that generate a given distribution $P(x, y, z)$. This motivates the following definition of identifiability:

Definition 2 *(identifiability) A query* q *is said to be* identifiable *in a model* M *iff, for any two theories* T_1 *and* T_2 *of* M,

$$q(T_1) = q(T_2), \text{ whenever } P_{T_1}(x) = P_{T_2}(x) \text{ and } P_{T_1}(x) > 0$$

A model M *is said to be identifiable relative to a set* Q *of queries if every member of* Q *is identifiable in* M. □

In other words, a model M is identifiable relative to a set Q of queries if every query in Q can be computed uniquely from the pair $\{M, P\}$, where P is any positive distribution over the observables that is compatible with M.

Technically, the reason for restricting the observed distributions to positive distributions is to avoid conditioning on events with zero probabilities. Conceptually, positivity ensures that each function is perturbed by some stochastic disturbance; these disturbances act like instrumental variables, or randomized experiments conducted by nature, in that they help reveal the strength of causal effects of some variables while others are kept constant.

It is clear, from the definition above, that every model is trivially identifiable relative to queries, such as q_3, which are addressed to the observed distribution $P(x)$. As discussed in the introduction, the focus of this paper is the set of *control queries*, like q_4, that are the primary (yet often forgotten) reason we use structural modeling [Haavelmo 1943].

2.1.3 Control Queries

Of special interest to us will be the set Q_2 of *pairwise* control queries, in which each query is of the type "Find the distribution of X_j given that X_i is held fixed at x_i", where i and j are arbitrary. Answers to such queries are the nonparametric analogs to the so-called "causal effects" or "total effects" in linear models, and in these models the answers can be computed directly from the structural coefficients. To answer such queries, we need to formalize the notion of "holding fixed" within the general framework of nonparametric structural models.

Given a model M and a subset S of variables, define a submodel M_s of M as the set of equations that results if the $|S|$ equations corresponding to the variables in S are deleted from M and $S = s$ is substituted in the remaining equations. For example, the model specified in Eqs. (19)-(20) is the submodel M_x of the model in Eqs. (1)-(3). The theories delineated by M_s will be denoted by T_s.

Definition 3 *(control queries) A control query $q = P(r|\hat{s})$ (read: the probability of $R = r$ given that S is* held fixed *at s) is a functional of the theories of M, defined by*

$$P_T(r|\hat{s}) = P_{T_s}(r).$$

In other words, the value of $P(r|\hat{s})$ in theory T is given by the probability $P(r)$ induced by the subtheory T_s of T. □

The notion of subtheories reflects the understanding that external interventions perturb the normal causal influences as represented by the structural equations. In particular, the primitive intervention "holding X fixed" has sharp, local effect on those mechanisms, that is, it totally neutralizes X from its normal influences and places it under a new influence (given by the intervention), while keeping all other influences unperturbed. This interpretation of control queries is an integral part of viewing structural equations as representing a set of *autonomous*, *stable*, or *invariant* mechanisms—a notion going back to [Frisch 1938, Haavelmo 1943, Marschak 1953] and later expanded by [Simon 1977] and [Goldberger 1973].[9] An explicit translation of interventions to "wiping out" equations from the model was first proposed by [Strotz & Wold 1960] and later used in [Fisher 1970] and [Sobel 1990] for defining effects decomposition (see Section 3.4). Formal graphical accounts of this notion are given in [Spirtes et al. 1993] and [Pearl 1993].

[9] As discussed briefly in Section 1, the notion of invariance, and its operational derivative of Definition 3 is, in fact, the defining feature of structural equations. We, therefore, depart from the views of [Sobel 1990] and others and do not refer to invariance as an assumption that requires further justification or judgment. In other words, invariance is what the investigator must already have in mind when he/she specifies the arguments of each function f_i in the model.

2.1.4 Graphs, Conditional Independence, and *d*-Separation

In this subsection, we review the properties of directed acyclic graphs (DAGs) as carriers of conditional independence information [Pearl 1988]. Readers familiar with this aspect of DAGs are advised to skip to Section 2.2.

Given a DAG G and a joint distribution P over a set $V = \{X_1, ..., X_n\}$ of variables, we say that G *represents* P if there is a one-to-one correspondence between the variables in X and the nodes of G, such that P admits the product decomposition

$$P(x_1, ..., x_n) = \prod_i P(x_i \mid pa_i) \tag{23}$$

where pa_i are the values of the direct predecessors (called *parents*), PA_i, of X_i in G. For example, the DAG in Figure 3 induces the decomposition

$$P(x_1, x_2, x_3, x_4, x_5) = P(x_1)\ P(x_2|x_1)\ P(x_3|x_1)\ P(x_4|x_2, x_3)\ P(x_5|x_4) \tag{24}$$

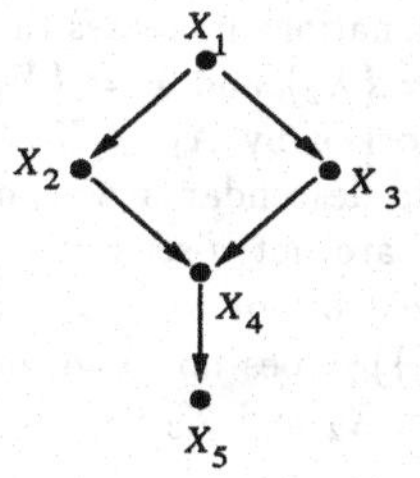

Figure 3:
A typical directed acyclic graph (DAG) representing the decomposition of Eq. (24).

A convenient way of characterizing the set of distributions represented by a DAG G is to list the set of (conditional) independencies that each such distribution must satisfy. Clearly, the decomposition in Eq. (23) implies (using the chain rule) that, given its parent set PA_i, each variable X_i is conditionally independent of all its other predecessors $\{X_1, X_2, ..., X_{i-1}\} \backslash PA_i$. We call this set of independencies *Markovian*, because it reflects the Markovian condition for state transitions: Each state is rendered independent of the past, given its immediately preceding state. However, the decomposition of Eq. (23) implies additional, less obvious independencies which can be read off the DAG by using a graphical criterion called *d*-separation [Pearl 1988]. To test whether X is independent of Y given Z in the distributions represented by G, we need to examine G and test whether the nodes corresponding to variables Z *d*-separate all paths from nodes in X to nodes in Y. By *path* we mean a sequence of consecutive edges (of any directionality) in the DAG.

Definition 4 (d-separation) *A path p is said to be d-separated (or blocked) by a set of nodes Z iff:*

(i) p contains a chain $i \longrightarrow j \longrightarrow k$ *or a fork* $i \longleftarrow j \longrightarrow k$ *such that the middle node* j *is in* Z*, or,*

(ii) p contains an inverted fork $i \longrightarrow j \longleftarrow k$ *such that neither the middle node* j *nor any of its descendants (in* G*) are in* Z*.*

If X, Y*, and* Z *are three disjoint subsets of nodes in a DAG* G*, then* Z *is said to d-separate* X *from* Y*, denoted* $(X \perp\!\!\!\perp Y)_G$*, iff* Z *d-separates every path from a node in* X *to a node in* Y *.*

The intuition behind *d*-separation is simple: In chains $X \rightarrow Z \rightarrow Y$ and forks $X \leftarrow Z \rightarrow Y$, the two extreme variables are dependent (marginally) but become independent of each other (i.e., blocked) once we know the middle variable. Inverted forks $X \rightarrow Z \leftarrow Y$ act the opposite way; the two extreme variables are independent (marginally) and become dependent (i.e., unblocked) once the value of the middle variable (i.e., the common effect) or any of its descendants is known. For example, finding that the pavement is wet or slippery (see Figure 1) renders Rain and Sprinkler dependent, because refuting one of these explanations increases the probability of the other.

In Figure 1, for example, $X = \{X_2\}$ and $Y = \{X_3\}$ are *d*-separated by $Z = \{X_1\}$; the path $X_2 \leftarrow X_1 \rightarrow X_3$ is blocked by $X_1 \in Z$, while the path $X_2 \rightarrow X_4 \leftarrow X_3$ is blocked because X_4 and all its descendants are outside Z. Thus $(X_2 \perp\!\!\!\perp X_3|X_1)_G$ holds in G. However, X and Y are not *d*-separated by $Z' = \{X_1, X_5\}$, because the path $X_2 \rightarrow X_4 \leftarrow X_3$ is unblocked by virtue of X_5, a descendant of X_4, being in Z'. Consequently, $(X_2 \perp\!\!\!\perp X_3|\{X_1, X_5\})_G$ does not hold; in words, learning the value of the consequence X_5 renders its causes X_2 and X_3 dependent, as if a pathway were opened along the arrows converging at X_4.

Theorem 1 [Verma & Pearl 1990, Geiger et al. 1990]. *For any three disjoint subsets of nodes* (X, Y, Z) *in a DAG* G*,* Z *d-separates* X *from* Y *in* G *implies that* X *is independent of* Y *conditional on* Z *in every probability distribution represented by* G*.*

Thus, a DAG can be viewed as an efficient scheme for representing Markovian independence assumptions and for deducing and displaying all the logical consequences of such assumptions. Note that the precise ordering of the nodes does not enter into the *d*-separation criterion; it is only the topology of the graph that determines the set of independencies that the probability P must satisfy.

An important property that follows from the *d*-separation characterization is a criterion for determining whether two given DAGs are observationally equivalent, that is, whether every probability distribution that is represented by one of the DAGs is also represented by the other.

Theorem 2 [Verma & Pearl 1990] *Two DAGs are observationally equivalent iff they have the same sets of edges and the same sets of v-structures, that is, two converging arrows whose tails are not connected by an arrow.*

Observational equivalence places a limit on our ability to infer the directionality of the links directionality from probabilities alone. For example, reversing the direction of the arrow between X_1 and X_2 in Figure 1 does not introduce any new v-structure. Therefore, this reversal yields an observationally equivalent DAGs, and the directionality of the link $X_1 \to X_2$ cannot be determined from probabilistic information. The arrows $X_2 \to X_4$ and $X_4 \to X_5$, however, are of different nature; there is no way of reversing their directionality without creating a new v-structure. Thus, we see that some probability functions P can constrain the directionality of some arrows in their DAG representation.

Additional properties of DAGs and their applications to evidential reasoning are discussed in [Geiger 1990, Lauritzen & Spiegelhalter 1988, Spiegelhalter et al. 1993, Pearl 1988, Pearl 1993, Pearl et al. 1990].

2.2 A Causal Calculus

This subsection establishes a set of sound (and possibly complete) inference rules by which probabilistic sentences involving actions and observations can be transformed to other such sentences, thus providing a syntactic method for deriving (or verifying) claims about actions and observations. Given the pair $< M, P >$, our main problem will be to facilitate the syntactic derivation of expressions of the form $P(x_j|set(x_i))$ from standard probability expressions.

Let X, Y, and Z be arbitrary disjoint sets of nodes in a DAG G. We denote by $(X \perp\!\!\!\perp Y|Z)_G$, the proposition "$Z$ d-separates X from Y in G" (see Definition 4). We denote by $G_{\overline{X}}$ ($G_{\underline{X}}$, respectively) the graph obtained by deleting from G all arrows pointing to (emerging from, respectively) nodes in X. In dealing with expressions involving both observed and fixed variables, we will use $P(y|\hat{x}, z)$, where the $\hat{}$ symbol identifies the variables that are kept constant externally. In words, the expression $P(y|\hat{x}, z)$ will stand for the probability of $Y = y$ given that $Z = z$ is observed and X is held constant at x.

Armed with this notation, we formulate the three basic inference rules of our calculus in the following theorem [Pearl 1995a]:

Theorem 3 *Let G be a DAG characterizing a structural model M, and let P be a distribution generated by some theory of M. Then, for any disjoint sets of variables X, Y, Z, and W, we have:*

Rule 1 *Insertion/deletion of observations*

$$P(y|\hat{x}, z, w) = P(y|\hat{x}, w) \text{ if } (Y \perp\!\!\!\perp Z|X, W)_{G_{\overline{X}}} \tag{25}$$

Rule 2 *Action/observation exchange*

$$P(y|\hat{x}, \hat{z}, w) = P(y|\hat{x}, z, w) \text{ if } (Y \perp\!\!\!\perp Z|X, W)_{G_{\overline{X}\underline{Z}}} \tag{26}$$

Rule 3 *Insertion/deletion of actions*

$$P(y|\hat{x}, \hat{z}, w) = P(y|\hat{x}, w) \text{ if } (Y \perp\!\!\!\perp Z|X, W)_{G_{\overline{X},\, \overline{Z(W)}}} \tag{27}$$

where $Z(W)$ is the set of Z-nodes that are not ancestors of any W-node in $G_{\overline{X}}$.

□

Each of the inference rules above can be proven [Pearl 1995a] from the basic interpretation of the "set(x)" operation as a replacement of the causal mechanism that connects X to its parents prior to the operation with a new mechanism $X = x$ introduced by the intervention (Definition 3). This results in a submodel M_x which is characterized by the subgraph $G_{\overline{X}}$ (named "manipulated graph" in [Spirtes et al. 1993]).

Rule 1 reaffirms d-separation as a valid test for Bayesian conditional independence in the distribution determined by the intervention $set(X = x)$, hence the graph $G_{\overline{X}}$.

Rule 2 provides conditions for an external intervention $set(Z = z)$ to have the same effect on Y as the passive observation $Z = z$. The condition amounts to $\{X \cup W\}$ blocking all back-door paths from Z to Y (in $G_{\overline{X}}$), since $G_{\overline{X}\underline{Z}}$ retains all (and only) such paths. Rule 2 is equivalent to the "back-door criterion"[10] of [Pearl 1993] and can also be derived from Theorem 7.1 in [Spirtes et al. 1993].

Rule 3 provides conditions for introducing (or deleting) an external intervention $set(Z = z)$ without affecting the probability of $Y = y$. The validity of this rule stems, again, from simulating the intervention $set(Z = z)$ by severing all relations between Z and its parents (hence the graph $G_{\overline{XZ}}$).

Corollary 1 *A query q: $P(y_1, ..., y_k|\hat{x}_1, ..., \hat{x}_m)$ is identifiable in model M if there exists a sequence of inference rules which transforms q into a standard (i.e., hat-free) probability expression.*

2.3 Computing Causal Effects: An Example

We will now demonstrate how these inference rules can be used to evaluate all control queries for the structural model specified in Eqs. (1)-(3). The graphical characterization of this model is given by the DAG G of Figure 4, which is identical to that of Figure 2 save for the explicit representation of the unobserved variable U. We will see that this structure permits us to quantify, using the causal calculus of Section 2.2, the effect of every action on every set of observed variables. Our task amounts to reducing expressions involving actions to those involving only observations, that is, to eliminating the "hat" symbol ($\hat{}$) from the query expressions.

The applicability of the inference rules in Theorem 3 requires that the d-separation conditions holds in various subgraphs of G, and the structure of each subgraph varies with the expressions to be manipulated. Figure 4 displays the subgraphs that will be needed for the derivations that follow.

Task-1, compute $P(z|\hat{x})$

This task can be accomplished in one step, since G satisfies the applicability condition

[10]The back-door criterion states that, if there exists a set S of observed variables which are nondescendants of X and which block every back-door path from X to Y (that is, paths ending with arrows pointing to X), then $P(y|\hat{x})$ is identifiable and is given by the formula

$$P(y|\hat{x}) = \sum_s P(y|x, s)P(s) \tag{28}$$

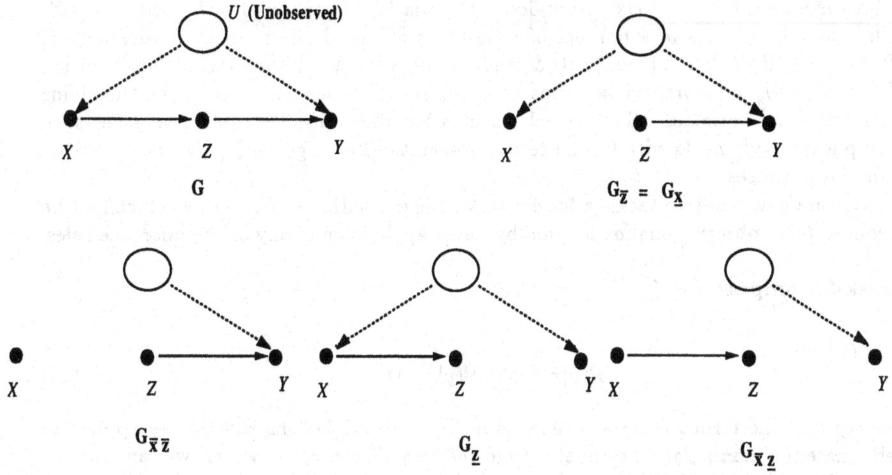

Figure 4:
Subgraphs of G used in the derivation of causal effects.

for Rule 2; namely, $X \perp\!\!\!\perp Z$ in $G_{\underline{X}}$ (because the path $X \leftarrow U \rightarrow Y \leftarrow Z$ is blocked by the collider at Y) and we can write

$$P(z|\hat{x}) = P(z|x) \tag{29}$$

Task-2, compute $P(y|\hat{z})$
Here we cannot apply Rule 2 to replace $\hat{z}$ with z, because $G_{\underline{Z}}$ contains a path from Z to Y (a so-called "back-door" path). Naturally, we would like to "block" this path by "adjusting for" covariates (such as X) that reside on that path. Symbolically, the "adjustment" operation involves conditioning and summing over all values of X, as follows:

$$P(y|\hat{z}) = \sum_x P(y|x,\hat{z})P(x|\hat{z}) \tag{30}$$

We now have to deal with two expressions involving $\hat{z}$, $P(y|x,\hat{z})$ and $P(x|\hat{z})$. The latter can readily be reduced to an observational quantity by applying Rule 3 for action deletion:

$$P(x|\hat{z}) = P(x) \text{ if } (Z \perp\!\!\!\perp X)_{G_{\overline{Z}}} \tag{31}$$

noting that, indeed, X and Z are d-separated in $G_{\overline{Z}}$. (This can also be seen immediately from G: manipulating Z will have no effect on X.) To reduce the former, $P(y|x,\hat{z})$, we apply Rule 2, which yields

$$P(y|x,\hat{z}) = P(y|x,z) \text{ if } (Z \perp\!\!\!\perp Y|X)_{G_{\underline{Z}}} \tag{32}$$

and note that X d-separates Z from Y in $G_{\underline{Z}}$. This allows us to write Eq. (30) as

$$P(y|\hat{z}) = \sum_x P(y|x,z)P(x) = E_x P(y|x,z) \tag{33}$$

which is a special case of the "back-door" formula (see footnote 10, (28)) with $S = X$. This formula appears in a number of treatments of causal effects (e.g., [Rosenbaum & Rubin 1983, Rosenbaum 1989, Pratt & Schlaifer 1988]) in which the legitimizing condition $(Z \perp\!\!\!\perp Y|X)_{G_{\underline{Z}}}$ is expressed in terms of conditional-independence judgments involving counterfactual variables. The causal calculus facilitated by Theorem 3 replaces such complicated judgments with formal tests (d-separation) on a graph (G) which represents familiar processes.

We are now ready to tackle a harder task—the evaluation of $P(y|\hat{x})$, which cannot be reduced to an observational expression by direct application of any of the inference rules.

Task-3, compute $P(y|\hat{x})$

Writing

$$P(y|\hat{x}) = \sum_z P(y|z,\hat{x})P(z|\hat{x}) \tag{34}$$

we see that the term $P(z|\hat{x})$ was reduced in Eq. (29) but that no rule can be applied to eliminate the manipulation symbol ˆ from the term $P(y|z,\hat{x})$. However, we can add a ˆ symbol to this term via Rule 2

$$P(y|z,\hat{x}) = P(y|\hat{z},\hat{x}) \tag{35}$$

since Figure 3 shows

$$(Y \perp\!\!\!\perp Z|X)_{G_{\overline{X}\underline{Z}}}$$

We can now delete the action $\hat{x}$ from $P(y|\hat{z},\hat{x})$ using Rule 3, since $Y \perp\!\!\!\perp X|Z$ holds in $G_{\overline{XZ}}$. Thus, we have

$$P(y|z,\hat{x}) = P(y|\hat{z}) \tag{36}$$

which was calculated in Eq. (33). Substituting Eqs. (33), (36), and (29) back into Eq. (34) yields

$$P(y|\hat{x}) = \sum_z P(z|x) \sum_{x'} P(y|x',z)P(x') \tag{37}$$

Eq. (37) was named the "front-door" formula in [Pearl 1995b], as it involves a (nonstandard) adjustment for a variable (Z) that stands between the cause (X) and the effect (Y).

Task-4, compute $P(y,z|\hat{x})$

$$P(y,z|\hat{x}) = P(y|z,\hat{x})P(z|\hat{x})$$

The two terms on the right-hand side were derived in Eqs. (29) and (36), from which we obtain

$$\begin{aligned} P(y,z|\hat{x}) &= P(y|\hat{z})P(z|x) \\ &= P(z|x)\textstyle\sum_{x'} P(y|x',z)P(x') \end{aligned} \tag{38}$$

Task-5, compute $P(x,y|\hat{z})$

$$\begin{aligned} P(x,y|\hat{z}) &= P(y|x,\hat{z})P(x|\hat{z}) \\ &= P(y|x,z)P(x) \end{aligned} \tag{39}$$

The first term on the right is obtained by Rule 2 (licensed by $G_{\underline{Z}}$) and the second term, by Rule 3 (as in Eq. (31)).

3 Graphical Tests of Identifiability

In the example above, we were able to compute all expressions of the form $P(r|\hat{s})$ where R and S are subsets of observed variables. In general, this will not be the case. For example, there is no general way of computing $P(y|\hat{x})$ from the observed distribution whenever the causal model contains the bow-pattern shown in Figure 5, in which X and Y are connected by both a causal link and a confounding arc. A confounding arc represents the existence in the diagram of a back-door path that contains only unobserved variables and has no converging arrows. A bow-pattern represents an equation

$$Y = f_Y(X, U)$$

where U is unobserved and dependent on X. Such an equation does not permit the identification of causal effects since any portion of the observed dependence between X and Y may always be attributed to spurious dependencies mediated by U.

The presence of a bow-pattern prevents the identification of $P(y|\hat{x})$ even when it is found in the context of a larger graph, as in Figure 5 (b). This is in contrast to linear models, where the addition of an arc to a bow-pattern can render $P(y|\hat{x})$ identifiable. For example, if Y is related to X via a linear relation $Y = bX + U$, where U is a zero-mean disturbance possibly correlated with X, then $b \triangleq \frac{\partial}{\partial x} E(Y|\hat{x})$ is not identifiable. However, adding an arc $Z \rightarrow X$ to the structure (that is, finding a variable Z that is correlated with X but not with U) would facilitate the computation of b via the instrumental-variable formula [Bowden & Turkington 1984, Bollen 1989]:

$$b \triangleq \frac{\partial}{\partial x} E(Y|\hat{x}) = \frac{E(Y|z)}{E(X|z)} = \frac{R_{yz}}{R_{xz}} \tag{40}$$

In nonparametric models, adding an instrumental variable Z to a bow-pattern (Figure 5(b)) does not permit the identification of $P(y|\hat{x})$. This is a familiar problem in the analysis of clinical trials in which treatment assignment (Z) is randomized (hence, no link enters Z), but compliance is imperfect [Pearl 1995b]. The confounding arc between X and Y in Figure 5(b) represents unmeasurable factors which influence both subjects' choice of treatment (X) and subjects' response to treatment (Y). In such trials, it is not possible to obtain an unbiased estimate of the treatment effect $P(y|\hat{x})$ without making additional assumptions on the dependence between compliance and response, as is done, for example, by Angrist et al. (1993) and Imbens & Angrist (1994). While the added arc $Z \rightarrow X$ permits us to calculate bounds on $P(y|\hat{x})$ [Robins 1989, Section 1g],[Manski 1990, Balke & Pearl 1994], and the upper and lower bounds may even coincide for certain types of distributions $P(x, y, z)$ [Balke & Pearl 1993, Pearl 1995b] there is no way of computing $P(y|\hat{x})$ for every positive distribution $P(x, y, z)$, as required by Definition 2. It is interesting to note that the noncompliance model of Figure 9(b) is falsifiable whenever X is discrete, but has no testable implications when X is continuous [Pearl 1995c] .

A general feature of nonparametric models is that the addition of arcs to a causal diagram can impede, but never assist, the identification of causal effects. This is because such addition reduces the set of d-separation conditions carried by the diagram and, hence, if a causal effect derivation fails in the original diagram, it is bound to fail in the augmented diagram as well. Conversely, any causal effect derivation that succeeds in the augmented diagram (by a sequence of symbolic transformations, as in Corollary 1) would succeed in the original diagram.

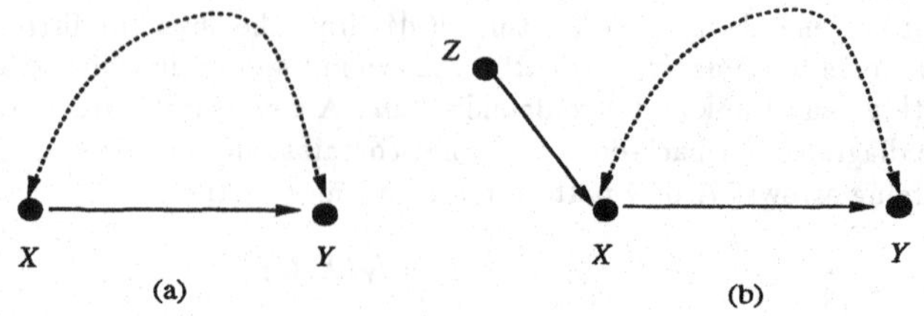

Figure 5:
(a) A bow-pattern: a confounding arc embracing a causal link $X \to Y$, thus preventing the identification of $P(y|\hat{x})$ even in the presence of an instrumental variable Z, as in (b).

Our ability to compute $P(y|\hat{x})$ for pairs (x, y) of singleton variables does not ensure our ability to compute joint distributions, such as $P(y_1, y_2|\hat{x})$. Figure 6, for example, shows a causal diagram where both $P(z_1|\hat{x})$ and $P(z_2|\hat{x})$ are computable, but $P(z_1, z_2|\hat{x})$ is not. Consequently, we cannot compute $P(y|\hat{x})$. Interestingly, the graph shown in Figure 6 is the smallest graph that does not contain the bow-pattern of Figure 5 and still presents an uncomputable causal effect.

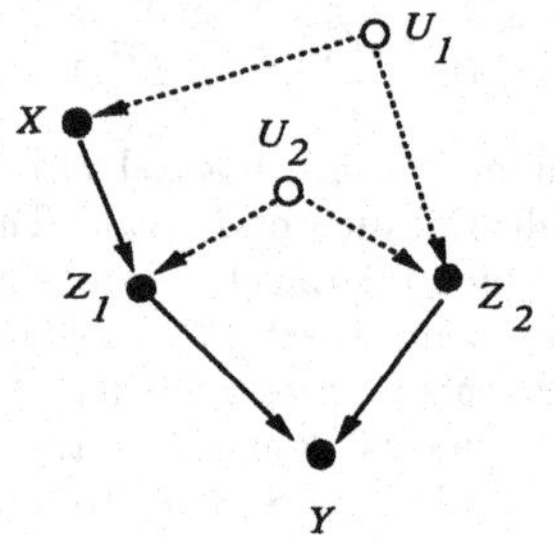

Figure 6:
A graph not containing a bow, but still prohibiting the identification of $P(y|\hat{x})$.

Another interesting feature demonstrated by Figure 6 is that computing the effect of a joint action is often easier than computing the effects of its constituent singleton actions.[11] Here, it is possible to compute $P(y|\hat{x}, \hat{z}_2)$ and $P(y|\hat{x}, \hat{z}_1)$, yet there is no way of computing $P(y|\hat{x})$. For example, the former can be evaluated by invoking Rule 2 in $G_{\overline{X}\underline{Z_2}}$, giving

$$P(y|\hat{x}, \hat{z}_2) \;=\; \sum_{z_1} P(y|z_1, \hat{x}, \hat{z}_2) P(z_1|\hat{x}, \hat{z}_2)$$

[11]This was brought to my attention by James Robins, who has worked out many of these computations in the context of sequential treatment management. Eq. (41) for example, can be obtained from Robin's G-computation formula [Robins 1989, Robins et al. 1992].

$$= \sum_{z_1} P(y|z_1, x, z_2)P(z_1|x) \quad (41)$$

The computation of $P(y|\hat{x})$, on the other hand, requires the conversion of $P(z_1|\hat{x}, z_2)$ into $P(z_1|x, z_2)$; Rule 2 is inapplicable because, when conditioned on Z_2, X and Z_1 are d-connected in $G_{\underline{X}}$ (through the dashed lines). A systematic procedure for identifying causal effects of multiple actions is provided in [Pearl & Robins 1995].

3.1 Identifying Models

Figure 7 shows simple diagrams in which the causal effect of X on Y, $P(y|\hat{x})$, is identifiable. Such structures are called identifying because their structures communicate a sufficient number of assumptions (missing links) to permit the identification of the target quantity $P(y|\hat{x})$. Unobserved (or latent) variables are not shown explicitly in these diagrams; rather, such variables are implicit in the confounding arcs (dashed lines). Every causal diagram with latent variables can be converted to an equivalent diagram involving measured variables interconnected by arrows and confounding arcs. This conversion corresponds to substituting out all unobserved variables from the structural equations of Eq. (21) and then constructing a new diagram by connecting any two variables X_i and X_j by (1) an arrow from X_j to X_i whenever X_j appears in the equation for X_i and (2) a confounding arc whenever the same U term appears in both f_i and f_j. The result is a diagram in which all unmeasured variables are exogenous and mutually independent.

Several features should be noted from examining the diagrams in Figure 7.

1. Since the removal of any arc or arrow from a causal diagram can only assist the identifiability of causal effects, $P(y|\hat{x})$ will still be identified in any edge-subgraph of the diagrams shown in Figure 7.

2. Likewise, the introduction of mediating observed variables onto any edge in a causal graph can assist, but never impede, the identifiability of any causal effect. Therefore, $P(y|\hat{x})$ will still be identified from any graph obtained by adding mediating nodes to the diagrams shown in Figure 7.

3. The diagrams in Figure 7 are maximal, in the sense that the introduction of any additional arc or arrow onto an existing pair of nodes would render $P(y|\hat{x})$ no longer identifiable.

4. Although most of the diagrams in Figure 7 contain bow-patterns, none of these patterns emanates from X (as is the case in Figure 8 (a) and (b) below). In general, a necessary condition for the identifiability of $P(y|\hat{x})$ is the absence of a confounding path between X and any of its children on any directed path from X to Y.

5. Diagrams (a) and (b) in Figure 7 contain no back-door paths between X and Y, and thus represent experimental designs in which there is no confounding bias between the treatment (X) and the response (Y) (i.e., X is strongly ignorable relative to Y [Rosenbaum & Rubin 1983]); hence, $P(y|\hat{x}) = P(y|x)$. Likewise, diagrams (c) and (d) in Figure 7 represent designs in which observed covariates, Z, block every back-door path between X and Y (i.e., X is conditionally ignorable given Z

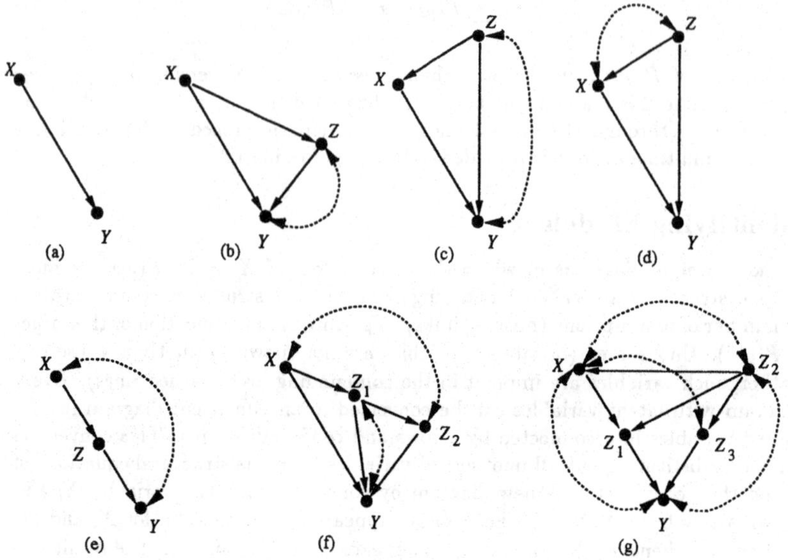

Figure 7:
Typical models in which the total effect of X on Y is identifiable. Dashed lines represent confounding paths, and Z represents observed covariates.

[Rosenbaum & Rubin 1983]); hence, $P(y|\hat{x})$ is obtained by a standard adjustment for Z (as in Eq. (28)):

$$P(y|\hat{x}) = \sum_z P(y|x,z)P(z) \tag{42}$$

6. For each of the diagrams in Figure 7, we can readily obtain a formula for $P(y|\hat{x})$, by using symbolic derivations patterned after those in Section 2.3. The derivation is often guided by the graph topology. For example, diagram (f) in Figure 7 dictates the following derivation. Writing

$$P(y|\hat{x}) = \sum_{z_1,z_2} P(y|z_1,z_2,\hat{x})P(z_1,z_2|\hat{x})$$

we see that the subgraph containing $\{X, Z_1, Z_2\}$ is identical in structure to that of diagram (e), with (Z_1, Z_2) replacing (Z, Y), respectively. Thus, $P(z_1, z_2|\hat{x})$ can be obtained from Eq. (38). Likewise, the term $P(y|z_1, z_2, \hat{x})$ can be reduced to $P(y|z_1, z_2, x)$ by Rule 2, since $(Y \perp\!\!\!\perp X|Z_1, Z_2)_{G_{\underline{X}}}$. Thus, we have

$$P(y|\hat{x}) = \sum_{z_1,z_2} P(y|z_1,z_2,x)\ P(z_1|x) \sum_{x'} P(z_2|z_1,x')\ P(x') \tag{43}$$

Applying a similar derivation to diagram (g) of Figure 7 yields

$$P(y|\hat{x}) = \sum_{z_1}\sum_{z_2}\sum_{x'} P(y|z_1,\ z_2,\ x')P(x')P(z_1|z_2,\ x)P(z_2) \tag{44}$$

Note that the variable Z_3 does not appear in the expression above, which means that Z_3 need not be measured if all one wants to learn is the causal effect of X on Y.

7. In diagrams (e), (f), and (g) of Figure 7, the identifiability of $P(y|\hat{x})$ is rendered feasible through observed covariates, Z, that are affected by the treatment X (i.e., Z being descendants of X). This stands contrary to the warning, repeated in most of the literature on statistical experimentation, to refrain from adjusting for concomitant observations that are affected by the treatment [Cox 1958, Rosenbaum 1984, Pratt & Schlaifer 1988]. It is commonly believed [Pratt & Schlaifer 1988] that if a concomitant Z is affected by the treatment, then it should be included in the analysis *only* if we want to learn the conditional effect *given* Z and must be excluded if we want to learn the unconditional total effects. The reason given for the exclusion is that the calculation of total effects often amounts to integrating out Z, which is functionally equivalent to omitting z to begin with.

   Diagrams (e), (f), and (g) show cases where one wants to learn the unconditional total effects of X and, still, the measurement of concomitants that are affected by X (e.g., Z, or Z_1) is necessary. However, the adjustment needed for such concomitants is nonstandard, involving two or more stages of the standard adjustment of Eq. (42) (see Eqs. (37), (43), and (44)).

8. Diagrams (b), (c), and (f) of Figure 7 deserve special attention. In each of these graphs, Y has a parent whose effect on Y is not identifiable yet the effect of X on Y is identifiable. This demonstrates that, contrary to linear analysis, local identifiability is not a necessary condition for global identifiability. In other words, to identify the effect of X on Y we need not insist on identifying each and every link of the paths from X to Y.

3.2 Nonidentifying Models

Figure 8 presents typical graphs in which the total effect of X on Y, $P(y|\hat{x})$, is not identifiable. Noteworthy features of these graphs are as follows.

1. All graphs in Figure 8 contain unblockable back-door paths between X and Y, that is, paths ending with arrows pointing to X which cannot be blocked by observed nondescendants of X. The presence of such a path in a graph is, indeed, a necessary test for nonidentifiability (see Theorem 3). It is not a sufficient test, though, as is demonstrated by Figure 7 (e), in which the back-door path (dashed) is unblockable and yet $P(y|\hat{x})$ is identifiable.

2. A sufficient condition for the nonidentifiability of $P(y|\hat{x})$ is the existence of a confounding path between X and any of its children on a path from X to Y, as shown in Figure 8 (b) and (c). A stronger sufficient condition is that the graph contain any of the patterns shown in Figure 8 as an edge-subgraph.

3. With the exception of (c), all the graphs in Figure 8 are minimal, that is, $P(y|\hat{x})$ is rendered identifiable by removing any arc or arrow from any of these graphs.

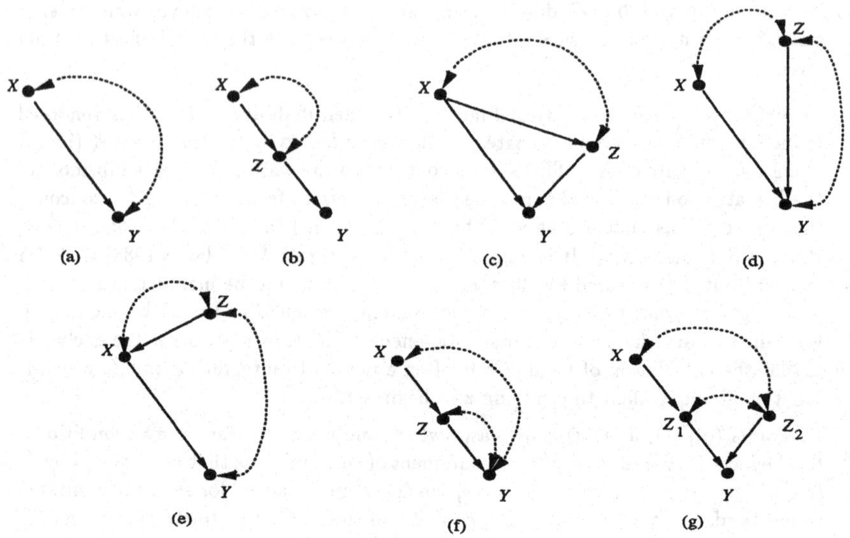

Figure 8:
Typical models in which $P(y|\hat{x})$ is not identifiable.

4. Graph (g) in Figure 8 does not have any bow-patterns and, moreover, every other causal effect is identifiable except that of X on Y. For example, we can identify $P(z_1|\hat{x})$, $P(z_2|\hat{x})$, $P(y,|\hat{z}_1)$, and $P(y|\hat{z}_2)$, but not $P(y|\hat{x})$. Thus, local identifiability is not sufficient for global identifiability. This is one of the main differences between nonparametric and linear models; in the latter, all causal effects can be determined from the structural coefficients, each coefficient representing the direct causal effect of one variable on its immediate successor (see Section 3.4).

3.3 Causal Inference by Surrogate Experiments

Suppose we wish to learn the causal effect of X on Y when X and Y are confounded and, for practical reasons of cost or ethics, we cannot control X by randomized experiment. In such situations, we naturally search for observed covariates that, if adjusted for, would eliminate the confounding effect between X and Y. Such covariates may not always be available, and the question arises whether $P(y|\hat{x})$ can be identified by randomizing a *surrogate* variable Z, which is easier to control than X. More generally, we are interested in a criterion by which a set Z of variables in the diagram can be identified and brought to the investigator's attention as potential surrogates for X.[12] Formally, this problem

[12]The main distinction between surrogate variables and *instrumental variables* as used in economics [Bowden & Turkington 1984], is that instrumental variables act as though they were randomized while surrogate variables are candidates for randomization. Additionally, the criterion for choosing surrogate variables need not be limited to the standard setting of instrumental variables depicted in Figure 5 (b); it includes any set of variables that would permit (if randomized) the identification of $P(y|\hat{x})$.

amounts to transforming $P(y|\hat{x})$ into expressions in which only members of Z obtain the hat symbol.

Diagram (e) in Figure 8 illustrate a simple structure which admits a surrogate experiment. The observed covariate Z is confounded with both X and Y, hence adjusting for Z does not permit the identification of $P(y|\hat{x})$ (i.e., X is not strongly ignorable conditional on Z, by the back-door criterion of Eq. (28)). However, if Z can be controlled by randomized trial, then we can measure $P(x, y|\hat{z})$, from which we can compute $P(y|\hat{x})$ using

$$P(y|\hat{x}) = P(y|x, \hat{z}) = P(y, x|\hat{z})/P(x|\hat{z}) \qquad (45)$$

The validity of Eq. (45) can be established by first applying Rule 3 to add $\hat{z}$,

$$P(y|\hat{x}) = P(y|\hat{x}, \hat{z}) \text{ because } (Y \perp\!\!\!\perp Z|X)_{G_{\overline{X}\overline{Z}}}$$

then applying Rule 2 to exchange $\hat{x}$ with x:

$$P(y|\hat{x}, \hat{z}) = P(y|x, \hat{z}) \text{ because } (Y \perp\!\!\!\perp X|Z)_{G_{\underline{X}\overline{Z}}}$$

The auxiliary diagrams permitting these steps are given in Figure 9.

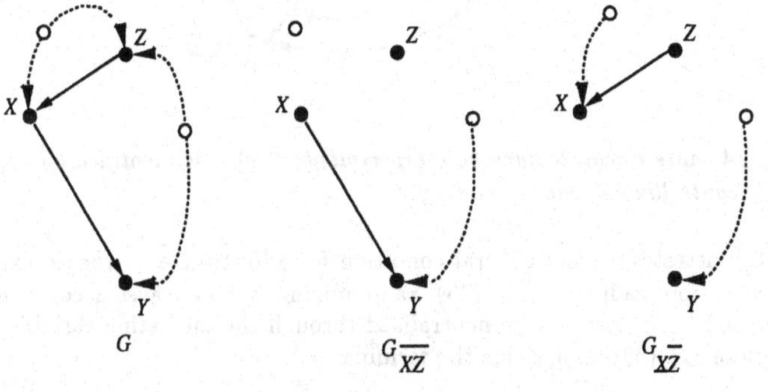

Figure 9:
A causal structure permitting the the identification of $P(y|\hat{x})$ by controlling Z, instead of X.

The use of surrogate experiments is not uncommon. For example, if we are interested in assessing the causal effect of cholesterol levels (X) on heart disease (Y), a reasonable experiment to conduct would be to control subjects' diet (Z), rather than exercising direct control over cholesterol levels in subjects' blood.

The derivation leading to Eq. (45) explicates a simple sufficient condition for qualifying a proposed variable Z as a surrogate for X: there must be no direct link from Z to Y and no confounding path between X and Y. Translated to our cholesterol example, this condition requires that there be no direct effect of diet on heart conditions and no confounding effect between cholesterol levels and heart disease.

Note that, according to Eq. (45), only one level of Z suffices for the identification of $P(y|\hat{x})$, for any values of y and x. In other words, Z need not be varied at all, just held constant by external force, and, if the assumptions embodied in G are valid, the r.h.s. of Eq. (45) should attain the same value regardless of the level at which Z is being held constant. In practice, however, several levels of Z will be needed to ensure that enough samples are obtained for each desired value of X. For example, if we are interested in the difference $E(Y|\hat{x}_1) - E(Y|\hat{x}_2)$, then we should choose two values z_1 and z_2 of Z which maximize the number of samples in x_1 and x_2, respectively, and write

$$E(Y|\hat{x}_1) - E(Y|\hat{x}_2) = E(Y|x_1, \hat{z}_1) - E(Y|x_2, \hat{z}_2)$$

Not surprisingly, this expression is equal to the instrumental-variable formula [Angrist et al. 1993]

$$E(Y|\hat{x}_1) - E(Y|\hat{x}_2) = \frac{E(Y|z_1) - E(Y|z_2)}{E(Y|x_1) - E(Y|x_2)}$$

when Z is randomized.

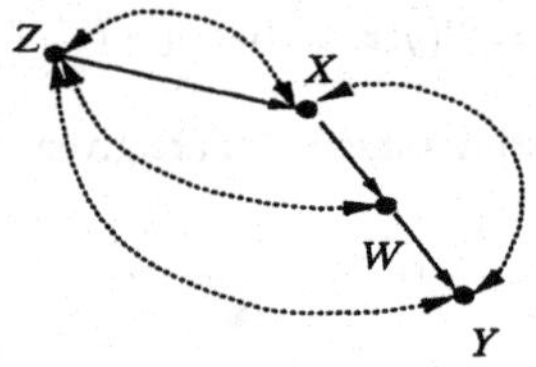

Figure 10:
A more elaborate surrogate experiment; $P(y|\hat{x})$ is identified by controlling Z and measuring W.

Figure 10 illustrates a more general condition for admitting a surrogate experiment. Unlike the condition leading to Eq. (43), randomizing Z now leaves a confounding arc between X and Y. This arc can be neutralized through the mediating variable W, as in the derivation of Eq. (36), and yields the formula

$$P(y|\hat{x}) = \sum_{w} P(w|x, \hat{z}) \sum_{x'} P(y|w, x', \hat{z}) P(x'|\hat{z})$$

Thus, the more general conditions for admitting a surrogate variable Z are:

1. X intercepts all directed paths from Z to Y, and,
2. $P(y|\hat{x})$ is identifiable in $G_{\overline{Z}}$.

3.4 Total, Direct, and Indirect Effects

Path analysis is noted for allowing researchers to decompose the influence of one variable on another into direct, indirect, and total effects [Bollen 1989, page 376]. Yet the path-analytic literature has not been successful in communicating these notions unambiguously to the rest of the scientific community. The standard definition of a total effect is expressed algebraically, in terms of a matrix B of "structural coefficients" [Bollen 1989], and these coefficients, circularly, are defined in terms of total effects when intervening variables are "held constant", [Alwin & Hauser 1975]. With the exception of [Sobel 1990], the notions of "intervening variables", "holding constant", and "structural coefficients" have not been given formal, operational definitions and have remained open to a variety of misinterpretations. Wermuth [1993], for example, interprets "holding X fixed" as "conditioning on X", and finds contradictions in the standard definition of structural equations. Freedman [1987] finds the notion of "fixing" an endogenous variable X to be "self-contradictory", as it conflicts with the assumption that the value of X is functionally determined by the explanatory variables in the equation for X. [Freedman 1987] summarizes the confusion in this area:

> a path model represents the analysis of observational data as if it were the result of an experiment. At points such as this, it would be helpful to know more about the structure of such hypothetical experiments: What is to be held constant, and what manipulated?

To explicate the structure of such hypothetical experiments we need a language in which the notion of "holding constant" is given both formal notation and operational interpretation. The mechanism-based interpretation of "holding constant" as an operation that deletes equations from the model (Definition 3), coupled with the $set(x)$ (or $\hat{x}$) notation introduced in Section 2, constitutes such a language and can be used to provide simple, unambiguous definitions of effect decomposition, for both parametric and nonparametric models.

We start with the general notion of causal effect $P(y|\hat{x})$, as in Definition 3, which applies to arbitrary sets of variables, X and Y. For singleton variables of interest, the notion of causal effect can be specialized to define total and direct effects, as follows.

Definition 5 (*total effect*) *The total effect of X on Y is given by $P(y|\hat{x})$, namely, the distribution of Y while X is held constant at x and all other variables are permitted to run their natural course.*

Definition 6 (*direct effect*) *The direct effect of X on Y is given by $P(y|\hat{x},\ \hat{s}_{XY})$ where S_{XY} is the set of all observed variables in the system, excluding X and Y.*

This definition ascribes to the direct effect the properties of an ideal laboratory; the scientist controls for all possible conditions S_{XY}. It is easy to show (e.g., by applying Rule 3) that there is no need to actually hold *all* other variables constant, since holding constant the direct parents of Y (excluding X) would have the same effect on Y. Thus, we obtain an equivalent definition for direct effect:

Corollary 2 *The direct effect of X on Y is given by $P(y|\hat{x},\ \hat{pa}_{Y\backslash X})$ where $pa_{Y\backslash X}$ stands for any realization of the variables appearing in the equation for Y, excluding X.*

Readers versed in linear analysis might find it a bit strange that the direct effect of X on Y involves other variables beside X and Y. However, considering that we are dealing with nonlinear interactions, the effect of X on Y should indeed depend on the levels at which we hold the other variables (in the equation for Y). Note also that causal effects are not defined in terms of differences between two expectations, or the relative change in Y

with a unit change in X. Such differences can always be determined from the probability distribution $P(y|\hat{x})$. In linear models, for example, the ratio

$$\frac{E(Y|\hat{x},\hat{pa}_{Y\backslash X}) - E(Y|\hat{x}',\hat{pa}_{Y\backslash X})}{x - x'} = \frac{\partial}{\partial x}E(Y|\hat{x}) = \text{const.}$$

reduces to the ordinary path coefficient between X and Y, regardless of the value taken by $pa_{Y\backslash X}$. In general, if X does not appear in the equation for Y, then $P(y|\hat{x},\hat{pa}_{Y\backslash X})$ defines a constant distribution on Y, independent of x, which matches our understanding of "having no direct effect". Note also that if PA_Y are not confounded with Y, we can remove the "hat" from the expressions above and define direct effects in terms of ordinary conditional probabilities $P(y|x, pa_{Y\backslash X})$.

The definitions above explicate the operational meaning of structural equations and path coefficients, and should end, I hope, an era of controversy and confusion regarding these entities. Specifically, if G is the graph associated with a set of structural equations, then the assumptions embodied in the equations can be read off G as follows: Every missing arrow, say between X and Y, represents the assumption that X has no causal effect on Y once we intervene and hold the parents of Y fixed. Every missing bi-directed link between X and Y represents the assumption that there are no common causes for X and Y, except those shown in G. Thus, the operational reading of the structural equation $Y = \beta X + \epsilon$ is: "In an ideal experiment where we control X to x and any other set Z of variables (not containing X or Y) to z, Y is independent of z and is given by $\beta x + \epsilon$." The meaning of β is simply $\frac{\partial}{\partial x}E(Y|\hat{x})$, namely, the rate of change (in x) of the expectation of Y in an experiment where X is held at x by external control This interpretation holds regardless of whether ϵ and X are correlated (e.g., via another equation $X = \alpha Y + \delta$.) Moreover, this interpretation provides an operational definition for the mystical error-term, ϵ, which is clearly a causal, rather than a statistical, entity.

In standard linear analysis, indirect effects are defined as the difference between the total and the direct effects [Bollen 1989]. In nonlinear analysis, differences lose their significance, and one must isolate the contribution of mediating paths in some alternative way. However, expressions of the form $P(y|\hat{x},\hat{z})$ cannot be used to isolate this contribution, because there is no physical means of selectively disabling a direct causal link from X to Y by holding some variables constant. This suggests that the notion of indirect effect indeed has no intrinsic operational meaning apart from providing a comparison between the direct and the total effects. In other words, a policy maker who asks for that part of the total effect transmitted by a particular intermediate variable or a group Z of such variables is really asking for a comparison of the effects of two policies, one in which Z is held constant, the other where it is not. The corresponding expressions for these two policies are $P(y|\hat{x},\ \hat{z})$ and $P(y|\hat{x})$, and this pair of distributions should therefore be taken as the most general representation of indirect effects. Similar conclusions are expressed in [Robins 1986] and [Robins & Greenland 1992].

3.5 Evaluating Conditional Policies

The interventions considered thus far were unconditional actions that merely force a variable or a group of variables X to take on some specified value x. In general, interventions may involve complex policies in which a variable X is made to respond in a specified way to some set Z of other variables, say through a functional relationship $X = g(Z)$ or through a stochastic relationship whereby X is set to x with probability $P^*(x|z)$. We will show that computing the effect of such policies is equivalent to computing the expression $P(y|\hat{x}, z)$.

Let $P(y|set(X = g(Z)))$ stand for the distribution (of Y) prevailing under the policy $(X = g(Z))$. To compute $P(y|set(X = g(Z)))$, we condition on Z and write

$$\begin{aligned} & P(y|set(X = g(Z))) \\ &= \sum_z P(y|set(X = g(z)), z) P(z|set(X = g(z))) \\ &= \sum_z P(y|\hat{x}, z)|_{x=g(z)} P(z) \\ &= E_z[P(y|\hat{x}, z)|_{x=g(z)}] \end{aligned}$$

where the equality

$$P(z|set(X = g(z))) = P(z)$$

stems from the fact that Z cannot be a descendant of X, hence, whatever control one exerts on X, it can have no effect on the distribution of Z. Thus, we see that the causal effect of a policy $X = g(Z)$ can be evaluated directly from the expression of $P(y|\hat{x}, z)$, simply by substituting $g(z)$ for x and taking the expectation over Z (using the observed distribution $P(z)$).

The identifiability condition for policy intervention is somewhat stricter than that for a simple intervention. Clearly, whenever a policy $set(X = g(Z))$ is identifiable, the simple intervention $set(X = x)$ is identifiable as well, as we can always get the latter by setting $g(Z) = X$. The converse, does not hold, however, because conditioning on Z might create dependencies that will prevent the successful reduction of $P(y|\hat{x}, z)$ to a hat-free expression.

A stochastic policy, which imposes a new conditional distribution $P^*(x|z)$ for x, can be handled in a similar manner. We regard the stochastic intervention as a random process in which the unconditional intervention $set(X = x)$ is enforced with probability $P^*(x|z)$. Thus, given $Z = z$, the intervention $set(X = x)$ will occur with probability $P^*(x|z)$ and will produce a causal effect given by $P(y|\hat{x}, z)$. Averaging over x and z gives

$$P(y|P^*(x|z)) = \sum_x \sum_z P(y|\hat{x}, z) P^*(x|z) P(z)$$

Since $P^*(x|z)$ is specified externally, we see again that the identifiability of $P(y|\hat{x}, z)$ is a necessary and sufficient condition for the identifiability of any stochastic policy that shapes the distribution of X by the outcome of Z.

It should be noted, however, that in planning applications the effect of an action may be to invalidate its preconditions. To represent such actions, temporally indexed causal networks may be necessary [Dean & Kanawaza 1989] or, if equilibrium conditions are required, cyclic graphs can be used [Balke & Pearl 1995].

4 Discussion

This paper demonstrates that:

1. The effect of intervening policies often be identified (from nonexperimental data) without resorting to parametric models.

2. The conditions under which such nonparametric identification is possible can be determined by simple graphical criteria.

3. When the effect of interventions is not identifiable, the causal graph may suggest non-trivial experiments which, if performed, would render the effect identifiable.

While the ability to assess the effect of interventions from nonexperimental data has many applications in the social and health sciences, perhaps the most practical result reported in this paper is the solution of the long standing problem of covariate adjustment. The reader might recognize Eq. (42) as the standard formula for covariate adjustment (also called "stratification"), which is used both for improving precision and for minimizing confounding bias. However, a formal, general criterion for deciding whether a set of covariates Z qualifies for adjustment has long been wanting. In the context of linear regression models, the problem amounts to deciding whether it is appropriate to add a set Z of variables to the regression of Y on X. Most of the statistical literature is satisfied with informal warnings that "Z should be quite unaffected by X" [Cox 1958, page 48], which is necessary but not sufficient (see Figure 8(d)) or that X should not precede Z [Shafer 1996, page 326], which is neither necessary nor sufficient. In some academic circles, a criterion called "ignorability" is invoked [Rosenbaum & Rubin 1983], which merely paraphrases the problem in the language of counterfactuals. Simplified, ignorability reads: Z is an admissible covariate relative to the effect of X on Y if, for every x, the value that Y would obtain had X been x is conditionally independent of X, given Z (see appendix II for further discussion of counterfactual analysis). In contrast, Eq. (26) provides an admissibility test which is both precise and meaningful, as it is applicable directly to the elementary processes (i.e., linkages in the graph) around which scientific knowledge is organized. This test (called the "back-door criterion" in [Pearl 1993]) reads: Z is an admissible set of covariates relative to the effect of X on Y if:

(i) no node in Z is a descendant of X, and

(ii) Z d-separates X from Y along any path containing an arrow into X (equivalently, $(Y \perp\!\!\!\perp X|Z)_{G_{\underline{X}}}$).

We see, for instance, that Z qualifies as admissible covariates relative the effect of X on Y in Figure 7(d) but not in Figure 8(d). The graphical definition of admissible covariates replaces statistical folklore with formal procedures, and should enable analysts to systematically select an optimal set of observations, namely, a set Z that minimizes measurement cost or sampling variability.

It is important to note several limitations and extensions of the method proposed in this paper. First, the structural models discussed so far consist only of behavioral equations; definitional equalities and equilibrium constraints are excluded, as these do not respond to intervention in the manner described in Definition 3. One way of handling mixtures of behavioral and equilibrium equations is to treat the latter as observational events, on which to condition the probabilities [Strotz & Wold 1960]. For example, the econometric equilibrium constraint $q_d = q_s$, equating quantity demanded and quantity supplied, would be treated by adding a "dummy" behavioral equation

$$S = q_s - q_d$$

(S connoting "stock growth") and, then, conditioning the resulting probabilities on the event $S = 0$. Such conditioning events tend to introduce new dependencies among the variables in the graphs, as dictated by the d-separation criterion. Consequently, in applying the inferences rule of Theorem 3, one would have to consult graphs in which the dummy variables have been permanently conditioned.

A second extension concerns the use of the causal calculus (Theorem 1) in cyclic models. The subtheory interpretation of control queries (Definition 3) carries over to cyclic systems [Strotz & Wold 1960, Sobel 1990], but then two issues must be addressed. First, the analysis of identification is meaningful only when the resulting system is stable. Therefore, we must modify the definition of identifiability by considering only the set of stable theories for each structural model and for each submodel [Fisher 1970]. Second, the d-separation criterion for DAGs must be extended to cover cyclic graphs as well. The validity of d-separation has been established for non-recursive linear models and extended, using an augmented graph to any arbitrary set of stable equations [Spirtes 1994]. However, the computation of control queries will be harder in cyclic networks, because complete reduction of control queries to hat-free expressions may require the solution of nonlinear equations.

Having obtained nonparametric formulas for causal effects does not imply, of course, that one should refrain from using parametric forms in the estimation phase of the study. When data are scarce, prior information about shapes of distributions and the nature of causal interactions can be extremely useful, and it can be incorporated into the analysis by limiting the distributions in the estimand formulas to whatever parametric family of functions are deemed plausible by the investigator. For example, if the assumptions of Gaussian, zero-mean disturbances and additive interactions are deemed reasonable in a given problem, then nonparametric formulas of the type (see Eq. (33))

$$P(y|\hat{x}) = \sum_z P(y|x, z)P(z) \tag{46}$$

will be converted to

$$E(Y|\hat{x}) = \int_y \int_z yf(y|x, z)f(z)dydz = R_{yx \cdot z}x \tag{47}$$

and the estimation problem reduces to that of estimating (e.g., by least-squares) the regression of Y on X and Z. Similarly, the estimand given in Eq. (37) can be converted to a product

$$E(Y|\hat{x}) = R_{xz}\beta_{zy \cdot x}x \tag{48}$$

where $\beta_{zy \cdot x}$ is the standardized regression coefficient. More sophisticated estimation techniques, tailored specifically for causal inference can be found in [Robins 1992].

Finally, a few comments regarding the notation introduced in this paper. Traditionally, statisticians have approved of only one method of combining subject-matter considerations with statistical data: the Bayesian method of assigning subjective priors to distributional parameters. To incorporate causal information within the Bayesian framework, plain causal statements such as "Y is affected by X" must be converted into sentences capable of receiving probability values, e.g., counterfactuals. Indeed, this is how Rubin's model has achieved statistical legitimacy: causal judgments are expressed as constraints on probability functions involving counterfactual variables (see Appendix II).

Causal diagrams offer an alternative language for combining data with causal information. This language simplifies the Bayesian route by accepting plain causal statements as its basic primitives. These statements, which merely identify whether a causal connection between two variables of interest exists, are commonly used in natural discourse and provide a natural way for scientists to communicate experience and organize knowledge. It can be anticipated, therefore, that by separating issues of identification and parametric form this article should serve to make the language of path analysis more accessible to the scientific community (see discussions following [Pearl 1995a]).

Acknowledgment

This investigation benefitted from discussions with Joshua Angrist, Peter Bentler, David Cox, Sander Greenland, Arthur Goldberger, David Hendry, Paul Holland, Guido Imbens, Ed Leamer, Rod McDonald, John Pratt, James Robins, Paul Rosenbaum, Donald Rubin, and Michael Sobel. The research was partially supported by Air Force grant #AFOSR 90 0136, NSF grant #IRI-9200918, and Northrop-Rockwell Micro grant #93-124.

APPENDIX I. Smoking and the Genotype Theory: An Illustration

To illustrate the usage of the causal effects computed in Subsection 2.3, we will associate the model of Figure 1 with a concrete example concerning the evaluation of the effect of smoking (X) on lung cancer (Y). According to many, the tobacco industry has managed to stay anti-smoking legislation by arguing that the observed correlation between smoking and lung cancer could be explained by some sort of carcinogenic genotype (U) which involves inborn craving for nicotine.[13]

The amount of tar (Z) deposited in a person's lungs is a variable that promises to meet the conditions specified by the structure of Figure 1. To justify the missing link between X and Y, we must assume that smoking cigarettes (X) has no effect on the production of lung cancer (Y) except that mediated through tar deposits. To justify the missing link between U and Z, we must assume that, even if a genotype is aggravating the production of lung cancer, it nevertheless has no effect on the amount of tar in the lungs except indirectly, through cigarette smoking.

To demonstrate how we can assess the degree to which cigarette smoking increases (or decreases) lung cancer risk, we will construct a hypothetical study in which the three variables, X, Y, and Z, were measured simultaneously on a large, randomly selected sample from the population. To simplify the exposition, we will further assume that all three variables are binary, taking on true (1) or false (0) values. A hypothetical data set from a study on the relations among tar, cancer, and cigarette smoking is presented in Table 1.

	Group Type	$P(x, z)$ Group Size (% of Population)	$P(Y = 1 \mid x, z)$ % of Cancer Cases in Group
$X = 0,\ Z = 0$	Non-smokers, No tar	47.5	10
$X = 1,\ Z = 0$	Smokers, No tar	2.5	90
$X = 0,\ Z = 1$	Non-smokers, Tar	2.5	5
$X = 1,\ Z = 1$	Smokers, Tar	47.5	85

Table 1

The table shows that 95% of smokers and 5% of non-smokers have developed high levels of tar in their lungs. Moreover, 81.51% of subjects with tar deposits have developed lung cancer, compared to only 14% among those with no tar deposits. Finally, within each of the two groups, tar and no tar, smokers show a much higher percentage of cancer than non-smokers do.

These results seem to prove that smoking is a major contributor to lung cancer. However, the tobacco industry might argue that the table tells a different story—that smoking actually decreases, not increases, one's risk of lung cancer. Their argument goes as follows. If you decide to smoke, then your chances of building up tar deposits are 95%, compared to 5% if you decide not to smoke. To evaluate the effect of tar deposits, we look separately at two groups, smokers and non-smokers. The table shows that tar deposits have a protective effect in both groups: in smokers, tar deposits lower cancer rates from 90% to 85%; in non-smokers, they lower cancer rates from 10% to 5%. Thus, regardless of

[13] For an excellent historical account of this debate, see [Spirtes et al. 1993, pp. 291–302].

whether I have a natural craving for nicotine, I should be seeking the protective effect of tar deposits in my lungs, and smoking offers a very effective means of acquiring them.

To settle the dispute between the two interpretations, we note that, while both arguments are based on stratification, the anti-smoking argument invokes an illegal stratification over a variable (Z) that is affected by the treatment (X). The tobacco industry's argument, on the the hand, is made up of two steps, neither of which involves stratification over treatment-affected variables: stratify over smoking to find the effect of tar deposit on lung cancer, then average (not stratify) over tar deposits when we consider each of the decision alternatives, smoking vs. non-smoking. This is indeed the intuition behind the formula in Eq. (32) and, given the causal assumptions of Figure 7, the tobacco industry's argument is the correct one (see [Pearl 1995a, Pearl 1994] for formal derivation).

To illustrate the use of Eq. (32), let us use the data in Table 1 to calculate the probability that a randomly selected person will develop cancer ($y_1 : Y = 1$) under each of the following two actions: smoking ($x_1 : X = 1$) or not smoking ($x_0 : X = 0$).

Substituting the appropriate values of $P(y|x)$, $P(y|x,z)$, and $P(x)$ gives

$$\begin{aligned} E[P(y_1|x_1,u)] &= .05(.10 \times .50 + .90 \times .50) + .95(.05 \times .50 + .85 \times .50) \\ &= .05 \times .50 + .95 \times .45 = .4525 \\ E[P(y_1|x_0,u)]) &= .95(.10 \times .50 + .90 \times .50) + .05(.05 \times .50 + .85 \times .50) \\ &= .95 \times .50 + .05 \times .45 = .4975 \qquad (49) \end{aligned}$$

Thus, contrary to expectation, the data prove smoking to be somewhat beneficial to one's health.

The data in Table 1 are obviously unrealistic and were deliberately crafted so as to support the genotype theory. However, this exercise was meant to demonstrate how reasonable qualitative assumptions about the workings of mechanisms can produce precise quantitative assessments of causal effects when coupled with nonexperimental data. In reality, we would expect observational studies involving mediating variables to refute the genotype theory by showing, for example, that the mediating consequences of smoking, such as tar deposits, tend to increase, not decrease, the risk of cancer in smokers and non-smokers alike. The estimand given in Eq. (32) could then be used for quantifying the causal effect of smoking on cancer.

APPENDIX II: Graphs, structural equations, and counterfactuals

This paper uses two representations of causal models: graphs and structural equations. By now, both representations have been considered controversial for almost a century. On the one hand, economists and social scientists have embraced these modeling tools, but they continue to debate the empirical content of the symbols they estimate and manipulate; as a result, the use of structural models in policy-making contexts is often viewed with suspicion. Statisticians, on the other hand, reject both representations as problematic (if not meaningless) and instead resort to the Neyman-Rubin counterfactual notation [Rubin 1990] whenever they are pressed to communicate causal information. This appendix presents an explication that unifies these three representation schemes in order to uncover commonalities, mediate differences, and make the causal-inference literature more generally accessible.

The primitive object of analysis in Rubin's counterfactual framework is the unit-based response variable, denoted $Y(x,u)$ or $Y_x(u)$, read: "the value that Y would obtain in unit u, had X been x". This variable has natural interpretation in structural equation models. Consider a set T of equations

$$X_i = f_i(PA_i, U_i) \quad i = 1, \ldots, n \tag{50}$$

where the U_i stand for latent exogenous variables (or disturbances), and the PA_i are the explanatory (observed) variables in the ith equation (pa_i is a realization of PA_i). (50) is similar to (14), except we no longer insist on the equations being recursive or on the U_i's being independent. Let U stand for the vector $(U_1, \ldots, U_n)$, let X and Y be two disjoint subsets of observed variables, and let T_x be the subtheory created by replacing the equations corresponding to variables in X with $X = x$, as in Definition 2. The structural interpretation of $Y(x,u)$ is given by

$$Y(x,u) \triangleq Y_{T_x}(u) \tag{51}$$

namely, $Y(x,u)$ is the (unique) solution of Y under the realization $U = u$ in the subtheory T_x of T. While the term *unit* in the counterfactual literature normally stands for the identity of a specific individual in a population, a unit may also be thought of as the set of attributes that characterize that individual, the experimental conditions under study, the time of day, and so on, which are represented as components of the vector u in structural modeling. Eq. (51) forms a connection between the opaque English phrase "the value that Y would obtain in unit u, had X been x" and the physical processes that transfer changes in X into changes in Y. The formation of the submodel T_x represents a minimal change in model T needed for making x and u compatible; such a change could result either from external intervention or from a natural yet unanticipated eventuality.

Given this interpretation of $Y(x,u)$, it is instructive to contrast the methodologies of causal inference in the counterfactual and the structural frameworks. If U is treated as a vector of random variable, then the value of the counterfactual $Y(x,u)$ becomes a random variable as well, denoted as $Y(x)$ or Y_x. The counterfactual analysis proceeds by imagining the observed distribution $P^*(x_1, \ldots, x_n)$ as the marginal distribution of an augmented probability function P^* defined over both observed and counterfactual variables. Queries about causal effects, written $P(y|\hat{x})$ in the structural analysis, are phrased as queries about the marginal distribution of the counterfactual variable of interest, written $P^*(Y(x) = y)$. The new entities $Y(x)$ are treated as ordinary random variables that are connected to the observed variables via consistency constraints (Robins, 1987) such as

$$X = x \Longrightarrow Y(x) = Y \tag{52}$$

and a set of conditional independence assumptions which the investigator must supply to endow the augmented probability, P^*, with causal knowledge, paralleling the knowledge that a structural analyst would encode in equations or in graphs.

For example, to communicate the understanding that in a randomized clinical trial (see Figure 5(b)) the way subjects react (Y) to treatments (X) is statistically independent of the treatment assignment (Z), the analyst would write $Y(x) \perp\!\!\!\perp Z$. Likewise, to convey the understanding that the assignment processes is randomized, hence independent of any variation in the treatment selection process, structurally written $U_X \perp\!\!\!\perp U_Z$, the analyst would use the independence constraint $X(z) \perp\!\!\!\perp Z$.

A collection of constraints of this type might sometimes be sufficient to permit a unique solution to the query of interest, for example, $P^*(Y(x) = y)$; in other cases, only bounds on the solution can be obtained. Section 4 explains why this approach is conceptually appealing to some statisticians, even though the process of eliciting judgments about counterfactual dependencies has so far not been systematized. When counterfactual variables are not viewed as by-products of a deeper, process-based model, it is hard to ascertain whether *all* relevant judgments have been articulated, whether the judgments articulated are redundant, or whether those judgments are self-consistent. The elicitation of such judgments can be systematized using the following translation from graphs.

Graphs provide qualitative information about the structure of both the equations in the model and the probability function $P(u)$, the former is encoded as missing arrows, the latter as missing dashed arcs. Each parent-child family (PA_i, X_i) in a causal diagram G corresponds to an equation in the model (50). Hence, missing arrows encode exclusion assumptions, that is, claims that adding excluded variables to an equation will not change the outcome of the hypothetical experiment described by that equation. Missing dashed arcs encode independencies among disturbance terms in two or more equations. For example, the absence of dashed arcs between a node Y and a set of nodes $Z_1, \ldots, Z_k$ implies that the corresponding error variables, $U_Y, U_{Z_1}, \ldots, U_{Z_k}$, are jointly independent in $P(u)$.

These assumptions can be translated into the counterfactual notation using two simple rules; the first interprets the missing arrows in the graph, the second, the missing dashed arcs.

1. Exclusion restrictions: For every variable Y having parents PA_Y, and for every set of variables S disjoint of PA_Y, we have
$$Y(pa_Y) = Y(pa_Y, s) \tag{53}$$
2. Independence restrictions: If $Z_1, \ldots, Z_k$ is any set of nodes not connected to Y via dashed arcs, we have
$$Y(pa_Y) \perp\!\!\!\perp \{Z_1(pa_{Z_1}), \ldots, Z_k(pa_{Z_k})\} \tag{54}$$

Given a sufficient number of such restrictions on P^*, it is possible to compute causal effects $P^*(Y(x) = y)$ using standard probability calculus together with the logical constraints (e.g., Eq. (52)) that couple counterfactual variables with their measurable counterparts. These constraints can be used as axioms, or rules of inference, in attempting to transform causal effect expressions, $P^*(Y(x) = y)$, into expressions involving only measurable variables. When such a transformation is found, the corresponding causal effect is identifiable, since P^* reduces then to P. The axioms needed for such transformation are:

$$\textbf{Degeneracy}: \quad Y(\emptyset) = Y \tag{55}$$
$$\textbf{Composition}: \quad Y(x) = Y(x, Z(x)) \text{ for any } Z \text{ disjoint of } \{X, Y\} \tag{56}$$
$$\textbf{Sure-thing}: \quad \text{If } Y(x,z) = Y(x',z)\ \forall\, x' \neq x, \text{ then } Y(x,z) = Y(z) \tag{57}$$

Degeneracy asserts that the observed value of Y is equivalent to a counterfactual variable $Y(x)$ in which the conditional part: "had X been x" is not enforced, that is, X is the empty set.

The Composition axiom[14] asserts:

$$\text{If } Y(x,z)=y \text{ and } Z(x)=z, \text{ then } Y(x)=y$$

and, conversely:

$$\text{If } Y(x)=y \text{ and } Z(x)=z, \text{ then } Y(x,z)=y$$

In words: "The value that Y would obtain had X been x is the same as that obtained had X been x and Z been z, where z is the value that Z would obtain had X been x".

The sure-thing axiom (named after Savage's "sure-thing principle") asserts that if $Y(x,z)=y$ for every value x of X, then the counterfactual antecedent $X=x$ is redundant, namely, we need not concern ourselves with the value that X actually obtains.

Properties (56)-(57) are theorems in the structural interpretation of $Y(x,u)$ as given in Eq. (51) [Galles & Pearl 1995a]. However, in the Neyman-Rubin model, where $Y(x,u)$ is taken as a primitive notion, these properties must be considered axioms which, together with other such properties, defines the abstract counterfactual conditioning operator "had X been x". It is easy to verify that composition and degeneracy imply the consistency rule of (52); substituting $X=\{\emptyset\}$ in (59) yields $Y=Y(z)$ if $Z=z$, which is equivalent to (52).

As an example, let us compute the causal effects associated with the model shown in Figure 2 (or Eqs. (1)-(3)). The parents sets a given by:

$$PA_X=\{\emptyset\},\ PA_Z=\{X\},\ PA_Y=\{Z\} \tag{58}$$

Consequently, the exclusion restrictions (53) translate into:

$$Z(x) = Z(y,x) \tag{59}$$
$$X(y) = X(z,y)=X(z)=X \tag{60}$$
$$Y(z) = Y(z,x) \tag{61}$$

The independence restrictions (54) translate into:

$$Z(x) \perp\!\!\!\perp \{Y(z),X\} \tag{62}$$

Task-1, compute $P^*(Z(x)=z)$ (Equivalently $P(z|\hat{x})$)
From (62) we have $Z(x) \perp\!\!\!\perp X$, hence

$$P^*(Z(x)=z)=P^*(Z(x)=z|x)=P^*(z|x)=P(z|x) \tag{63}$$

Task-2, compute $P^*(Y(z)=y)$ (Equivalently $P^*(y|\hat{z})$)

$$P^*(Y(z)=y)=\sum_x P^*(Y(z)=y|x)P^*(x) \tag{64}$$

[14] This axiom was communicated to me by James Robins (1995, in conversation) as a property needed for defining a structure he calls "finest fully randomized causal graphs" [Robins 1986, pp. 1419–1423]. In Robins' analysis, $Y(x,z)$ and $Z(x)$ may not be defined.

From (62) we have

$$Y(z) \perp\!\!\!\perp Z(x)|X \tag{65}$$

hence

$$\begin{aligned} P^*(Y(z)=y|x) &= P^*(Y(z)=y|x, Z(x)=z) && \text{by (52)} \\ &= P^*(Y(z)=y|x, z) && \text{by (40)} \\ &= P^*(y|x, z) && \text{by (40)} \\ &= P(y|x, z) \end{aligned} \tag{66}$$

Substituting (66) in (64), gives

$$P^*(Y(z)=y) = \sum_x P(y|x,z)P(x) \tag{67}$$

which is the celebrated covariate-adjustment formula for causal effect, as in Eq. (42).

Task-3, compute $P^*(Y(x)=y)$ (Equivalently $P(y|\hat{x})$)
For any arbitrary variable Z, we have (by composition)

$$Y(x) = Y(x, Z(x))$$

In particular, since $Y(x,z) = Y(z)$ (from (61)), we have

$$Y(x) = Y(x, Z(x)) = Y(Z(x))$$

and

$$\begin{aligned} P^*(Y(x)=y) &= P^*(Y(Z(x))=y) \\ &= \sum_z P^*(Y(Z(x))=y)|Z(x)=z)\ P^*(Z(x)=z) \\ &= \sum_z P^*(Y(z)=y)|Z(x)=z)\ P^*(Z(x)=z) \\ &= \sum_z P^*(Y(z)=y)\ P^*(Z(x)=z) \end{aligned}$$

since $Y(z) \perp\!\!\!\perp Z(x)$.

$P^*(Y(z)=y)$ and $P^*(Z(x)=z)$ were computed in (67) and (63), respectively, hence

$$P^*(Y(x)=y) = \sum_z P(z|x) \sum_{x'} P(y|z,x')P(z')$$

In summary, the structural and counterfactual frameworks are complementary of each other. Structural analysts can interpret counterfactual sentences as constraints over the solution set of a given system of equations (51) and, conversely, counterfactual analysts can use the constraints (over P^*) given by Eqs. (53) and (54) as a definition of graphs, structural equations and the physical processes which they represent.

References

[Alwin & Hauser 1975] Alwin, D.F., and Hauser, R.M., "The decomposition of effects in path analysis," *American Sociological Review*, 40, 37-47, 1975.

[Angrist et al. 1993] Angrist, J.D., Imbens, G.W., and Rubin, D.B., "Identification of causal effects using instrumental variables," Department of Economics, Harvard University, Cambridge, MA, Technical Report No. 136, June 1993. To appear in *JASA*.

[Balke & Pearl 1993] Balke, A., and Pearl, J., "Nonparametric bounds on causal effects from partial compliance data," Department of Computer Science, University of California, Los Angeles, Technical Report R-199, 1993. Submitted to the *Journal of the American Statistical Association.*

[Balke & Pearl 1994] Balke, A., and Pearl, J., "Counterfactual Probabilities: Computational Methods, Bounds, and Applications," In R. Lopez de Mantaras and D. Poole (Eds.), *Uncertainty in Artificial Intelligence - 10*, Morgan Kaufmann, San Mateo, 46-54, 1994.

[Balke & Pearl 1995] Balke, A., and Pearl, J., "Counterfactuals and policy analysis in structural models," In P. Besnard and S. Hanks (Eds.), *Uncertainty in Artificial Intelligence 11*, Morgan Kaufmann, San Francisco, 11-18, 1995.

[Bollen 1989] Bollen, K.A., *Structural Equations with Latent Variables*, John Wiley and Sons, New York, 1989.

[Bowden & Turkington 1984] Bowden, R.J., and Turkington, D.A., *Instrumental Variables*, Cambridge University Press, Cambridge, 1984.

[Cox 1958] Cox, D.R., *Planning of Experiments*, John Wiley and Sons, New York, 1958.

[Dean & Kanawaza 1989] Dean, T., and Kanazawa, K., "A Model for Reasoning about Persistence and Causation," *Computational Intelligence*, **5**, 142-150, 1989.

[Fisher 1970] Fisher, F.M., "A correspondence principle for simultaneous equation models," *Econometrica*, 38, 73-92, 1970.

[Freedman 1987] Freedman, D., "As others see us: A case study in path analysis" (with discussion), *Journal of Educational Statistics*, 12, 101-223, 1987.

[Frisch 1938] Frisch, R., "Statistical versus theoretical relations in economic macrodynamics," League of Nations Memorandum, 1938. (Reproduced, with Tinbergen's comments, in *Autonomy of Economic Relations*, Oslo: Universitetets Socialokonomiske Institutt, 1948). Cited in M.S. Morgan, *The History of Econometric Ideas*, Cambridge University Press, Cambridge, 1990.

[Galles & Pearl 1995a] Galles, D., and Pearl, J., "Axioms of causal relevance," Department of Computer Science, University of California, Los Angeles, Technical Report R-240, January 1996. Submitted to AAAI-96.

[Galles & Pearl 1995b] Galles, D., and Pearl, J., "Testing identifiability of causal effects," In P. Besnard and S. Hanks (Eds.), *Uncertainty in Artificial Intelligence 11*, Morgan Kaufmann, San Francisco, CA, 185-195, 1995.

[Geiger 1990] Geiger, D., "Graphoids: A qualitative framework for probabilistic inference," UCLA, Ph.D. Dissertation, Computer Science Department, 1990.

[Geiger & Pearl 1988] Geiger, D., and Pearl, J., "On the logic of causal models," *Proceedings of the 4th Workshop on Uncertainty in Artificial Intelligence*, St. Paul, MN, 1988, pp. 136-147. Also in L. Kanal et al. (Eds.), *Uncertainty in Artificial Intelligence*, 4, Elsevier Science Publishers, North-Holland, Amsterdam, pp. 3-14, 1990.

[Geiger et al. 1990] Geiger, D., Verma, T.S., and Pearl, J., "Identifying independence in Bayesian networks." *Networks*, volume 20, pages 507-534. John Wiley and Sons, Sussex, England, 1990.

[Goldberger 1973] Goldberger, A.S., *Structural Equation Models in the Social Sciences*, Seminar Press, New York, 1973.

[Haavelmo 1943] Haavelmo, T., "The statistical implications of a system of simultaneous equations," *Econometrica*, 11, 1-12, 1943.

[Lauritzen & Spiegelhalter 1988] Lauritzen, S.L., and Spiegelhalter, D.J., "Local computations with probabilities on graphical structures and their applications to expert systems," *Proceedings of the Royal Statistical Society, Series B*, 50, 154-227, 1988.

[Marschak 1953] Marschak, J., "Economic measurements for policy and prediction," in T. Koopmans and W. Hood (Eds.), *Studies in Econometric Method, Cowles Commission Monograph 14*, Chapter 1, Yale University Press, New Haven, 1953.

[Manski 1990] Manski, C.F., "Nonparametric bounds on treatment effects," *American Economic Review, Papers and Proceedings,* 80, 319-323, 1990.

[Pearl 1988] Pearl, J., *Probabilistic Reasoning in Intelligence Systems*, Morgan Kaufmann, San Mateo, 1988.

[Pearl 1993] Pearl, J., "Graphical models, causality and intervention," *Statistical Science*, 8(3) 266-269, 1993.

[Pearl 1994] Pearl, J., "A probabilistic calculus of actions," in R. Lopez de Mantaras and D. Poole (Eds.), *Uncertainty in Artificial Intelligence* 10, Morgan Kaufmann, San Mateo, pp. 454-462, 1994.

[Pearl 1995a] Pearl, J., "Causal diagrams for experimental research," *Biometrika*, 82(4) 669-710, 1995.

[Pearl 1995b] Pearl, J., "Causal inference from indirect experiments," *Artificial Intelligence in Medicine Journal*, 7, 561-582, 1995.

[Pearl 1995c] Pearl, J., "On the testability of causal models with latent and instrumental variables," In P. Besnard and S. Hanks (Eds.), *Uncertainty in Artificial Intelligence 11*, Morgan Kaufmann Publishers, San Francisco, 435-443, 1995.

[Pearl & Robins 1995] Pearl, J., and Robins, J., "Probabilistic evaluation of sequential plans from causal models with hidden variables," In P. Besnard and S. Hanks (Eds.), *Uncertainty in Artificial Intelligence 11*, Morgan Kaufmann, San Francisco, 444-453, 1995.

[Pearl & Verma 1991] Pearl, J., and Verma, T., "A theory of inferred causation," in J.A. Allen, R. Fikes, and E. Sandewall (Eds.), *Principles of Knowledge Representation and Reasoning: Proceeding of the Second International Conference*, Morgan Kaufmann, San Mateo, 1991, pp. 441-452.

[Pearl et al. 1990] Pearl, J., Geiger, D., and Verma, T., "The logic of influence diagrams, in R.M. Oliver and J.Q. Smith (Eds.), *Influence Diagrams, Belief Nets and Decision Analysis*, John Wiley and Sons, New York, 1990, pp. 67-87.

[Pratt & Schlaifer 1988] Pratt, J., and Schlaifer, R., "On the interpretation and observation of laws," *Journal of Economics*, 39, 23-52, 1988.

[Robins 1986] Robins, J.M., "A new approach to causal inference in mortality studies with a sustained exposure period – applications to control of the healthy workers survivor effect," *Mathematical Modeling*, Vol. 7, 1393-1512, 1986.

[Robins 1989] Robins, J.M., "The analysis of randomized and non-randomized AIDS treatment trials using a new approach to causal inference in longitudinal studies," in L. Sechrest, H. Freeman, and A. Mulley (Eds.), *Health Service Research Methodology: A Focus on AIDS*, NCHSR, U.S. Public Health Service, 1989, pp. 113-159.

[Robins 1992] Robins, J., "Estimation of the time-dependent accelerated failure time model in the presence of confounding factors," *Biometrika*, 79(2), 321-334, 1992.

[Robins & Greenland 1992] Robins, J.M., and Greenland, S., "Identifiability and exchangeability for direct and indirect effects," *Epidemiology*, 3(2), 143-155, 1992.

[Robins et al. 1992] Robins, J.M., Blevins, D., Ritter, G., and Wulfsohn, M., "*G*-Estimation of the Effect of Prophylaxis Therapy for Pneumocystis carinii Pneumonia on the Survival of AIDS Patients," *Epidemiology*, Vol. 3, Number 4, 319-336, 1992.

[Rosenbaum 1984] Rosenbaum, P.R., "The consequences of adjustment for a concomitant variable that has been affected by the treatment," *Journal of the Royal Statistical Society, Series A (General)*, 147(5), 656-666, 1984.

[Rosenbaum 1989] Rosenbaum, P.R., "The role of known effects in observational studies," *Biometrics*, 45, 557-569, 1989.

[Rosenbaum & Rubin 1983] Rosenbaum, P., and Rubin, D., "The central role of propensity score in observational studies for causal effects," *Biometrica*, 70, 41-55, 1983.

[Rubin 1990] Rubin, D.B., "Formal modes of statistical inference for causal effects," *Journal of Statistical Planning and Inference*, 25, 279-292, 1990.

[Shafer 1996] Shafer, G., *The Art of Causal Conjecture*. MIT Press, Cambridge, 1996. Forthcoming.

[Simon 1977] Simon, H.A., *Models of Discovery: and Other Topics in the Methods of Science*, D. Reidel, Dordrecht, Holland, 1977.

[Spirtes 1994] Spirtes, P., "Conditional independence in directed cyclic graphical models for feedback," Department of Philosophy, Carnegie-Mellon University, Pittsburg, PA, Technical Report CMU-PHIL-53, May 1994.

[Spirtes et al. 1993] Spirtes, P., Glymour, C., and Schienes, R., *Causation, Prediction, and Search*, Springer-Verlag, New York, 1993.

[Sobel 1990] Sobel, M.E., "Effect analysis and causation in linear structural equation models," *Psychometrika*, 55(3), 495-515, 1990.

[Spiegelhalter et al. 1993] Spiegelhalter, D.J., Lauritzen, S.L., Dawid, P.A., and Cowell, R.G., "Bayesian analysis in expert systems," *Statistical Science*, 8(3), 219-247, 1993.

[Strotz & Wold 1960] Strotz, R.H., and Wold, H.O.A., "Recursive versus nonrecursive systems: An attempt at synthesis," *Econometrica*, 28, 417-427, 1960.

[Verma 1990] Verma, T.S., "Causal networks: Semantics and expressiveness," in R. Shachter et al. (Eds.), *Uncertainty in Artificial Intelligence*, 4, Elsevier Science Publishers, North-Holland, Amsterdam, 1990, 69-76.

[Verma & Pearl 1990] Verma, T.S. and Pearl, J., "Equivalence and synthesis of causal models." In *Uncertainty in Artificial Intelligence*, Elsevier Science Publishers, Cambridge, pages 6, 220-227, 1990.

[Wermuth 1993] Wermuth, N., "On block-recursive regression equations" (with discussion), *Brazilian Journal of Probability and Statistics*, 6, 1-56, 1992.

Causal Inference from Complex Longitudinal Data

James M. Robins
Departments of Epidemiology and Biostatistics
Harvard School of Public Health
Boston, MA 02115
email: robins@hsph.harvard.edu

1. Introduction

The subject-specific data from a longitudinal study consist of a string of numbers. These numbers represent a series of empirical measurements. Calculations are performed on these strings of numbers and causal inferences are drawn. For example, an investigator might conclude that the analysis provides strong evidence for "a direct effect of AZT on the survival of AIDS patients controlling for the intermediate variable – therapy with aerosolized pentamidine." The nature of the relationship between the sentence expressing these causal conclusions and the computer calculations performed on the strings of numbers has been obscure. Since the computer algorithms are well-defined mathematical objects, it is important to provide formal mathematical definitions for the English sentences expressing the investigator's causal inferences.

I proposed a formal theory of counterfactual causal inference (Robins, 1986, 1987) that extended Rubin's (1978) "point treatment" theory to longitudinal studies with direct and indirect effects and time-varying treatments, confounders, and concomitants. The purpose of this paper is to provide a summary and unification of this theory.

In my theory, causal questions that could be asked concerning the direct and indirect effects of the measured variables on an outcome became mathematical conjectures about the causal parameters of event trees that I called causally interpreted structured tree graphs (CISTGs). I defined randomized CISTGs (RCISTGs) and showed that for RCISTGs, a subset of the population causal parameters were non-parametrically identified by the G-computation algorithm formula. A RCISTG is an event tree in which the treatment received at time t is randomly allocated (ignorable) conditional on past treatment, outcome, and covariate history. In the absence of physical randomization, as in an observational study, the assumption that a CISTG is an RCISTG is non-identifiable.

Pearl (1995), and Spirtes, Glymour, and Scheines (SGS) (1993) recently developed a formal theory of causal inference based on causal directed acyclic graphs (DAGs). I show that these causal DAGs are the same mathematical objects as particular RCISTGs and thus a theorem in one causal theory is a theorem in the other (Robins, 1995b).

The standard approach to the estimation of a time-varying treatment on an outcome of interest is to model the probability of the outcome at time t as a function of past treatment history. This approach may be biased, whether or not one further adjusts for the past history of time dependent confounding covariates, when these covariates predict subsequent outcome and treatment history and are themselves influenced by past treatment. In this setting, I have proposed several methods that can provide unbiased estimates of the causal effect of a time-varying treatment in the presence of time varying confounding factors. In this paper, I describe two of these methods of estimation: estimation of the conditional probabilities in the G-computation algorithm formula (Robins 1986, 1987, 1989), and G-estimation of structural nested models (Robins 1989, 1992, 1993, 1994, 1995a, 1996). The G-computation algorithm formula is equivalent to the marginal distribution of the outcome in the manipulated subgraph of a DAG in which all arrows into the treatment variables are removed and the vector of treatment random variables X is set to a treatment regime or plan x of interest. This marginal distribution has a causal interpretation as the effect of treatment regime x on the outcome if the graph is a RCISTG, i.e., treatment received at each time t was randomly allocated (i.e., ignorable) conditional on past treatment and covariate history.

However, estimation of causal effects based on the G-computation algorithm is seriously non-robust since the inevitable misspecification of statistical models for the conditional law of the DAG variables given their parents results in bias under the causal null hypothesis of no treatment effect. In contrast, an approach based on estimation of structural nested models can often avoid this bias. Mathematically, structural nested models reparameterize the joint law of the DAG in terms of parameters that represent contrasts between the marginal distributions of the outcome in the manipulated graph with treatment X set to different values of x. Causally the parameters of a structural nested model represent the causal effect of a final brief blip of treatment on the outcome of interest. However, in Sec. 8.2, we show that structural nested models, like models for the conditional laws of the DAG variables given their parents, are non-robust for testing the null hypothesis of no direct effect of a time-dependent treatment X_1 controlling for a time-dependent intermediate variable X_2. Therefore, in Appendix 3, we introduce an extension of structural nested models, the "direct effect" structural nested models, that are suitable for testing the null hypothesis of no direct effect.

In Section 9, we discuss how one can use G-estimation to estimate the parameters of a structural nested model even when treatment at time t is not allocated at random given the past (i.e., the CISTG is not a RCISTG), provided that one can correctly specify a parametric or semiparametric model for the probability of treatment at t conditional on past covariate and treatment history and on the (possibly unobserved) value of a subject's counterfactual outcome (Robins, 1996).

In Sections 1-9, we assume that treatment and covariate processes change (jump) at the pre-specified times. In Section 10 we relax this assumption by introducing continuous-time structural nested models whose parameter ψ reflects the instantaneous causal effect of the treatment rate (Robins, 1996). In Sec. 11, I argue that the faithfulness assumption of Pearl and Verma (1991) and SGS (1993) should not be used to draw causal conclusions from observational data.

2. Standard Analysis of Sequential Randomized Trials

The following type of example originally motivated the development of the methods described in this article. The graph in Figure 1 represents the data obtained from a hypothetical (oversimplified) sequential randomized trial of the joint effects of AZT (A_0) and aerosolized pentamidine (A_1) on the survival of AIDS patients. AZT inhibits the AIDS virus. Aerosolized pentamidine prevents pneumocystis pneumonia (PCP), a common opportunistic infection of AIDS patients. The trial was conducted as follows. Each of 32,000 subjects was randomized with probability .5 to AZT ($A_0 = 1$) or placebo ($A_0 = 0$) at time t_0. All subjects survived to time t_1. At time t_1, it was determined whether a subject had had an episode of PCP ($L_1 = 1$) or had been free of PCP ($L_1 = 0$) in the interval $(t_0, t_1]$. Since PCP is a potential life-threatening illness, all subjects with $L_1 = 1$ were treated with aerosolized pentamidine (AP) therapy ($A_1 = 1$) at time t_1. Among subjects who were free of PCP ($L_1 = 0$), one-half were randomized to receive AP at t_1 and one half were randomized to placebo ($A_1 = 0$). At time t_2, the vital status was recorded for each subject with $Y = 1$ if alive and $Y = 0$ if deceased. We view A_0, L_1, A_1, Y as random variables with realizations a_0, l_1, a_1, y. All investigators agreed that the data supported a beneficial effect of treatment with AP ($A_1 = 1$) because, among the 8,000 subjects with $A_0 = 1$, $L_1 = 0$, AP was assigned at random and the survival rates were greater among those given AP, since

$$P[Y = 1 \mid A_1 = 1, L_1 = 0, A_0 = 1] - P[Y = 1 \mid A_1 = 0, L_1 = 0, A_0 = 1] = 3/4 - 1/4 = 1/2 \tag{2.1}$$

The remaining question was whether, given that subjects were to be treated with AP, should or should not they also be treated with AZT. That is, we wish to determine whether the direct effect of AZT on survival controlling for (the potential intermediate variable) AP is beneficial or harmful (when all subjects receive AP). The most straightforward way to examine this question is to compare the survival rates in groups with a common AP treatment who differ on their AZT treatment. Reading from Figure 1 we observe, after collapsing over the data on L_1-status, that

$$P[Y = 1 \mid A_0 = 1, A_1 = 1] - P[Y = 1 \mid A_0 = 0, A_1 = 1] = 7/12 - 10/16 = -1/24 \tag{2.2}$$

suggesting a harmful effect of AZT. However, the analysis in (2.2) fails to account for the possible confounding effects of the extraneous variable PCP(L_1). [We refer to PCP here as an "extraneous variable" because the causal question of interest, i.e., the question of whether AZT has a direct effect on survival controlling for AP, makes no reference to PCP. Thus adjustment for PCP is necessary only insofar as PCP is a confounding factor.] It is commonly accepted that PCP is a confounding factor and must be adjusted for in the analysis if PCP is *(a)* an independent risk (i.e., prognostic) factor for the outcome and *(b)* an independent risk factor for (predictor of) future treatment. By "independent" risk factor in *(a)* and *(b)* above, we mean a variable that is a predictor conditional upon all other measured variables occurring earlier than the event being predicted. Hence, to check condition *(a)*, we must adjust for A_0 and A_1; to check condition *(b)*, we must adjust for A_0.

Reading from Figure 1, we find that conditions *(a)* and *(b)* are both true, i.e.,

$$.5 = P[Y = 1 \mid L_1 = 1, A_0 = 1, A_1 = 1] \neq P[Y = 1 \mid L_1 = 0, A_0 = 1, A_1 = 1] = .75 \tag{2.3}$$

and

$$1 = P[A_1 = 1 \mid L_1 = 1, A_0 = 1] \neq P[A_1 = 1 \mid L_1 = 0, A_0 = 1] = .5 \tag{2.4}$$

The standard approach to the estimation of the direct effect of AZT controlling for AP in the presence of a confounding factor (PCP) is to compare survival rates among

Figure 1: Data From a Hypothetical Study

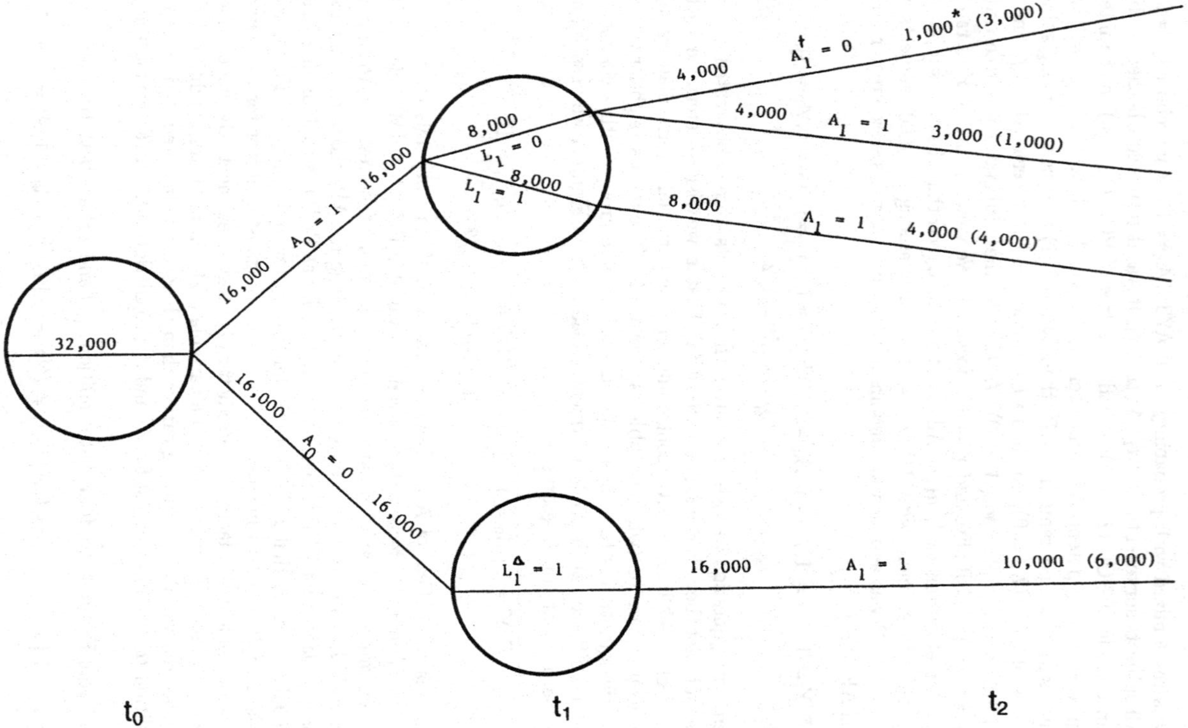

* Survivors (Y=1) at t_2. [Deaths (Y= 0) at t_2 in parentheses.]

† A_k measured just after time t_k, k=0,1.

Δ L_1 measured at time t_1.

groups with common AP and confounder history (e.g., $L_1 = 1$, $A_1 = 1$) but who differ in AZT treatment. Reading from Figure 1, we obtain

$$\begin{aligned} &P[Y=1 \mid A_0=1, L_1=1, A_1=1] - P[Y=1 \mid A_0=0, L_1=1, A_1=1] \\ &= 4{,}000/8{,}000 - 10{,}000/16{,}000 = -1/8 \end{aligned} \tag{2.5}$$

Hence the analysis adjusted for PCP also suggests an adverse direct effect of AZT on survival controlling for AP.

However, the analysis adjusted for PCP is also problematic, since *(a)* Rosenbaum (1984) and Robins (1986, 1987) argue that it is inappropriate to adjust (by stratification) for an extraneous risk factor that is itself affected by treatment, and *(b)*, reading from Figure 1, we observe that PCP is affected by previous treatment, i.e.,

$$.5 = P[L_1 = 1 \mid A_0 = 1] \neq P[L_1 = 1 \mid A_0 = 0] = 1 \tag{2.6}$$

Thus, according to standard rules for the estimation of causal effects, *(a)* one cannot adjust for the extraneous risk factor PCP, since it is affected by a previous treatment (AZT); yet one must adjust for PCP since it is a confounder for a later treatment (AP). Thus it may be that, in line with the adage that association need not be causation, neither (2.2) nor (2.5) may represent the direct causal effect of AZT controlling for AP. Since both treatments (AZT and AP) were randomized, one would expect that there should exist a "correct" analysis of the data such that the association observed in the data under that analysis has a causal interpretation as the direct effect of AZT controlling for AP. In the next section, we derive such a "correct" analysis based on the G-computation algorithm of Robins (1986). We show that there is, in fact, no direct causal effect of AZT controlling for AP. That is, given that all subjects take AP, it is immaterial to the survival rate in the study population whether or not AZT is also taken.

3. A Formal Theory of Causal Inference

We use this section to derive the G-computation algorithm formula that will allow a correct analysis of the trial in Section 2.

3.1. The Observed Data

Let $V = \overline{V}_J \equiv (V_0, \ldots, V_J)$ be a $J+1$ random vector of temporally ordered variables with associated distribution function $F_V(v)$ and density $f_V(v)$ with respect to a measure μ where, for any $Z = (Z_0, \ldots, Z_M), \overline{Z}_m \equiv (Z_0, Z_1, \ldots, Z_m)$ is the history through $m, m \leq M$. We assume each component V_j of V is either a real-valued continuous random variable or discrete. The measure μ is the product measure of Lebesgue and counting measures corresponding to the continuous and discrete components of V. Now let $\overline{A}_K = (A_0, A_1, \ldots, A_K)$ be a temporally ordered $K+1$ subvector of V consisting of the treatment variables of interest. Denote by t_k the time at which treatment A_k is received. Let L_k be the vector of all variables whose temporal occurrence is between treatments A_{k-1} and A_k with L_1 being the variables preceding A_1 and L_{K+1} being the variables succeeding A_K. Hence, $\overline{L}_{K+1} = (L_0, \ldots, L_{K+1})$ is the vector of all non-treatment variables.

For notational convenience, define $\overline{A}_{-1}, \overline{L}_{-1}, \overline{V}_{-1}$ be identically 0 for all subjects. We view $\overline{A}_K$ as a sequence of treatment (control, exposure) variables whose causal effect on $\overline{L}_{K+1}$ we wish to evaluate. To do so, we must define a feasible treatment regime. Before doing this, in Figure 2, with $K = 1$ we use the trial of Section 2 to clarify our notation.

<u>Figure 2</u>

	V_0	V_1	V_2	V_3	V_4
	L_0	A_0	L_1	A_1	$Y \equiv L_2$
Variable Name:	0	AZT at t_0	PCP at t_1	AP at t_1	survival to t_2

Note that since no data has been collected prior to A_0, L_0 can be set to 0 for all subjects. Also, since all subjects are alive at t_1, we do not require a variable recording survival to t_1.

<u>Remark</u>: Our results do not actually require that the temporal ordering of the univariate components of the vector variable L_k need be known. [Pearl and Robins (1995) discuss conditions under which one may further relax the assumption that the temporal ordering of the variables in V is completely known.]

3.2. Treatment Regimes and Counterfactual Data

We adopt the convention that if two covariate histories, e.g., $\overline{\ell}_m$ and $\overline{\ell}_k$, are used in the same expression with $k > m$ then $\overline{\ell}_m$ is the initial segment of $\overline{\ell}_k$ through t_m. Also $\mathbf{Z}$ will denote the support of a random variable Z. We define a feasible treatment regime g to be a function $g\left(t_k, \overline{\ell}_k\right)$ that assigns to each t_k and $\overline{\ell}_k \in \overline{\mathbf{L}}_k$ a treatment $a_k \in \mathbf{A}_k$. Let $\mathbf{g}$ be the set of all feasible regimes. Given the function $g\left(t_k, \overline{\ell}_k\right)$ of two arguments, we define the function $g\left(\overline{\ell}_k\right)$ of one argument by the relation $g\left(\overline{\ell}_k\right) = \left\{g\left(t_m, \overline{\ell}_m\right); 0 \le m \le k\right\}$. Since there is a one-to-one relationship between the functions, we shall identify the regime g with both functions. If $g\left(t_k, \overline{\ell}_k\right)$ is a constant, say a_k^*, not depending on $\overline{\ell}_k$ for each k, we say regime g is non-dynamic and denote it by $g = \overline{a}^*$, $\overline{a}^* = (a_0^*, \ldots, a_K^*)$. Otherwise we say the regime g is dynamic.

<u>Example (3.1)</u>: In the hypothetical trial in Sec. 2, the non-dynamic regime in which subjects are forced to take AZT and then AP is $g = (1, 1)$. The regime in which subjects are forced not to take AZT but then to take AP is $g = (0, 1)$. The dynamic regime g^* in which subjects take AZT and then take AP only if they develop PCP is given by $g^*\left(\overline{\ell}_0\right) = 1$, $g^*\left(\overline{\ell}_1 = 1\right) = (1, 1)$ and $g^*\left(\overline{\ell}_1 = 0\right) = (1, 0)$.

<u>Remark</u>: In medical trials, subjects are usually assigned to dynamic regimes since, if toxicity develops at t_k (with toxicity recorded in L_k), the treatment must be modified accordingly.

It will now be convenient to consider the realizations of random variables for particular subjects i. We shall say that, in the observed study, subject i has followed a treatment history consistent with a particular regime g through t_k if $g\left(\overline{\ell}_{ki}\right) = \overline{a}_{ki}$.

We reserve the i subscript for study subjects and write $\overline{\ell}_{(K+1)i} \equiv \overline{\ell}_i$ and $\overline{a}_{Ki} \equiv \overline{a}_i$. For each subject i and regime g, we shall assume there exists (possibly counterfactual data) $\overline{\ell}_{(K+1)ig} \equiv \overline{\ell}_{ig}$ and $\overline{a}_{Kig} \equiv \overline{a}_{ig}$ representing the ℓ-history and a-history that would be observed in the closest possible world to this world (Lewis, 1973) in which a subject followed a treatment history consistent with regime g. Here $\overline{\ell}_{ig}$ determines $\overline{a}_{ig}$ through the relationship $g\left(t_k, \overline{\ell}_{kig}\right) = a_{kig}$; this formalizes the idea that subject i would have followed regime g.

Remark (3.1): For certain variables (such as gender) we may not understand what it means to speak of the closest possible world to this one in which one's gender were different. Such variables will not be regarded as potential treatments. (See Robins (1986, 1995b) and Rubin (1978).) In contrast, the literature on causal theories based on *DAGs* (Pearl, 1994; SGS, 1993) appears to suggest that the causal effect of any variable (including gender) is potentially well-defined (but see Heckerman and Shachter, 1995).

We define the counterfactual data on subject i to be $\left\{\overline{\ell}_{ig}, \overline{a}_{ig}; g \in \mathbf{g}\right\}$. Note that, by assumption, the counterfactual data on subject i do not depend on the observed or counterfactual data for any other subject (Rubin, 1978). We shall make a consistency assumption that serves to link the counterfactual data with the observed data. This assumption states that if subject i has an observed treatment history through t_k equal to that prescribed by regime g, then his observed outcome history through t_{k+1} will equal his counterfactual outcome history under regime g through that time.

Consistency Assumption: For $k = 0, \ldots, K$,

$$g\left(\overline{\ell}_{ki}\right) = \overline{a}_{ki} \text{ implies } \overline{\ell}_{(k+1)ig} = \overline{\ell}_{(k+1)i} \tag{3.1}$$

Remark: Interestingly, our main theorems [Theorems (3.1) and (3.2)] only require that (3.1) holds for $k = K$.

We shall assume $\left(\overline{\ell}_i, \overline{a}_i, \left\{\overline{\ell}_{ig}, \overline{a}_{ig}; g \in \mathbf{g}\right\}\right), i = 1, \ldots, n$, are realizations of independent identically distributed random vectors $\left(\overline{L}_i, \overline{A}_i, \left\{\overline{L}_{ig}, \overline{A}_{ig}; g \in \mathbf{g}\right\}\right)$. This will be the case if, as discussed in Robins (1995b), we can regard the n study subjects as randomly sampled without replacement from a large superpopulation of N subjects and our interest is the causal effect of treatment in the superpopulation. For notational convenience, we shall often suppress the i subscript. Robins (1986, 1987) represented the observed data and counterfactual data for the n subjects in "event trees" called causally interpreted structured tree graphs (CISTGs), whose topology is illustrated by Figure 1. These CISTGs are mathematically equivalent to the observed and counterfactual data for the n subjects plus the consistency assumption (3.1). The structure, developed in this section, is sufficient to provide a formal mathematical definition for essentially any English sentence expressing a statement about causal effects. For example, in the context of the trial of Section 2, the null hypothesis that there is no direct effect of AZT on any subject's survival through t_2 controlling for AP (when all subjects receive AP) can, in the notation of Ex. 3.1, be formally expressed as $Y_{ig=(1,1)} = Y_{ig=(0,1)}$ with probability 1.

3.3. Sequential Randomization, the G-computation Algorithm, and a Correct Analysis of the Trial of Sec. 2

We now suppose that, as is usually the case, the outcome of interest is a particular function $Y = y\left(\overline{L}_{K+1}\right)$ of $\overline{L}_{K+1}$ rather than the entire $\overline{L}_{K+1}$. For example, in the trial in Sec. 2, $Y = L_2 \equiv L_{K+1}$ denoted survival through time t_2. The purpose of this section is to give sufficient conditions to identify the law of $Y_g = y\left(\overline{L}_{(K+1)g}\right)$ from the observed data and to provide an explicit computational formula, the G-computation algorithm formula, for the density $f_{Y_g}(y)$ of Y_g under these conditions. If the A_k had been assigned at random by the flip of a coin, then for each regime g we would have, for $k = 0, 1, \ldots, K$, and all $\overline{\ell}_k$

$$\overline{L}_g \coprod A_k \mid \overline{L}_k = \overline{\ell}_k, \overline{A}_{k-1} = g\left(\overline{\ell}_{k-1}\right) \tag{3.2}$$

even if (as in Sec. 2) the probability that the coin landed heads depended on $\overline{L}_k$ and $\overline{A}_{k-1}$. Eq. (3.2) states that among subjects with covariate history $\overline{\ell}_k$ through t_k, and treatment history $g\left(\overline{\ell}_{k-1}\right)$ through t_{k-1}, treatment A_k at time t_k is independent of the counterfactual outcomes $\overline{L}_g \equiv \overline{L}_{K+1,g}$. This is because $\overline{L}_g$, like gender or age at enrollment, is a fixed characteristic of a subject unaffected by the randomized treatments a_k actually received. Note that such physical sequential randomization of the a_k would imply that (3.2) would hold for all $\bar{A}_{k-1}$ histories. However, we shall only require that it hold for the $\bar{A}_{k-1}$ histories $g\left(\overline{L}_{k-1}\right)$ consistent with regime g. Theorem 1 below states that, given (3.1) and (3.2), the density $f_{Y_g}(y)$ is identified provided that the probability density that a subject follows regime g through t_K is non-zero.

Our results also apply to observational studies in which (3.2) is true. However, in observational studies, (3.2) cannot be guaranteed to hold and is not subject to an empirical test. Therefore, it is a primary goal of epidemiologists conducting an observational study to collect in $\overline{L}_k$, data on a sufficient number of covariates to try to make Eq. (3.2) at least approximately true. For example, suppose the data in Figure 1 was from an observational study rather than a sequentially randomized study. Because physicians tend to initiate AP preferentially for subjects who have had a recent bout of PCP, and because recent bouts of PCP signify poor prognosis (i.e., patients with small values of Y_g), Eq. (3.2) would be false if PCP at t_1 were not a component of L_1. It is because physical randomization guarantees Eq. (3.2) that most people accept that valid causal inferences can be obtained from randomized studies (Rubin, 1978).

For any random Z, let $S_Z(z) = P(Z \geq z)$ be the survivor probability where for a j vector Z, we define $Z \geq z \Leftrightarrow (Z_1 \geq z_1, \ldots, Z_j \geq z_j)$. For reasons that will become clear in Sec. 7, it will often be more convenient to state some results in terms of survivor probabilities or cumulative distribution functions [i.e., $1 - S_Z(z)$] than in terms of densities. Adopt the convention that the conditional density and survival probability of an observable random variable given other observable random variables will be denoted by $f(\cdot \mid \cdot)$ and $S(\cdot \mid \cdot)$. The conditional density or survival probability of a counterfactual random variable with respect to regime g given another counterfactual random variable with respect to that regime will be denoted by $f^g(\cdot \mid \cdot)$ or $S^g(\cdot \mid \cdot)$. Finally, the conditional density or survival probability of a counterfactual random variable with respect to regime g given an observed random variable will be denoted by $f^{g0}(\cdot \mid \cdot)$ or $S^{g0}(\cdot \mid \cdot)$. Further we adopt the convention that $g\left(\overline{\ell}_k\right)$ will always represent a realization of the variable $\overline{A}_k$. Thus, for example, $f^{g0}\left(\ell_k \mid \overline{\ell}_{k-1}, g\left(\overline{\ell}_{k-1}\right)\right) \equiv$ $f_{L_{kg}|\overline{L}_{k-1},\overline{A}_{k-1}}\left(\ell_k \mid \overline{\ell}_{k-1}, g\left(\overline{\ell}_{k-1}\right)\right)$, $f\left\{g\left(\overline{\ell}_k\right) \mid \overline{\ell}_k, g\left(\overline{\ell}_{k-1}\right)\right\} \equiv$ $f_{\overline{A}_k|\overline{L}_{k-1},\overline{A}_{k-1}}\left(g\left(\overline{\ell}_k\right) \mid \overline{\ell}_{k-1}, g\left(\overline{\ell}_{k-1}\right)\right)$, $f^g\left(\ell_k \mid \overline{\ell}_{k-1}\right) \equiv f_{L_{kg}|\overline{L}_{(k-1)g}}\left(\ell_k \mid \overline{\ell}_{k-1}\right)$.

Using this notation, the condition that for $k = (0, \ldots, K)$

$$f\left\{\overline{\ell}_k, g\left(\overline{\ell}_{k-1}\right)\right\} \neq 0 \Rightarrow f\left\{g\left(\overline{\ell}_k\right) \mid \overline{\ell}_k, g\left(\overline{\ell}_{k-1}\right)\right\} \neq 0 \tag{3.3}$$

states that subjects whose history is consistent with regime g prior to t_k have a positive density of "remaining" on regime g at t_k.

In the following theorem due to Robins (1986, 1987), we adopt the convention that $\bar{\ell}_{-1} \equiv 0, g\left(\bar{\ell}_{-1}\right) \equiv 0$.

Theorem 3.1: If (3.1) - (3.3) hold for regime g, then Eqs. (3.4) - (3.9) are true.

$$f^{g}\left(\ell_k \mid \bar{\ell}_{k-1}\right) = f^{g0}\left(\ell_k \mid \bar{\ell}_{k-1}, g\left(\bar{\ell}_{k-2}\right)\right) = f\left(\ell_k \mid \bar{\ell}_{k-1}, g\left(\bar{\ell}_{k-1}\right)\right) \tag{3.4}$$

$$S^{g}\left(y \mid \bar{\ell}_K\right) = S^{go}\left(y \mid \bar{\ell}_K, g\left(\bar{\ell}_{K-1}\right)\right) = S\left(y \mid \bar{\ell}_K, g\left(\bar{\ell}_K\right)\right) \tag{3.5}$$

Remark: Eq. (3.4) says that one time-step ahead innovations in the L-process in the counterfactual world where all subjects followed regime g equal the one step ahead innovations in the observed world among subjects whose previous treatment is consistent with g. Eq. (3.5) states that the conditional distribution of Y given $\bar{L}_K$ in the counterfactual world is equal to the same distribution in the observed world among the subjects whose previous treatment is consistent with g.

$$f^{g}\left(\bar{\ell}_m \mid \bar{\ell}_k\right) = f^{go}\left(\bar{\ell}_m \mid \bar{\ell}_k, g\left(\bar{\ell}_{k-1}\right)\right) = \prod_{j=k+1}^{m} f\left(\ell_j \mid \bar{\ell}_{j-1}, g\left(\bar{\ell}_{j-1}\right)\right), m > k \tag{3.6}$$

$$\begin{aligned} f^{g}\left(y, \bar{\ell}_K \mid \bar{\ell}_k\right) &= f^{go}\left(y, \bar{\ell}_K \mid \bar{\ell}_k, g\left(\bar{\ell}_{k-1}\right)\right) = \\ & f\left(y \mid \bar{\ell}_K, g\left(\bar{\ell}_K\right)\right) \prod_{j=k+1}^{K} f\left(\ell_j \mid \bar{\ell}_{j-1}, g\left(\bar{\ell}_{j-1}\right)\right) \end{aligned} \tag{3.7}$$

Remark: Eq. (3.6) and (3.7) state that multi-step ahead innovations in the counterfactual world can be built up from one step innovations in the observed world. Note (3.6) and (3.7) depend on the law of the observed data only through the densities $f\left(\ell_j \mid \bar{\ell}_{j-1}, \bar{a}_{j-1}\right)$.

$$\begin{aligned} S^{g}\left(y \mid \bar{\ell}_k\right) &= S^{go}\left(y \mid \bar{\ell}_k, g\left(\bar{\ell}_{k-1}\right)\right) = \int \cdots \iint S\left(y \mid \bar{\ell}_K, g\left(\bar{\ell}_K\right)\right) \\ & \prod_{j=k+1}^{K} f\left(\ell_j \mid \bar{\ell}_{j-1}, g\left(\bar{\ell}_{j-1}\right)\right) d\mu\left(\ell_j\right) \end{aligned} \tag{3.8}$$

$$S^{g}(y) = \int \cdots \iint S\left(y \mid \bar{\ell}_K, g\left(\bar{\ell}_K\right)\right) \prod_{j=0}^{K} f\left(\ell_j \mid \bar{\ell}_{j-1}, g\left(\bar{\ell}_{j-1}\right)\right) d\mu\left(\ell_j\right) \tag{3.9}$$

Remark: Eq. (3.9) states that the marginal survival probability of Y_g is obtained by a weighted average of the $S\left(y \mid \bar{\ell}_K, g\left(\bar{\ell}_K\right)\right)$ with weights proportional to $\omega\left(\bar{\ell}_K\right) \equiv \prod_{j=0}^{K} f\left[\ell_j \mid \bar{\ell}_{j-1}, g\left(\bar{\ell}_{j-1}\right)\right]$. Eq. (3.8) has a similar interpretation except that it conditions on the covariate history $\bar{\ell}_k$.

It will be convenient to let $b\left(y, \overline{\ell}_k, g\right)$ denote the rightmost side of (3.8). Note $b\left(y, \overline{\ell}_k, g\right)$ depends only on the law of the observables $F_V(v)$, the regime g, y, and $\overline{\ell}_k$. In fact, the dependence on $F_V(v)$ is only through the densities $f\left(\ell_j \mid \overline{\ell}_{j-1}, \overline{a}_{j-1}\right)$ since Y is a function of $\overline{L}_{K+1}$. Similarly, let $b(y, g)$ denote the right hand side of (3.9).

Example: A Correct Analysis of the Trial of Sec. 2: In the sequential randomized trial of Figure 1, we let $K = 1, L_0 \equiv 0, Y = L_2$. Then the probability a subject would survive to t_2 $(Y = 1)$ if all subjects were treated with AZT at t_0 and aerosolized pentamidine at t_1 is $S^g(y = 1)$ with g the non-dynamic regime $g = (1, 1)$ and equals, by Eq. (3.9),

$$\sum_{\overline{\ell}_1} S\left(y = 1 \mid \overline{\ell}_1, g\left(\overline{\ell}_1\right)\right) f\left(\overline{\ell}_1 \mid \ell_0, g\left(\overline{\ell}_0\right)\right) = S\left(y = 1 \mid \ell_1 = 1, a_1 = 1, a_0 = 1\right)$$

$$f\left(\ell_1 = 1 \mid a_0 = 1\right) + S\left(y = 1 \mid \ell_1 = 0, a_0 = 1, a_1 = 1\right) f\left(\ell_1 = 0 \mid a_0 = 1\right) =$$

$$(4,000/8,000)(8,000/16,000) + (3,000/4,000)(8,000/16,000) = 10,000/16,000$$

Similarly $S^g(y = 1)$ for regime $g = (0, 1)$ is $10,000/16,000$. Hence, there is, by definition, no direct effect of AZT on survival controlling for AP (when all subjects take AP) since $S^{g=(1,1)}_{(1)} = S^{g=(0,1)}_{(1)}$.

Remark: Note, $f^{g=(1,0)}(y = 1, \ell_1 = 1)$ is not identified, since evaluating (3.7) with $k = 0$ we obtain $f(y = 1 \mid \ell_1 = 1, a_0 = 1, a_1 = 0) f(\ell_1 = 1 \mid a_0 = 1)$, but the first factor is unidentified on account of the conditioning event having probability zero. This reflects the fact that (3.3) fails, since $f\left[g\left(t_1, \overline{\ell}_1 = 1\right) \mid \overline{\ell}_1 = 1, g\left(\overline{\ell}_0\right)\right] = f\left[a_1 = 0 \mid \ell_1 = 1, a_0 = 1\right] = 0$ even though $f\left(\overline{\ell}_1 = 1, g\left(\overline{\ell}_0\right)\right) = f\left(\ell_1 = 1, a_0 = 1\right) = \frac{1}{4} \neq 0$.

Remark: If L_k or A_k have continuous components then, in order to avoid measure theoretic difficulties due to the existence of different versions of conditional distributions, we shall impose the assumption that $f\left(\ell_k \mid \overline{\ell}_{k-1}, \overline{a}_{k-1}\right)$ is continuous in all arguments that represent realizations of random variables with continuous distributions, so that the right hand sides of (3.4)-(3.9) are unique. Gill and Robins (1996) discuss measure theoretic issues in greater depth.

3.4. Sequential Randomization w.r.t. Y Only

Suppose in place of (3.2) we imposed only the weaker condition that, for all k,

$$Y_g \coprod A_k \mid \overline{L}_k = \overline{\ell}_k, \overline{A}_{k-1} = g\left(\overline{\ell}_{k-1}\right). \tag{3.10}$$

When (3.10) holds, we will say that given $\overline{L}$, treatment is (sequentially) randomized with respect to Y_g. When (3.2) holds, we say that, given $\overline{L}$, treatment is (sequentially) randomized w.r.t. regime g. Robins (1992) refers to (3.10) as the assumption of no unmeasured confounders. Robins (1986, 1987) encoded (3.10) and (3.2) in event tree representations called randomized CISTGs w.r.t. Y and randomized CISTGs (RCISTGs). Although (3.10) does not logically imply (3.2), nevertheless, since physical randomization of treatment implies the stronger condition (3.2), one might wonder whether it would ever make substantive sense to impose (3.10) but not (3.2). Interestingly, Robins (1987, 1993) discussed such substantive settings. See Sec. 4 below for an example. For the moment, we restrict ourselves to determining which, if any, parts of Theorem 3.1 remain true under the weaker assumption (3.10) of sequential randomization w.r.t. Y_g.

Theorem 3.2 (Robins, 1987): If (3.1), (3.3), and (3.10) hold for regime g, then Eq. (3.9) and the right-most equalities in Eqs. (3.5) and (3.8) are true. However, $f^g\left(\ell_k \mid \overline{\ell}_{k-1}\right)$, $f^{g0}\left(\ell_k \mid \overline{\ell}_{k-1}, g\left(\overline{\ell}_{k-1}\right)\right)$, and $S^g\left(y \mid \overline{\ell}_k\right)$ are not identified.

Theorem 3.2 states that we can not identify densities that involve the counterfactual random variable $\overline{L}_g$. However, we can identify the marginal distribution of Y_g and the conditional distribution of Y_g given the observed data $\overline{L}_k, \overline{A}_{k-1} = g\left(\overline{L}_{k-1}\right)$. For completeness, a proof of Theorem (3.2) is given in Appendix 1.

4. Confounding

Suppose we believe (3.10) holds for all $g \in \mathbf{g}$ but that a subset $\overline{U}_{K+1}$ of $\overline{L}_{K+1}$ is not observed. The observed subset $\overline{O}_{K+1}$ of $\overline{L}_{K+1}$ includes the outcome variables Y. Define $\overline{U}_k = \overline{L}_k \bigcap \overline{U}_{K+1}$ and $\overline{O}_k = \overline{L}_k \bigcap \overline{O}_{K+1}$ to be the unobserved and observed non-treatment variables through time t_k. The goal of this section is to define restrictions on the joint distribution of $V = \left(\overline{L}_{K+1}, \overline{A}_K\right) \equiv \left(\overline{L}, \overline{A}\right)$ such that (3.10) will imply that for a given $g \in \mathbf{g}(O) \equiv \left\{g \in \mathbf{g}; g\left(\overline{\ell}_k\right) = g\left(\overline{o}_k\right) \text{ for all } k\right\}$

$$Y_g \coprod A_k \mid \overline{O}_k, \overline{A}_{k-1} = g\left(\overline{O}_{k-1}\right) \text{ for all } k \tag{4.1}$$

since, then, for that $g \in \mathbf{g}(O)$, by Theorem (3.2), $S^g(y)$ is identified from data $\left(\overline{A}_K, \overline{O}_{K+1}\right)$ and can be computed by the g-computation algorithm formula (3.9) with o substituted for ℓ. $\mathbf{g}(O)$ is exactly the set of treatment regimes that a person can follow without having data on $\overline{U}_{K+1}$. Denote by $b\left(y, \overline{\ell}_k, g\right)$ the right hand side of Eq. (3.8). Note $b\left(y, \overline{\ell}_k, g\right)$ is a functional of the law $F_V(v)$ of V.

Theorem: If, for a regime $g \in \mathbf{g}(\mathbf{O})$ and each y, k,

$$E\left[b\left(y, \overline{L}_k, g\right) \mid \overline{O}_k, A_k, \overline{A}_{k-1} = g\left(\overline{O}_{k-1}\right)\right]$$
$$\text{does not depend on } A_k \tag{4.2}$$

then Eq. (3.10) implies Eq. (4.1).

Proof: Given (3.10), by Theorem (3.2) we have that Eq. (3.8) implies that (4.2) is equivalent to $E\left[I\left(Y_g > y\right) \mid \overline{O}_k, A_k, \overline{A}_{k-1} = g\left(\overline{O}_{k-1}\right)\right]$ does not depend on A_k, which implies (4.1).

When (3.10) and (4.1) are true, we say that $\overline{U}_{K+1}$ is a non-confounder for the effect of $\overline{A}_K$ on Y given data on $\overline{O}_{K+1}$ (Robins, 1986).

Corollary 4.1: If for regime $g \in \mathbf{g}(\mathbf{O})$ and each k, there exists $U_{bk} \subseteq \overline{U}_k$ such that

$$b\left(y, \overline{L}_k, g\right) = b\left(y, \left(\overline{O}_k, U_{bk}\right), g\right) \tag{4.3}$$

and

$$U_{bk} \coprod A_k \mid \overline{O}_k, \overline{A}_{k-1} = g\left(\overline{O}_{k-1}\right), \tag{4.4}$$

then (3.10) implies (4.1).

Proof: The supposition of the Corollary implies Eq. (4.2).

The suppositions of Theorem (4.1) and Corollary (4.1) are difficult to check because $b\left(y, \overline{L}_k, g\right)$ is the high dimensional integral given by the RHS of (3.8). The following Corollary provides restrictions on the law $F_V(v)$ of $V = \left(\overline{L}, \overline{A}\right)$ that can be checked without integration.

<u>Corollary 4.2</u>: If for a regime $g \in \mathbf{g}(\mathbf{O})$ and each k, $\overline{U}_k = (U_{bk}, U_{ak})$, U_{bk} satisfies (4.4), and $O_k \amalg U_{ak} \mid \overline{O}_{k-1}, \overline{A}_{k-1} = g\left(\overline{O}_{k-1}\right), U_{bk}$, then (3.10) implies (4.1).

Proof: The supposition of the Corollary (4.2) imply Eqs. (4.3) and (4.4).

4.1. Confounding in DAGs

Pearl and Robins (1995) and Robins and Pearl (1996) have recently developed a graphical approach [based on representing $F_V(v)$ by a directed acyclic graph (DAG)] for determining other sufficient non-integral conditions on $F_V(v)$ under which Eq. (3.10) implies (4.1). Specifically, without loss of generality, we represent V by a complete DAG T^* consistent with the ordering of the variables $V = (V_0, \ldots, V_J)$ (Pearl, 1995). That is, V are the vertices (nodes) of T^* and $F_V(v) = \prod_{j=0}^{J} f_{V_j|Pa_j}(v_j \mid pa_j)$ where Pa_j are the parents of V_j on T^* and pa_j and v_j are realizations. By the completeness of the DAG T^*, $Pa_j = \bar{V}_{j-1}$. We can remove arrows from V_m to V_j, $m < j$, on T^* if and only if $f_{V_j|\bar{V}_{j-1}}(v_j \mid \overline{v}_{j-1})$ does not depend on v_m. Henceforth, we denote by T the DAG in which all such arrows that the investigator knows *a priori* can be removed have been removed. Thus, the resulting DAG T may no longer be complete. Now given a regime $g \in \mathbf{g}(O)$, let $\overline{Y}_k^{\dagger}$ be the smallest subset of $\overline{O}_k$ such that $g\left(\overline{O}_k\right) = g\left(\overline{Y}_k^{\dagger}\right)$, i.e., the $\bar{Y}_k^{\dagger}$ are the variables actually used to assign treatment A_k at t_k.

Let T^{gm} be the DAG T modified so that for $k \geq m$ *(i)* all directed edges from $\bar{Y}_k^{\dagger}$ to A_k are included and represent the deterministic dependence of $A_k = g\left(\overline{Y}_k^{\dagger}\right)$ on $\overline{Y}_k^{\dagger}$ under regime g, and *(ii)* there are no other edges into A_k. Abbreviate T^{g0} as T^g. We call T^g the g-manipulated graph. T^g is the manipulated graph of SGS (1993). The G-computation algorithm formula $b(y, g)$ is the marginal survival distribution of Y based on the manipulated graph (Robins, 1995b).

Let T_k^g be the DAG that has edges into $A_1, \ldots, A_k$ as in T, edges into $A_{k+1}, \ldots, A_K$ as in T^g, no edges out of A_k and is elsewhere identical to T. That is, T_k^g is $T^{g(k+1)}$ with all edges out of A_k removed. Suppose Y is a subset of the variables in $\overline{O}_{K+1}$.

We then have

<u>Theorem 4.2</u>: (Robins and Pearl, 1996). If

$$\left(Y \amalg A_k \mid \bar{O}_k, \overline{A}_{k-1}\right)_{T_k^g}, k \leq K, \tag{4.5}$$

then (3.10) implies (4.1). Here $(A \amalg B \mid C)_{T_k^g}$ stands for d-separation of A and B given C in T_k^g (Pearl, 1995). Note that checking d-separation is a purely graphical (i.e., visual) procedure that avoids any integration.

Theorem 4.3: (Robins and Pearl, 1996). Eq. (4.5) is true if and only if, for $k \leq K$, $\bar{U}_k = (U_{ak}, U_{bk})$ for possibly empty mutually exclusive sets U_{ak}, U_{bk} satisfying *(i)* $\left(U_{bk} \amalg A_k \mid \overline{O}_k, \overline{A}_{k-1}\right)_{T^{g(k+1)}}$ and *(ii)* $\left(U_{ak} \amalg Y \setminus \overline{O}_k \mid \overline{O}_k, \overline{A}_k, U_{bk}\right)_{T^{g(k+1)}}$. Here $A \setminus B$ is the set of elements in A but not in B.

Remark: U_{ak} need not be contained in $U_{a(k+1)}$, and similarly for U_{bk} and $U_{b(k+1)}$.

Example (4.1): Suppose underlying the observed variables in Figures 1 and 2 are the variables in Figure 3, where each column represents alternative but equivalent names for a particular variable.

Figure 3

V_0	V_1	V_2	V_3	V_4	V_5
U_0	A_0	U_1	O_1	A_1	$Y = O_2$
L_0	A_0	L_{11}	L_{12}	A_1	L_2
PCP	AZT	Im	PCP	AP	Survival

with $K = 1$ and $L_1 \equiv (L_{11}, L_{12})$ in Fig. 3, $O_1 \equiv L_{12}$ from Fig. 3 is L_1 of Fig. 2, and $O_2 = L_2 = Y$ of Fig. 3 is $L_2 = Y$ of Fig. 2. In Fig. 3, $U_0 = L_0$ is an unmeasured variable that represents whether a subject had PCP prior to the time t_0 of AZT treatment, and $I_m \equiv L_{11}$ is a subject's unrecorded underlying immune status between t_0 and t_1. We now assume that the data in Fig. 1 are from an observational study rather than a sequential randomized trial. Without physical sequential randomization, we can no longer know *a priori* that (3.2) holds for the variables in Figures 1 and 2. However, we shall assume that in our observational study, Eq. (3.2) holds for the variables in Fig. 3. Further, we suppose that the joint distribution of $(V_0, \ldots, V_5)$ can be represented by the DAG T in Fig. 4.

Figure 4

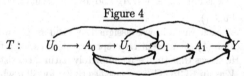

$T:$ 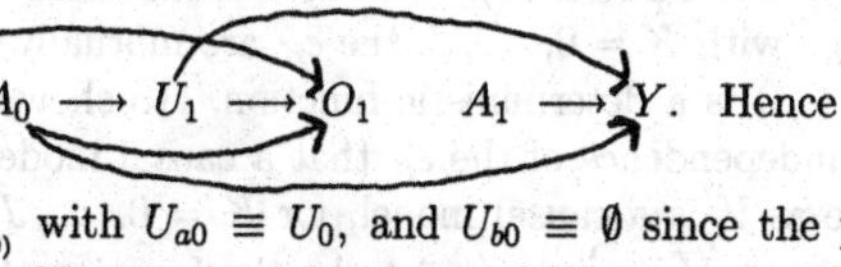

Our goal is to determine whether (4.1) holds for the variables in Fig. 3 (or, equivalently, whether (3.10) holds for the variables in Fig. 2) for the regimes $g = (a_0, a_1) = (1, 1)$ and $g = (0, 1)$. To do so, we use Theorem 4.3.

$T^{g(1+0)}$: 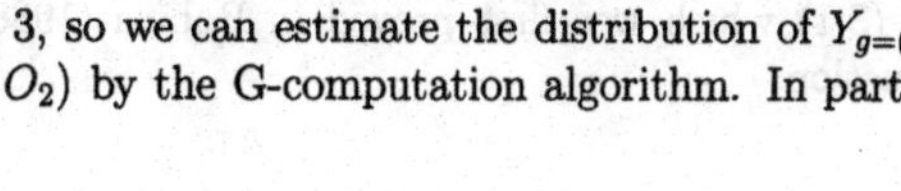 . Hence $T^{g(1+0)}$ satisfies $\left(U_{a0} \amalg Y \mid \bar{A}_0, \overline{O}_0 = \emptyset\right)_{T^{g(1+0)}}$ with $U_{a0} \equiv U_0$, and $U_{b0} \equiv \emptyset$ since the path $U_0 O_1 U_1 Y$ is blocked by the collider O_1. Hence, the hypothesis of Theorem (4.3) holds for $k = 0$. Furthermore, since $T^{g(1+1)} = T$, the hypothesis of Theorem (4.3) holds for $k = 1$ with $U_{b1} = (U_0, U_1), U_{a1} = \emptyset$. Hence, we conclude that, by Theorem (4.3), (4.1) holds for the variables in Fig. 3, so we can estimate the distribution of $Y_{g=(1,1)}$ and $Y_{g=(0,1)}$ from the data (A_0, O_1, A_1, O_2) by the G-computation algorithm. In particular, if the distribution of the observed variables in Fig. 3 (i.e., of the variables in Figs. 1 and 2) is given by Fig. 1, then, by our calculations in Sec. 3, $S^{g=(1,1)}(y=1) = S^{g=(0,1)}(y=1) = 10,000/16,000$. [However, neither the hypothesis of Theorem (4.2) nor Eq. (4.1) holds for the regime g^* of Example (3.1) in which subjects take AZT and then take AP only if they develop PCP between t_0 and t_1. In fact, $S^{g^*}(y)$ is not identified.]

Remark: As we have seen in the above example, the graphical conditions in Theorems (4.2) and (4.3) can be checked visually without performing integration. However, they require that we understand enough about the joint law $F_V(v)$ of V so that we can correctly construct an (incomplete) DAG T with appropriate arrows removed from the complete DAG T^*. Since $\bar{U}_{K+1}$ is unobserved, this understanding cannot be based on empirical associations in our data and thus must be based on prior subject matter knowledge or beliefs. In many (but not all) contexts, our prior beliefs concerning causal effects will be sharper than our beliefs concerning the magnitude of non-causal associations (since causal effects tend to be more stable across time and place). Thus, we might have sharper beliefs concerning which arrows can be removed from the complete DAG T^* to form the DAG T if each arrow on T^* represents a direct causal effect. Thus, we might like to include enough variables in $\bar{U}_{K+1}$ and thus in V such that any arrow from V_m to $V_j, j > m$, on T^* represents the causal effect of V_m on V_j controlling for the other variables in $\overline{V}_{j-1}$. Then the absence of an arrow from V_m to V_j on T would represent the absence of a direct causal effect. That is, we would like $f\left[v_j \mid \overline{v}_{j-1}\right]$ not to depend on v_m if and only if V_m has no direct effect on V_j controlling for the other variables in $\bar{V}_{j-1}$. This, of course, requires that each V_m can be regarded as a potential treatment which, as discussed in Remark (3.1), may not always be the case.

We will have accomplished this if we include a sufficient number of variables in $V \equiv \overline{V}_J$ such that,

$$\text{For each } K, K = 0, \ldots, J, \text{ Eq. (3.2) is true with } \overline{A}_K \equiv (V_0, V_1, \ldots, V_{K-1}) \qquad (*)$$

for each **g**. In that case, V_K is in $\overline{L}_{K+1} \equiv \overline{V}_J / \overline{A}_K$ and, by Theorem (3.1), $f_{V_{K,g=(\overline{v}_{K-1})}}(v_K) \equiv f^{g=(\overline{v}_{K-1})}(v_K) = f(v_K \mid \overline{v}_{K-1})$.

Remark: (*) says that each variable was assigned at random given the past. Robins (1986) encodes assumption (*) in an event tree representation called a full RCISTG (FRCISTG) as fine as V. He shows that (under a completely natural consistency assumption) a FRCISTG as fine as V is an RCISTG (i.e., satisfies (3.2)) for all g when $\overline{A}_K$ now represents any subset of the variables V. A FRCISTG as fine as V (i.e., *) is mathematically equivalent to a causal model (i.e. a non-parametric structural equations model) in the sense of Pearl (1995) and SGS (1993). A DAG T with nodes $V = \overline{V}_J$ is a causal model if $V_K = r_K(Pa_K, \epsilon_K)$ with $K = 0, \ldots, J$, the ϵ_K are mutually independent random variables, and each $r_K(\cdot,\cdot)$ is a deterministic function. To show the equivalence, it is easy to check using the independence of the ϵ_K that a causal model satisfies (*). To see that a FRCISTG as fine as V is a causal model, for $K = 0, \ldots, J$, let $Pa_K = \overline{V}_{K-1}$, define $\epsilon_K = \left\{V_{K,g=\overline{v}_{K-1}}; \overline{v}_{K-1} \in \overline{\mathbf{V}}_{K-1}\right\}$, $r_K(\cdot,\cdot)$ to be the function that selects the appropriate component of ϵ_K, and check that (*) implies mutual independence of the ϵ_K. Using this equivalence, it immediately follows that the manipulation theorem of Spirtes et al. (1993) is a corollary of Theorem (3.1) which was first proved in Robins (1986). See Robins (1995b) for additional discussion.

Example (4.2): Suppose now that, as discussed above, we included the variables U_0 and U_1 in Fig. 3 precisely so that we would believe that condition (*) holds for the entire set of 6 variables. Then (assuming our beliefs are correct) the arrows in DAG T of Fig. 4 have direct causal interpretations. We immediately see from Fig. 4 that, since PCP status prior to time 0 (U_0) determined AZT treatment in A_0, the study cannot possibly represent the sequential randomized trial described in Sec. 2 (since, in that trial, AZT treatment was determined at random by the flip of a fair coin). Further, other than the treatments A_0 and A_1, only the underlying immune status U_1 has a direct causal effect on survival (O_2) since there are no arrows from either U_0 or O_1 to O_2. That is, PCP is not a direct causal risk factor for death controlling for A_0, A_1, and U_1. Further, we observe that the AP treatment (A_1) at t_1 is only influenced by treatment A_0 and measured PCP (O_1). In particular, the decision whether to treat with AP (A_1) at t_1 did not depend on the unmeasured PCP status U_0 or the underlying immune status U_1.

Relationship of Figure 4 to Figure 1:

In the data given in Fig. 1, we have noted in Sec., 2 that PCP status ($L_1 = O_1$) is an independent risk factor for death controlling for the other measured variables A_0 and A_1. From the observed data in Fig. 1 alone, we have no way to determine whether this association is causal or not. However, our assumption (*) for the variables in Fig. 3 implies that Fig. 4 represents the underlying causal relationships. Hence the association between PCP and survival controlling for A_0 and A_1 in Fig. 1 represents the fact that the true underlying causal risk factor U_1 of Fig. 4 is not measured and thus, since U_1 causes PCP (O_1), an association between O_1 and survival (conditional on A_0 and A_1) is induced in Fig. 1.

Relationship of Figure 4 to the Causal Hypothesis of Interest:

Our hypothesis of interest, as discussed in Sec. 2, is whether there is a direct causal effect of A_0 on $Y \equiv O_2$ controlling for $A_1 = 1$. Our prior beliefs included in DAG T of Fig. 4 were not sufficient to accept or reject our hypothesis *a priori*. However, our prior beliefs were sufficiently strong that we were able to empirically test our hypothesis by applying the G-computation algorithm to the observed data in Fig. 1. We discovered that there was no direct effect of A_0 on Y controlling for A_1. This absence of a direct effect of A_0 controlling for A_1 could be due to the fact that *(a)* the arrow from A_0 to Y in Fig. 4 is (as a fact of nature) missing and, *(b)* the arrow from A_0 to U_1 is missing. [The arrow from U_1 to Y_2 cannot be missing since, in the data, $Y \not\!\perp\!\!\!\perp O_1 \mid A_0, A_1$.] Alternatively, both these arrows could be present but the magnitude of the effect of A_0 on Y controlling for A_1 represented by the arrow from A_0 to Y is exactly balanced by an equal in magnitude (but opposite in sign) effect of A_0 on Y controlling for A_1 determined by the arrows from A_0 to U_1 and from U_1 to Y. Based solely on our assumptions encoded in DAG T in Fig. 4, the assumption (*), and the data in Fig. 1, we cannot empirically discriminate between these alternative explanations. Note that the direct arrow from A_0 to Y in Fig. 4 represents precisely that part of the possible direct effect of A_0 on Y controlling for A_1 that is not through (mediated by) underlying immune status U_1.

Finally we note that Fig. 4 is an example of the phenomenon discussed in Sec. 3 in which, for the observed data in Fig. 2, (3.2) is false but (3.10) is true for the regime $g = (1, 1)$. That is, for regime $g = (1, 1)$, even if (*) holds for the variables in Fig. 4, we have randomization [in the absence of data on (U_0, U_1)] w.r.t. $Y_g \equiv O_{2g}$ but not w.r.t. O_{1g} and O_{2g} jointly. This is because the association between A_0 (AZT) and O_1 (PCP after t_0) is confounded by the unmeasured factor PCP prior to t_0 (U_0). Thus we can estimate the effect of regime $g = (1,1)$ on survival but not its effect on PCP at t_1 (O_1); as a consequence, as noted above, we cannot estimate the effect of the dynamic regime g^* of Example (3.1) on survival [i.e., Eq. (3.10) is false for Y_{g^*}].

5. The g-Null Hypothesis

In many settings, the treatments $\overline{A}_K = (A_0, \ldots, A_K)$ represent a single type of treatment given at different times. For example, all the A_k may represent AP doses at various times t_k. Often, in such settings, an important question is whether the g-null hypothesis of no effect of treatment on Y is true, i.e.,

g-Null Hypothesis:

$$S^{g_1}(y) = S^{g_2}(y) \text{ for all } y \text{ and } g_1, g_2 \in \mathbf{g}. \tag{5.1}$$

This reflects the fact that if the **g**-null hypothesis is true, the distribution of Y will be the same under any choice of regime g and thus it does not matter whether the treatment is given or withheld at any time. Now when (3.1), (3.3), and (3.10) are true for all $g \in \mathbf{g}$, Theorem (3.2) implies that the g-null hypothesis is true if and only if the following "g"-null hypothesis is true.

"g"-Null Hypothesis:

$$b(y, g_1) = b(y, g_2) \text{ for all } y \text{ and } g_1, g_2 \in \mathbf{g} \tag{5.2}$$

where $b(y, g)$ is given by the rightmost side of Eq. (3.9) and depends only on the joint distribution $F_V(v)$ of the observables.

<u>**Remark 5a**</u>: The "g"-null hypothesis is not implied by the weaker condition that for all $g \in \mathbf{g}$, (3.1), (3.3), and (3.10) are true, and $b(y, g = (\overline{a}_1)) = b(y, g = (\overline{a}_2))$ for all non-dynamic regimes $\overline{a}_1 \equiv \overline{a}_{1K}$ and $\overline{a}_2 \equiv \overline{a}_{2K}$. However, the following Lemma is true.

<u>Lemma</u>: The "g"-null hypothesis is true if and only if

$$b\left(y, \overline{\ell}_k, g = (\overline{a}_1)\right) = b\left(y, \overline{\ell}_k, g = (\overline{a}_2)\right) \text{ for each } y, k, \overline{\ell}_k, \overline{a}_1, \overline{a}_2 \text{ with } \overline{a}_1 \text{ and } \overline{a}_2 \text{ agreeing through } t_{k-1}, \text{ i.e., } \overline{a}_{1(k-1)} = \overline{a}_{2(k-1)}.$$

<u>**Corollary**</u>: If (3.1), (3.3), and (3.10) hold for all non-dynamic regimes $g = (\overline{a})$, and, for each $y, \overline{\ell}_k, \overline{a}_{k-1}$, $S^{g_1 0}\left(y \mid \overline{\ell}_k, \overline{a}_{k-1}\right) = S^{g_2 0}\left(y \mid \overline{\ell}_k, \overline{a}_{k-1}\right)$ for any $g_1 = (\overline{a}_1)$, and $g_2 = (\overline{a}_2)$ with $\overline{a}_{1(k-1)} = \overline{a}_{2(k-1)} = \overline{a}_{(k-1)}$, then the "g"-null hypothesis holds. The usefulness of this Corollary is that we only require randomization with respect to Y_g for non-dynamic regimes.

<u>**Remark 5b**</u>: The above results can be generalized to the case where (3.3) may not hold for all $g \in \mathbf{g}$. Let $\mathbf{g}^*$ be the subset of $\mathbf{g}$ on which $b(y, g)$ is defined. $\mathbf{g}^*$ may be strictly contained in $\mathbf{g}$ if (3.3) is not always true. Define the $\mathbf{g}^*$-null hypothesis and the "$\mathbf{g}^*$"-null hypothesis as in (5.1) and (5.2), except with $\mathbf{g}^*$ replacing $\mathbf{g}$. Then we have that the $\mathbf{g}^*$-null hypothesis is true if and only if the "$\mathbf{g}^*$"-null hypothesis is true.

<u>**Remark 5c**</u>: Difficulties in testing the "$\mathbf{g}^*$"-null hypothesis: The following difficulties arise when attempting to construct tests of the "g"-null hypothesis or the "$\mathbf{g}^*$"-null hypothesis.

First each $b(y,g)$ is a high dimensional integral that cannot be computed analytically and thus must be evaluated by a Monte Carlo approximation - the Monte Carlo G-computation algorithm described by Robins (1987, 1989). Second the cardinality of **g** is enormous. However, as we now discuss, the greatest difficulty is attributable to the fact that the "**g**"-null hypothesis does not imply either that the component $f\left(\ell_j \mid \overline{\ell}_{j-1}, \overline{a}_{j-1}\right)$ of $b(y,g)$ does not depend on $\overline{a}_{j-1}$ or that the component $S\left(y \mid \overline{\ell}_K, \bar{a}_K\right)$ does not depend on $\bar{a}_K$. Rather, it only implies that the entire integral $b(y,g)$ does not depend on g. In many settings where the "g"-null or "**g***"-null hypothesis holds, each of the above components of $b(y,g)$ will depend on the treatment history.

Example (5.1): In a modified Fig. 1 in which the survival rate in the 4,000 subjects with history $(A_0 = 1, L_1 = 0, A_1 = 0)$ is .75 rather than .25, the "**g***-null" hypothesis is true [since $b(y,g)$ equal 10,000/16,000 for all three g for which $b(y,g)$ can be computed, i.e., the regimes $g = (1,1)$, $g = (0,1)$, and g^* of example (3.1)]. Yet both of the above components of $b(y,g)$ depend on treatment history $\overline{a}$. Further, we have seen in Example (4.2) that the data in Fig. 1 or the modified version of Fig. 1 could be obtained from a fairly realistic setting in which the underlying causal theory is given by DAG T of Fig. 4 under assumption (*).

This example is a particular case of the following general result.

Lemma: If *(i)* $Y = L_{K+1}$; *(ii)* $\overline{L}_K$ is an independent predictor of Y, i.e. $f\left(y \mid \overline{\ell}_K, \overline{a}_K\right)$ depends on $\overline{\ell}_K$; *(iii)* $\overline{A}_{j-1}$ is the independent predictor of L_j, i.e., $f\left(\ell_j \mid \overline{\ell}_{j-1}, \overline{a}_{j-1}\right)$ depends on $\overline{a}_{j-1}$; *(iv)* there is no effect of treatment on the outcome Y so the **g**-null hypothesis holds; and *(v)* (3.1), (3.3), and (3.10) hold as in a sequential randomized trial, so the "**g**"-null hypothesis is true; then $f\left(y \mid \overline{\ell}_K, \overline{a}_K\right)$ must also depend on $\overline{a}_K$. The suppositions of the Lemma will hold in a sequential randomized trial in which the time-varying treatment has no effect on the outcome of interest Y, but the treatment does effect an intermediate outcome that (like PCP) *(i)* is not itself a direct cause of Y, but *(ii)* is caused by and thus correlated with an unmeasured direct causal risk factor for Y (such as immune status).

When both $f\left(y \mid \overline{\ell}_K, \overline{a}_K\right)$ and $f\left(\ell_j \mid \overline{\ell}_{j-1}, \overline{a}_{j-1}\right)$ depend on a-history and ℓ_j is multivariate with continuous components (so, given realistic sample sizes, parametric or semiparametric models are required), the most straightforward approach to testing the "**g**"-null hypothesis will fail.

Specifically, the most straighforward parametric likelihood-based approach to estimation of $b(y,g)$ is to specify parametric models $f\left(\ell_j \mid \overline{\ell}_{j-1}, \overline{a}_{j-1}; \Delta\right)$ and $S\left(y \mid \overline{\ell}_K, \overline{a}_K; \theta\right)$. This parameterization corresponds to the standard parameterization of a DAG model, since we are parameterizing the conditional distribution of each variable given its parents. The maximum likelihood estimators $\widehat{\Delta}$ and $\widehat{\theta}$ of Δ and θ are obtained by maximizing $\prod_i f\left(Y_i \mid \overline{L}_{Ki}, \overline{A}_{Ki}; \theta\right) \prod_{j=0}^{K} f\left(L_{ji} \mid \overline{L}_{(j-1)i}, \overline{A}_{(j-1)i}; \Delta\right)$. $b(y,g)$ is then estimated (using Monte Carlo integration) with model derived estimators substituted for the unknown conditional laws in the rightmost side of (3.9). Robins (1986) provided a worked examples of this approach in the context of a survival outcome Y. However, it is clear from inspecting the right-hand side of (3.9) that there will in general be no parametric subvector, say ψ, of (θ, Δ) that will take a fixed value, say 0, whenever the "g"-null hypothesis (5.2) holds. In particular, this problem will arise whenever, as in Example 5.1, the "g"-null hypothesis (5.2) holds but both $f\left(\ell_j \mid \overline{\ell}_{j-1}, \overline{a}_{j-1}; \Delta\right)$ and $S\left(y \mid \overline{\ell}_K, \overline{a}_K; \theta\right)$ depend on $\overline{a}$ history. In fact, I believe that, for non-linear models, it is essentially impossible to specify an unsaturated parametric or semiparametric model for which there exists a parameter vector, (Δ^*, θ^*) say, such that $f\left(\ell_j \mid \overline{\ell}_{j-1}, \overline{a}_{j-1}; \Delta^*\right)$ depends on $\overline{a}_{j-1}$ and $f\left(y \mid \overline{\ell}_K, \overline{a}_K; \theta^*\right)$ depends on both $\overline{\ell}_K$ and $\overline{a}_K$ and yet the "g"-null hypothesis holds. As a consequence, in settings like that of Example (5.1), the "g"-null hypothesis, even when true, will almost certainly be rejected whatever data are obtained! Specifically, suppose each L_k is discrete with p-levels, each A_k is discrete with p^*-levels and $Y = L_{K+1}$. Then it follows from Eq. (6.1) below that the "g"-null hypothesis implies $\Omega \equiv (p^* - 1)(p - 1)\left\{(p^* p)^{K+1} - 1\right\} p / (p^* p - 1)$ independent equality constraints on the joint distribution of the observables so, in general, most non-linear models with fewer than Ω parameters will reject the "g"-null hypothesis before any data are obtained. If either L_k or A_k has continuous components, an uncountable number of equality constraints must hold and the above difficulty will be even worse. In contrast, linear models need not imply pre-data rejection of the "g"-null hypothesis. For example, suppose $V = \left(\overline{A}_1, \overline{L}_2\right)$, $Y = L_2$, $L_0 \equiv 0$ and all variables are continuous. Suppose we specify the linear models $Y \mid \overline{A}_1, \overline{L}_1 \sim N\left(\beta_0 + \beta_1 A_0 + \beta_2 A_1 + \beta_3 L_1, \sigma_y^2\right)$, $L_1 \mid A_0 \sim N\left(\alpha_0 + \alpha_1 A_0, \sigma_{\ell_1}^2\right)$. Then a sufficient condition for the "g"-null hypothesis to hold is easily shown to be $\beta_2 = 0$ and $\beta_1 + \beta_3 \alpha_1 = 0$. The above difficulty in testing the "g"-null hypothesis under the standard DAG parameterization motivated the development of the structural nested models described in Sec. 7. Further motivation for these models is now provided by describing a proper approach to testing the "g"-null hypothesis.

6. g-Null Theorem and g-Null Tests

A proper approach to testing the "g"-Null Hypothesis (5.2) is based on the following g-null theorem:

g-null theorem: The "g"-Null Hypothesis (5.2) is true $\Leftrightarrow$

$$Y \coprod A_k \mid \overline{L}_k, \overline{A}_{k-1}, \quad k = (0, \ldots, K) \tag{6.1}$$

Example: In the modified version of Fig. 1 described in Example (5.1),

$$f(y = 1 \mid a_0 = 0) = f(y = 1 \mid a_0 = 1) = 10/16$$

and

$$f(y = 1 \mid a_0 = 1, \ell_1 = 0, a_1 = 1) = f(y = 1 \mid a_0 = 1, \ell_1 = 0, a_1 = 0) = 3/4 \tag{6.2}$$

so (6.1) is true (as it must be) since we showed in Example (5.1) above that (5.2) is true. However, in the unmodified Fig. 1, (6.2) is false, which implies (5.2) must be false; in particular, $10/16 = b(y, g = (1,1)) = b(y, g = (0,1)) \neq b(y, g^*)$ with regime g^* as defined in Example (3.1).

Remark: If L_k is discrete with p-levels, A_k is discrete with p^*-levels, and $Y = L_{K+1}$, by writing (6.1) as $f\left(A_k \mid \overline{L}_k, \overline{A}_{k-1}, Y\right) = f\left(A_k \mid \overline{L}_k, \overline{A}_{k-1}\right)$, $k = (0, \ldots, K)$, it is straightforward to show that the joint distribution of the observables satisfies
$(p - 1)(p^* - 1)\left\{(p^* p)^{K+1} - 1\right\} p / (p^* p - 1)$ independent equality constraints under the "g"-null hypothesis.

Remark: The g-null theorem may appear less surprising upon realizing that, given (3.10) is true for all $g \in \mathbf{g}$, the following sharp null hypothesis implies both (5.2) and (6.1).

$$\text{Sharp g-null hypothesis} : Y = Y_{g_1} = Y_{g_2} \text{ with probability 1 for all } g_1, g_2 \in \mathbf{g} \tag{6.3}$$

Specifically (3.10) plus (6.3) implies (6.1). Further, (6.3) implies the g-null hypothesis (5.1) which, given (3.10), implies (5.2).

We use (6.1) to construct tests of the "g"-null hypothesis. Consider first the case in which A_k is dichotomous, Y is univariate, and L_k is discrete with a small number of levels. Let $Num\left(\overline{\ell}_k, \overline{a}_{k-1}\right)$ and $V\left(\overline{\ell}_k, \overline{a}_{k-1}\right)$ be the numerator and its estimated variance of a two sample test (e.g., t-test, log-rank or Mantel extension test, Wilcoxon test) of the hypothesis that subjects with history $a_k = 1, \overline{\ell}_k, \overline{a}_{k-1}$ have the same distribution of Y as subjects with history $a_k = 0, \overline{\ell}_k, \overline{a}_{k-1}$, satisfying, under (6.1), $E\left[Num\left(\overline{\ell}_k, \overline{a}_{k-1}\right)\right] = 0$ and $E\left\{V\left(\overline{\ell}_k, \overline{a}_{k-1}\right)\right\} = Var\left\{Num\left(\overline{\ell}_k, \overline{a}_{k-1}\right)\right\}$. This relationship will be satisfied for most two sample tests. Let $Num = \sum_{k=0}^{K} \sum_{\overline{\ell}_k} \sum_{\overline{a}_{k-1}} Num\left(\overline{\ell}_k, \overline{a}_{k-1}\right)$ and $V = \sum_{k=0}^{K} \sum_{\overline{\ell}_k} \sum_{\overline{a}_{k-1}} V\left(\overline{\ell}_k, \overline{a}_{k-1}\right)$. It can be shown that, under (6.1), the terms in Num are all uncorrelated (Robins, 1986). Thus, under regularity conditions, $Num/\sqrt{V}$ will have an asymptotic standard normal distribution (Robins, 1986) and will thus constitute a non-parametric test of the "g"-null hypothesis (5.2). This will be a non-parametric test of the g-null hypothesis (5.1) if Eq. (3.10) holds for all $g \in \mathbf{g}$. However, the power of this test may be poor. This reflects the fact that for k greater than 2 or 3, $Num\left(\overline{\ell}_k, \overline{a}_{k-1}\right)$ will usually be zero because it would be rare (with the sample sizes occurring in practice) for 2 subjects to have the same $\overline{\ell}_k, \overline{a}_{k-1}$ history but differ on a_k.

To obtain greater power for discrete a_k and ℓ_k and to handle a_k and ℓ_k with continuous components, tests of (6.1) can be based on parametric or semiparametric models. For instance, suppose A_k is dichotomous and we can correctly specify a logistic model

$$f\left(A_m = 1 \mid \overline{L}_m, \overline{A}_{m-1}\right) = \left\{1 + \exp\left(-\alpha_0' W_m\right)\right\}^{-1} \tag{6.4}$$

where W_m is a p-dimensional function of $\overline{L}_m, \overline{A}_{m-1}$.

Let $Q_m \equiv q\left(Y, \overline{L}_m, \overline{A}_{m-1}\right)$ where $q(\cdot,\cdot,\cdot)$ is any known real-valued function chosen by the data analyst. Let θ be the coefficient of Q_m when θQ_m is added to the regressors $\alpha_0' W_m$ in (6.4). If, for each m, (6.4) is true for some α_0, then hypothesis (6.1) is equivalent to the hypothesis the true value θ_0 of θ is zero. A score test of the hypothesis $\theta_0 = 0$ can then be computed using logistic regression software where, when fitting the logistic regression model, each subject is regarded as contributing $K+1$ independent observations - one at each treatment time $t_0, t_1, \ldots, t_K$. Robins (1992) provides mathematical justification. This is a semiparametric test since it only requires we specify a parametric model for a L.H.S. of (6.4) rather than for the entire joint distribution of V. As discussed later, the choice of the function $q(\cdot,\cdot,\cdot)$ will effect the power of the test but not its α-level.

Robins (1992) refers to these tests as the g-tests. They are extensions of the propensity score methods of Rosenbaum and Rubin (1983) and Rosenbaum (1984) to time-dependent treatments and covariates.

7. Structural Nested Models:

In this Section we motivate, define, and characterize the class of structural nested models. In this Section, we shall only consider the simplest structural nested model - a structural nested distribution model for a univariate continuous outcome measured after the final treatment time t_K. Robins (1989, 1992, 1994, 1995) considers generalizations to discrete outcomes, multivariate outcomes, and failure time outcomes. Discrete outcomes are also treated in Sec. 8 below.

7.1. Motivation for Structural Nested Models

We assume L_{K+1} is a univariate continuous-valued random variable with a continuous distribution function and denote it by Y. To motivate SNMs, we first note that none of the tests of the "g"-null hypothesis discussed in Sec. 6 was a parametric likelihood-based test. Further, our g-test of the "g"-null hypothesis is unlinked to any estimator of $b(y,g)$ based on the G-computation algorithm. Our first goal in this subsection will be to derive a complete reparameterization of the joint distribution of $V \equiv \left(\overline{L}, \overline{A}\right)$ that will offer a unified likelihood-based approach to testing the "g"-null hypothesis (5.2) and estimating the function $b(y,g)$. Then in Sec. 8 we will develop a unified approach to testing (5.2) and estimating $b(y,g)$ based on the semiparametric g-test of Sec. 6.

7.2. New Characterizations of the "g"-Null and g-Null Hypotheses

Denote by $b\left(y, \overline{\ell}_k, g, F\right)$ the right-hand side of (3.8) where the dependence on the law F of V is made explicit. The first step in constructing a reparameterization of the likelihood is a new characterization of the "g"-null hypothesis (5.2). We assume the conditional distribution of Y given$\left(\overline{\ell}_m, \overline{a}_m\right)$ has a continuous density with respect to Lebesgue measure. Given any treatment history $\overline{a} = \overline{a}_K$, adopt the convention that $(\overline{a}_m, 0)$ will be the treatment history $\overline{a}_K^{(1)}$ characterized by $a_k^{(1)} = a_k$ if $k \leq m$ and $a_k^{(1)} = 0$ if $k > m$.

Let $\gamma\left(y, \overline{\ell}_m, \overline{a}_m, F\right)$ be the unique function satisfying

$$b\left[\gamma\left(y, \overline{\ell}_m, \overline{a}_m, F\right), \overline{\ell}_m, g = (\overline{a}_{m-1}, 0), F\right] = b\left(y, \overline{\ell}_m, g = (\overline{a}_m, 0), F\right) \tag{7.1}$$

It follows from its definition that: *(a)* $\gamma\left(y, \overline{\ell}_m, \overline{a}_m, F\right) = y$ if $a_m = 0$; *(b)* $\gamma\left(y, \overline{\ell}_m, \overline{a}_m, F\right)$ is increasing in y; and *(c)* the derivative of $\gamma\left(y, \overline{\ell}_m, \overline{a}_m, F\right)$ w.r.t. y is continuous. Examples of such functions are

$$\gamma\left(y, \overline{\ell}_m, \overline{a}_m, F\right) = y + 2a_m + 3a_m a_{m-1} + 4a_m w_m \tag{7.2}$$

where w_m is a given univariate function of $\overline{\ell}_m$ and

$$\gamma\left(y, \overline{\ell}_m, \overline{a}_m, F\right) = y \exp\left\{2a_m + 3a_m a_{m-1} + 4a_m w_m\right\} \tag{7.3}$$

Our interest in $\gamma\left(y, \overline{\ell}_m, \overline{a}_m, F\right)$ is based on the following theorem proved in Robins (1989, 1995a).

Theorem (7.1): $\gamma\left(y, \overline{\ell}_m, \overline{a}_m, F\right) = y$ for all $y, m, \overline{\ell}_m, \overline{a}_m$ if and only if the "g"-null hypothesis (5.2) holds.

Interpretation in terms of counterfactuals: Although Theorem (7.1) is a theorem referring only to the joint distribution of the observables, it is of interest to examine its implications for the distribution of the counterfactual variables under (3.1) and (3.10). Let $\gamma^{\dagger}\left(y, \overline{\ell}_m, \overline{a}_m\right)$ be the unique function such that, conditional on the observed data $\overline{\ell}_m, \overline{a}_m$, $\gamma^{\dagger}\left(Y_{g=(\overline{a}_m,0)}, \overline{\ell}_m, \overline{a}_m\right)$ has the same distribution as $Y_{g=(\overline{a}_{m-1},0)}$. If conditional on $\overline{\ell}_m, \overline{a}_m$, $\gamma^{\dagger}\left(Y_{g=(\overline{a}_m,0)}, \overline{\ell}_m, \overline{a}_m\right) = Y_{g=(\overline{a}_{m-1},0)}$ with probability 1, we say that we have local rank preservation since then, conditionally, $y_{ig=(\overline{a}_m,0)} > y_{jg=(\overline{a}_m,0)} \Leftrightarrow y_{ig=(\overline{a}_{m-1},0)} > y_{jg=(\overline{a}_{m-1},0)}$ so the ranks (orderings) of any two subjects i's and j's Y values will be preserved under these alternative treatment regimes. It is clear that, if we have local rank preservation, $\gamma^{\dagger}\left(y, \overline{\ell}_m, \overline{a}_m\right)$ applied to $Y_{g=(\overline{a}_m,0)}$ removes the effect on Y of a final brief blip of treatment of magnitude a_m at t_m leaving us with $Y_{g=(\overline{a}_{m-1},0)}$. We thus call $\gamma^{\dagger}\left(y, \overline{\ell}_m, \overline{a}_m\right)$ the "blip-down" function. Local rank preservation is a strong untestable assumption that we would rarely expect to hold, but even without local rank preservation we still view $\gamma^{\dagger}\left(y, \overline{\ell}_m, \overline{a}_m\right)$ as representing the effect of a final blip of treatment of size a_m since its maps (conditional on $\overline{\ell}_m, \overline{a}_m$) quantiles of $Y_{g=(\overline{a}_m,0)}$ into those of $Y_{g=(\overline{a}_{m-1},0)}$: from its definition $\gamma^{\dagger}\left(y, \overline{\ell}_m, \overline{a}_m\right)$ satisfies

$$S^{g=(\overline{a}_{m-1},0),0}\left\{\gamma^{\dagger}\left(y, \overline{\ell}_m, \overline{a}_m\right) \mid \overline{\ell}_m, \overline{a}_m\right\} = S^{g=(\overline{a}_m,0),0}\left(y \mid \overline{\ell}_m, \overline{a}_m\right) \tag{7.4}$$

But, given (3.1), (3.3), and the sequential randomization assumption (3.10) hold for all $g \in \mathbf{g}$, we have that Theorem (3.2) and Eq. (3.8) imply $\gamma^{\dagger}\left(y, \overline{\ell}_m, \overline{a}_m\right) = \gamma\left(y, \overline{\ell}_m, \overline{a}_m\right)$. Hence, Theorem (7.1) states:

> Given (3.1), (3.3), and (3.10) hold for all $g \in \mathbf{g}$, there is no effect of a final blip of treatment of size a_m among subjects with each observed history $\overline{a}_m, \overline{\ell}_m$ [i.e., $\gamma^{\dagger}\left(y, \overline{\ell}_m, \overline{a}_m\right) = y$] if and only if the g-null hypothesis (5.1) is true.

Remark: If the randomization assumption (3.10) is false, $\gamma^{\dagger}\left(y, \overline{\ell}_m, \overline{a}_m\right) = y$ does not imply the g-null hypothesis: for example, if $K = 2$ then, for subjects with observed history $(A_0 = 0, A_1 = 1)$, there may be an effect on $Y = L_2$ of the treatment a_1 when following the non-dynamic regime $g = (a_0 = 1, a_1 = 1)$.

Remark: If (3.2) is also true for all $g \in \mathbf{g}$, it follows from Theorem (3.1) and Eq. (3.8) that $\gamma^{\dagger}\left(y, \overline{\ell}_m, \overline{a}_m\right)$ also satisfies the following relationship among strictly counterfactual variables.

$$S^{g=(\overline{a}_{m-1},0)}\left\{\gamma^{\dagger}\left(y, \overline{\ell}_m, \overline{a}_m\right) \mid \overline{\ell}_m\right\} = S^{g=(\overline{a}_m,0)}\left(y \mid \overline{\ell}_m\right) \tag{7.5}$$

7.3. Pseudo-Structural and Structural Nested Distribution Models

In view of theorem (7.1), our approach will be to construct a parametric model for $\gamma(y, \overline{\ell}_m, \overline{a}_m, F)$ depending on a parameter ψ such that $\gamma(y, \overline{\ell}_m, \overline{a}_m, F) = y$ if and only if the true value ψ_0 of the parameter is 0. We will then reparameterize the likelihood of the observables $(\overline{L}, \overline{A})$ in terms of a random variable which is a function of the observables and the function $\gamma(y, \overline{\ell}_m, \overline{a}_m, F)$. As a consequence, likelihood-based tests of the hypothesis $\psi_0 = 0$ will produce likelihood-based tests of the "g"-null hypothesis (5.2).

Definition: The distribution $F \equiv F_V$ of $V \equiv \left(\overline{L}, \overline{A}\right)$ follows a pseudo-structural nested distribution model (PSNDM) $\gamma^*\left(y, \overline{\ell}_m, \overline{a}_m, \psi_0\right)$ if $\gamma\left(y, \overline{\ell}_m, \overline{a}_m, F\right) = \gamma^*(y, \overline{\ell}_m, \overline{a}_m, \psi_0)$ where

$\gamma^*(\cdot,\cdot,\cdot,\cdot)$ is a known function; (2) ψ_0 is a finite vector of unknown parameters to be estimated taking values in R^p; (3) for each value of $\psi \epsilon R^p$, $\gamma^*\left(y,\bar{\ell}_m,\bar{a}_m,\psi\right)$ satisfies the conditions *(a)*, *(b)*, *(c)* and *(d)* that were satisfied by $\gamma\left(y,\bar{\ell}_m,\bar{a}_m,F\right)$; (4) $\partial\gamma^*\left(y,\bar{\ell}_m,\bar{a}_m,\psi\right)/\partial\psi'$ and $\partial^2\gamma^*\left(y,\bar{\ell}_m,\bar{a}_m,\psi\right)/\partial\psi'\partial t$ are continuous for all ψ; and (5) $\gamma^*\left(y,\bar{\ell}_m,\bar{a}_m,\psi\right)=y$ if $\psi=0$ so that $\psi_0=0$ represents the "g-" null hypothesis (5.2).

We say the law of $\left(\overline{L},\overline{A},\left\{Y_g,\overline{L}_g,\overline{A}_g; g\in \mathbf{g}\right\}\right)$ follows a structural nested distribution model (SNDM) $\gamma^*\left(y,\bar{\ell}_m,\bar{a}_m,\psi_0\right)$ if the definition of a pseudo-structural nested model holds when we replace $\gamma\left(y,\bar{\ell}_m,\bar{a}_m,F\right)$ by $\gamma^\dagger\left(y,\bar{\ell}_m,\bar{a}_m\right)$. Theorem (3.2) implies that, given (3.1), (3.3), and (3.10) hold for all $g\in\mathbf{g}$, the function $\gamma^*\left(y,\bar{\ell}_m,\bar{a}_m,\psi_0\right)$ is a structural nested model if and only if it is a pseudo-structural nested model.

Examples of appropriate functions $\gamma^*\left(y,\bar{\ell}_m,\bar{a}_m,\psi\right)$ can be obtained from Eqs. (7.2) and (7.3) by replacing the quantities 2, 3 and 4 by the components of $\psi=(\psi_1,\psi_2,\psi_3)$. We call models for $\gamma\left(y,\bar{\ell}_m,\bar{a}_m,F\right)$ pseudo-structural because, although they mimic structural nested models, pseudo-SNDM are models for the observable rather than the counterfactual random variables.

7.4. The Reparameterization of the Likelihood

Next we recursively define random variables $H_K(\gamma),\ldots,H_0(\gamma)$ that depend on the observables $V\equiv\left(\overline{L},\overline{A}\right)$ as follows. $H_K(\gamma)\equiv\gamma\left(Y,\overline{L}_K,\overline{A}_K,F\right)$, $H_m(\gamma)\equiv h_m\left(Y,\overline{L}_K,\overline{A}_K,\gamma\right)\equiv \gamma\left(H_{m+1}(\gamma),\overline{L}_m,\overline{A}_m,F\right)$, and $H(\gamma)\equiv H_0(\gamma)$. Note by Theorem (7.1) if the "g"-null hypothesis (5.2) is true, then $H(\gamma)=Y$.

<u>Example:</u> If $\gamma\left(y,\bar{\ell}_m,\bar{a}_m,F\right)$ is given by Eq. (7.2), then $H_m(\gamma)=$ $Y+\sum_{k=m}^{K}(2A_k+3A_kA_{k-1}+4A_kW_k)$.

Interpretation in terms of Counterfactuals: Let $H\left(\gamma^\dagger\right)$ be defined like $H(\gamma)$ but with the (counterfactual) blip-down function $\gamma^\dagger$ replacing the function γ. Robins (1993) proved that $H\left(\gamma^\dagger\right)$ has, under consistency assumption (3.1), the same marginal distribution as $Y_{g=(0)}$, the counterfactual variable when treatment is always withheld. Following Robins et al. (1992), we say $H\left(\gamma^\dagger\right)$ is the value of Y when "blipped all the way down."

Since $\gamma\left(Y,\overline{L}_m,\overline{A}_m,F\right)$ is increasing in Y, the map from $V\equiv\left(Y,\overline{L}_K,\overline{A}_K\right)$ to $\left(H(\gamma),\overline{L}_K,\overline{A}_K\right)$ is 1-1 with a strictly positive Jacobian determinant. Therefore, $f_{Y,\overline{L}_K,\overline{A}_k}\left(Y,\overline{L}_K,\overline{A}_K\right)=\{\partial H(\gamma)/\partial Y\}\,f_{H(\gamma),\overline{L}_K,\overline{A}_K}\left(H(\gamma),\overline{L}_K,\overline{A}_K\right)$. However, Robins (1989, 1995) proves that

$$A_m \amalg H(\gamma)\mid \overline{L}_m,\overline{A}_{m-1} \tag{7.6}$$

It follows that

$$f_{Y,\overline{L}_K,\overline{A}_K}\left(Y,\overline{L}_K,\overline{A}_K\right)=\{\partial H(\gamma)/\partial Y\}\,f\{H(\gamma)\}$$

$$\prod_{m=0}^{K} f\left(L_m\mid \overline{L}_{m-1},\overline{A}_{m-1},H(\gamma)\right)f\left[A_m\mid \bar{L}_m,\bar{A}_{m-1}\right] \tag{7.7}$$

(7.7) is the aforementioned reparameterization of the density of the observables.

Thus we have succeeded in reparameterizing the joint density of the observables in terms of the function $\gamma\left(y,\overline{\ell}_m,\overline{a}_m,F\right)$, its derivative with respect to y, and the densities $f\left[\ell_m \mid \overline{\ell}_{m-1},\overline{a}_{m-1},h\left(\gamma\right)\right]$, $f\left(h\left(\gamma\right)\right)$, and $f\left[a_m \mid \overline{\ell}_m,\overline{a}_{m-1}\right]$.

We can then specify a fully parametric model for the joint distribution of the observables by specifying *(a)* a pseudo-SNFTM $\gamma^*\left(y,\overline{\ell}_m,\overline{a}_m,\psi_0\right)$ for $\gamma(y,\overline{\ell}_m,\overline{a}_m,F)$, and *(b)* parametric models $f\left[\ell_m \mid \overline{\ell}_{m-1},\overline{a}_{m-1},h\left(\psi_0\right);\phi_0\right]$, $f\left(h\left(\psi_0\right);\eta_0\right)$, and $f\left[a_m \mid \overline{\ell}_m,\overline{a}_{m-1};\alpha_0\right]$ for the above densities, where $h(\psi)$ is defined like $h(\gamma)$ except $\gamma^*\left(y,\overline{\ell}_m,\overline{a}_m,\psi\right)$ replaces $\gamma(y,\overline{\ell}_m,\overline{a}_m,F)$. It follows from (7.7) that the maximum likelihood estimates of (ϕ_0,η_0,ψ_0) are the values of $\left(\widehat{\phi},\widehat{\eta},\widehat{\psi}\right)$ that maximize

$$\prod_i \left\{\left[\frac{\partial h_i\left(\psi\right)}{\partial y_i}\right] \bullet f\left(h_i\left(\psi\right);\eta\right)\right\} \bullet \prod_{m=0}^{m=K} f\left(\ell_{m,i} \mid \overline{\ell}_{m-1,i},\overline{a}_{m-1,i},h_i\left(\psi\right);\phi\right) \qquad (7.8)$$

Since the "g"-null hypothesis (5.2) is equivalent to the hypothesis that $\psi_0 = 0$, the reparameterization (7.7) has allowed us to construct fully parametric likelihood-based tests of the "g"-null hypothesis (5.2) based on the Wald, score, or likelihood ratio test for $\psi_0 = 0$.

7.5. Estimation of $b\left(y,g\right)$

If our fully parametric likelihood-based test of the null hypothesis $\psi_0 = 0$ rejects, we would wish to employ these same parametric models to estimate $b\left(y,g\right)$ for each $g \in \mathbf{g}$. We shall accomplish this goal in two steps. First we provide a Monte Carlo algorithm which produces independent realizations of a random variable whose survivor function is $b\left(y,g\right)$.

In order to accomplish this, we shall use the function $q(h,\overline{\ell}_m,\overline{a}_m,F)$ defined recursively for all $\overline{\ell}_m$, $\overline{a}_m$ and h as follows. $q(h,\overline{\ell}_0,\overline{a}_0,F) = \gamma^{-1}\left(h,\overline{\ell}_0,\overline{a}_0,F\right)$ where the "blip-up function" $\gamma^{-1}\left(h,\overline{\ell}_k,\overline{a}_k,F\right) \equiv y$ if $\gamma\left(y,\overline{\ell}_k,\overline{a}_k,F\right) = h$. For $1 \leq k \leq m$,

$$q(h,\overline{\ell}_k,\overline{a}_k,F) = \gamma^{-1}(q(h,\overline{\ell}_{k-1},\overline{a}_{k-1},F),\overline{\ell}_k,\overline{a}_k,F).$$

Example: If $\gamma\left(y,\overline{\ell}_m,\overline{a}_m,F\right)$ is given by Eq. (7.2), $q\left(h,\overline{\ell}_k,\overline{a}_k,F\right) = h - \sum_{m=0}^{k}\left(2a_m + 3a_m a_{m-1} + 4a_m w_m\right)$.

MC Algorithm: Given a regime g:

Step (1): Set $v = 1$

Step (2): Draw h_v from $f_{H(\gamma)}(h)$

Step (3): Draw $\ell_{o,v}$ from $f[\ell_o \mid h_v]$

Step (4): Do for $m = 1,\ldots,K$

Step (5): Draw $\ell_{m,v}$ from $f[\ell_m \mid \overline{\ell}_{m-1,v},g(\overline{\ell}_{m-1,v}),h_v]$.

Step (6): Compute $y_{v,g} = q\left(h_v,\overline{\ell}_{K,v},g\left(\overline{\ell}_{K,v}\right),F\right)$ and return to Step (1).

Robins (1989, 1995) shows that the $(y_{1,g},\ldots,y_{v,g},\ldots)$ are independent realizations of a random variable with survivor function $b\left(y,g\right)$.

Second, since $f_{H(\gamma)}(h)$ and $f[\ell_m \mid \overline{\ell}_{m-1,v}, g(\overline{\ell}_{m-1,v}), h_v]$ are unknown, in practice, we will draw h_v from $f(h;\widehat{\eta})$ and $\ell_{m,v}$ from $f[\ell_m \mid \overline{\ell}_{m-1,v}, g(\overline{\ell}_{m-1,v}), h_v; \widehat{\phi}]$. Also, will we use $q^*[h_v, \overline{\ell}_{m-1,v}, g(\overline{\ell}_{m-1,v}), \widehat{\psi}]$ in place of the unknown $q[h_v, \overline{\ell}_{m-1,v}, g(\overline{\ell}_{m-1,v}), F]$ where $q^*[h_v, \overline{\ell}_{m-1,v}, g(\overline{\ell}_{m-1,v}), \widehat{\psi}]$ is defined like $q[h_v, \overline{\ell}_{m-1,v}, g(\overline{\ell}_{m-1,v}), F]$ except with $\gamma^{*-1}\left(h, \overline{\ell}_k, \overline{a}_k, \widehat{\psi}\right)$ replacing $\gamma^{-1}\left(h, \overline{\ell}_k, \overline{a}_k, F\right)$ and $\gamma^{*-1}\left(h, \overline{\ell}_k, \overline{a}_k, \widehat{\psi}\right) \equiv y$ if $\gamma^*\left(y, \overline{\ell}_k, \overline{a}_k, \widehat{\psi}\right) = h$.

Remark 7.5.1: If $\gamma\left(y, \overline{\ell}_m, \overline{a}_m, F\right) \equiv \gamma\left(y, \overline{a}_m, F\right)$ does not depend on $\overline{\ell}_m$ (i.e., there is no treatment-covariate interaction in the blip function), then, in the above algorithm for a non-dynamic regime $g = (\overline{a})$, steps (3)-(5) can be eliminated since $q\left(h, \overline{\ell}_K, \overline{a}_K, F\right) \equiv q\left(h, \overline{a}_K, F\right)$ does not depend on $\overline{\ell}_K$; as a consequence, to draw from the law of $Y_{g=(\overline{a})}$ one does not need to model the conditional density of the variables L_m. In fact, $S^{g=(\overline{a})}(y) = pr\left\{q\left[H, \overline{a}_K, F\right] > y\right\}$.

Example: If $\gamma\left(y, \overline{\ell}_m, \overline{a}_m, F\right) = y + 2a_m + 3a_m a_{m-1}$ then $q\left(h, \overline{a}_K, F\right) = h - \sum_{m=0}^{K} 2a_m + 3a_m a_{m-1}$.

Remark: Eq. (7.7) is only a reparameterization. In particular, Eqs. (7.6) and (7.7) do not translate into restrictions on the joint distribution of $V = \left(Y, \overline{L}_K, \overline{A}_K\right)$ since any law for V satisfies (7.6) and (7.7). Conversely, *(i)* any function $\gamma\left(y, \overline{\ell}_m, \overline{a}_m\right)$ satisfying $\gamma\left(y, \overline{\ell}_m, \overline{a}_m\right) = 0$ if $a_m = 0$, $\partial\gamma\left(y, \overline{\ell}_m, \overline{a}_m\right)/\partial y$ is positive and continuous and *(ii)* densities $f_H(h)$, $f_{L_m}\left(\ell_m \mid \overline{\ell}_{m-1}, \overline{a}_{m-1}, h\right)$ and $f_{A_m}\left(a_m \mid \overline{\ell}_m, \overline{a}_{m-1}\right)$ together induce a unique law for $V = \left(Y, \overline{L}_K, \overline{A}_K\right)$ by (a) using $f_H(\bullet) f_{L_m}(\bullet \mid \bullet)$ and $f_{A_m}(\bullet \mid \bullet)$ to determine a joint distribution for $\left(H, \overline{L}_K, \overline{A}_K\right)$ satisfying $H \amalg A_m \mid \overline{A}_{m-1}, \overline{L}_m$ and then (b) defining Y to be $q\left(H, \overline{L}_K, \overline{A}_K\right)$ with $q(\cdot,\cdot,\cdot)$ defined in terms of $\gamma(\cdot,\cdot,\cdot)$ in the obvious manner. The joint distribution of V satisfies (7.7).

8. Semiparametric Inference in SNMs

8.1. g-Estimation of ψ_0

In this Section, we assume A_m is dichotomous. Robins (1992) discusses generalizations to multivariate A_m with (possibly) continuous components. Robins (1992) argues that one will have better prior knowledge about, and thus can more accurately model, the densities $f\left(A_m = 1 \mid \overline{L}_m, \overline{A}_{m-1}\right)$ [as in Eq. (6.4)] than the densities occurring in Eq. (7.8). Indeed, with some loss of efficiency, if A_k and L_k are discrete, we can use a saturated model in Eq. (6.4), thus eliminating all possibility of misspecification. It is for this reason we prefer to test the "g"-null hypothesis $\psi_0 = 0$ using the g-test of Sec. 6 rather than the likelihood-based test of Sec. 7.4. Here, we describe how to obtain $n^{\frac{1}{2}}$-consistent g-estimates $\widetilde{\psi}$ of ψ_0 which are based on model (6.4) and thus will be consistent with the g-test of ψ_0 in the sense that 95% confidence intervals for ψ_0 will fail to cover 0 if and only if the g-test of $\psi_0 = 0$ rejects. Specifically, we *(a)* add $\theta' Q_m(\psi)$ $\left[\text{rather than } \theta Q_m\right]$ to the regressors $\alpha_0' W_m$ in (6.4) where $Q_m(\psi) = q^*\left\{H(\psi), \overline{L}_m, \overline{A}_{m-1}\right\}$, $q^*()$ is a known vector-valued function of $\dim \psi$ chosen by the data analyst, θ is a $\dim \psi$ valued parameter;

(b) define the G-estimate $\widetilde{\psi}$ to be the value of ψ for which the logistic regression score test of $\theta = 0$ is precisely zero; and *(c)* a 95% large sample confidence set for ψ_0 is the set of ψ for which the score test (which I call a G-test) of the hypothesis $\theta = 0$ fails to reject. (Robins, 1992). The parameter ψ is treated as a fixed constant when calculating the score test. The optimal choice of the function $q^*()$ is considered in Sec. 9.

Given $\widetilde{\psi}$, we estimate $b(y,g)$ by *(i)* finding $\widetilde{\phi}$ that maximizes (7.8) with the expression in set braces set to 1 and with ψ fixed at $\widetilde{\psi}$, and *(ii)* using the empirical distribution of $H_i\left(\widetilde{\psi}\right)$ as an estimate for the distribution of $H(\gamma)$ and *(iii)* use the MC algorithm of Sec. (7.5) to estimate $b(y,g)$ based on $\left(\widetilde{\psi}, \widetilde{\phi}\right)$ and the empirical law of $H_i\left(\widetilde{\psi}\right)$.

8.2. Subtleties in the Estimation of Direct Effect Using SNDMs

Consider again an AIDS study and suppose subjects taking aerosolized pentamidine (AP) for pneumocystis pneumonia (PCP) are less likely to take AZT treatment than other subjects because of the concern that the prophylaxis therapy may exacerbate AZT-related toxicities. In this case, the net (that is, overall) effect of AP on the outcome Y will underestimate its direct effect. In this section, we consider the estimation of the direct effect of AP using SNDMs. We redefine $A_k = (A_{Pk}, A_{Zk})$ with A_{Pk} and A_{Zk} respectively the aerosolized pentamidine and AZT dosage rates in the interval $(t_k, t_{k+1}]$. For simplicity, we will suppose that both treatments are dichotomous: $A_{Pk} = 1$ if the subject is taking prophylaxis therapy in the interval and $A_{Pk} = 0$ otherwise, with A_{Zk} similarly defined.

Let g_1, g_2, g_3, g_4 represent, respectively, the non-dynamic regimes (1) always withhold AZT and AP; (2) always withhold AZT, always take AP; (3) always take AZT, always withhold AP; and (4) always take both AZT and AP. The contrasts *(a)* $C_{12}(y) = S^{g_2}(y) - S^{g_1}(y)$ and *(b)* $C_{34}(y) = S^{g_4}(y) - S^{g_3}(y)$ represent, respectively, the direct effect AP on the distribution (survivorship function) of Y with AZT *(a)* always withheld and *(b)* always taken. We now show that even if the causal blip function $\gamma^\dagger\left(y, \overline{\ell}_m, \overline{a}_m\right)$ does not depend on AP history $\overline{a}_{Pm}$, nonetheless, AP may have a direct effect on Y when AZT is always taken (i.e., $C_{34}(y) \neq 0$).

Suppose that we have a correctly specified SNDM

$$\gamma^*\left(y, \overline{\ell}_m, \overline{a}_m, \psi\right) = y + \psi_1 a_{Pm} + \psi_2 a_{Zm} + \psi_3 a_{Pm} w_m + \psi_4 a_{Zm} w_m, \tag{8.1}$$

say, for the causal blip function $\gamma^\dagger\left(y, \overline{\ell}_m, \overline{a}_m\right)$ and w_m is a univariate function of $\overline{\ell}_m$ (such as the indicator of whether a subject has experienced a bout of PCP through t_m). Then if (3.1), (3.3), and assumption (3.10) of sequential randomization with respect to Y hold, we can estimate $S^g(y)$ for all $g \in \mathbf{g}$ and thus the contrasts $C_{12}(y)$ and $C_{34}(y)$ using the methods described in Secs. (7.5) or (8.1).

Furthermore, if $0 = \psi_{0,1} = \psi_{0,2} = \psi_{0,3} = \psi_{0,4}$, the g-null hypothesis holds by Theorem 7.1, and thus, $C_{12}(y) = C_{34}(y) = 0$ for all y. Here $\psi_{0,k}$ is the true value of the parameter ψ_k. Suppose next that $\psi_{0,1} = \psi_{0,3} = \psi_{0,4} = 0$, but $\psi_{0,2} \neq 0$. Then, by Remark 7.5.1, $C_{12}(y) = C_{34}(y) = 0$ remains true since $q(h, \overline{a}_K, F)$ depends on $\overline{a}_K$ only through $\overline{a}_{ZK}$.

More interestingly, suppose that

$$\psi_{0,1} = \psi_{0,3} = 0,\ \psi_{0,2} \neq 0,\ \psi_{0,4} \neq 0 \tag{8.2}$$

so that the blip function $\gamma^\dagger\left(y,\overline{\ell}_m,\overline{a}_m\right)$ still does not depend on the AP history $\overline{a}_{Pm}$. One might wrongly suspect that if the blip function does not depend on AP history, then AP has no direct effect on survival. It is clear, however, from the MC algorithm of Sec. 7.5 that if the true values of the parameters of model (8.1) are given by (8.2), then $C_{12}(y)=0$ for all y, but $C_{34}(y)\neq 0$ for some y, unless $f\left(\ell_m \mid \overline{\ell}_{m-1},\overline{a}_{m-1},h\right)$ does not depend on AP history $\overline{a}_{P(m-1)}$ for all m (i.e., $L_m \amalg \overline{A}_{P(m-1)} \mid \overline{A}_{Z(m-1)},\overline{L}_{m-1},H(\psi_0)$).

Remark: Heuristically, this reflects the fact that, if subjects are always on AZT, then AP may affect Y by directly affecting L_m and then L_m affects Y through the term $\psi_4 a_{Zm} w_m$ in model (8.1). We say "heuristically" because, under our assumption (3.10) of randomization with respect to Y, the "effects" of AP on L_m and of L_m on Y may not have causal interpretations.

Thus, given a SNDM that includes a non-zero A_{Zm}-$\overline{L}_m$ interaction (such as (8.1)), we can test the hypothesis $C_{12}(y)=0$ of no direct affect of A_P on Y when A_Z is always withheld by testing whether the blip function depends on $\overline{A}_{Pm}$. However, we cannot test the hypothesis $C_{34}(y)=0$ of no direct affect of A_P on Y when A_Z is always taken without modelling the conditional distribution of the confounders L_m given $\overline{L}_{m-1},\overline{A}_{m-1}$ and $H(\psi_0)$.

Hence, if we were particularly interested in testing the null hypothesis $C_{34}(y)=0$, then, in specifying our blip function, we should redefine the baseline level of AZT to be "zero" when a subject is actively taking AZT. That is, $A_{Zk}=0$ if a subject is taking AZT in $(t_k,t_{k+1}]$ and $A_{Zk}=1$ otherwise. If, with this redefinition, the blip function $\gamma^\dagger\left(y,\overline{\ell}_m,\overline{a}_m\right)$ does not depend on AP history $\overline{a}_{Pm}$, then $C_{34}(y)$ will be 0, although, now, $C_{12}(y)$ may be non-zero.

However, if interest lies in the joint null hypothesis $C_{12}(y)=C_{34}(y)=0$ of no direct effect of AP on Y controlling for any level of AZT treatment, use of structural nested models is inadequate since (a) $\gamma^\dagger\left(y,\overline{\ell}_m,\overline{a}_m\right)$ may depend on both $\overline{a}_{Pm}$ and $\overline{\ell}_m$, (b) $f\left(\ell_m \mid \overline{\ell}_{m-1},\overline{a}_{m-1},h\right)$ may depend on $\overline{a}_{P(m-1)}$, and yet (c) the joint null hypothesis may be true. In such a case, for essentially all non-linear SNDMs $\gamma^*\left(y,\overline{\ell}_m,\overline{a}_m,\psi\right)$ and parametric models $f\left(\ell_m \mid \overline{\ell}_{m-1},\overline{a}_{m-1},h;\phi\right)$, there would exist no parameter vector (ψ,ϕ) for which (a), (b) and (c) are true for data generated under the law governed by (ψ,ϕ). As a consequence, the true joint null hypothesis would almost certainly be rejected whatever the data. In Appendix 3, we offer a new extension of structural nested models, the direct-effect structural nested models, which overcome this difficulty by reparameterizing the joint distribution of the observables in terms of a parameter ψ_P that, when zero, implies $C_{12}(y)=C_{34}(y)=0$.

8.3. Structural Nested Mean Models

SNDMs only apply to outcomes Y with continuous distributions. In contrast, the structural nested mean models (SNMMs) discussed in this section apply to both continuous and discrete Y. We continue to assume that $Y\equiv L_{K+1}$ is univariate. We now let $\gamma^\dagger\left(\overline{\ell}_m,\overline{a}_m\right)=E\left[Y_{g=(\overline{a}_m,0)}-Y_{(\overline{a}_{m-1},0)} \mid \overline{L}_m=\overline{\ell}_m,\overline{A}_m=\overline{a}_m\right]$ be the effect on the mean of Y of a final blip of treatment a_m on subjects with observed history $\left(\overline{\ell}_m,\overline{a}_m\right)$. Note

$\gamma^{\dagger}\left(\overline{\ell}_m, \overline{a}_m\right) = 0$ if $a_m = 0$. Let $H_m = Y - \sum_{k=m}^{K} \gamma^{\dagger}\left(\overline{L}_k, \overline{A}_k\right)$ and redefine $H \equiv H_0$. Robins (1994) shows that, under the consistency assumption (3.1),

$$E\left[H_m \mid \overline{L}_m, \overline{A}_m\right] = E\left[Y_{g=(\overline{a}_{m-1},0)} \mid \overline{L}_m, \overline{A}_m\right] . \tag{8.3}$$

In particular, $E(H) = E\left[Y_{g=(0)}\right]$. Further if Eqs. (3.1), (3.3), and the sequential randomization assumption (3.10) hold, then by Theorem (3.2) and Eq. (8.3), $E\left[H_m \mid \overline{L}_m, \overline{A}_m\right] = E\left[H_m \mid \overline{L}_m, \overline{A}_{m-1}\right]$ so that H is mean independent of A_m given $\left(\overline{L}_m, \overline{A}_{m-1}\right)$, i.e.,

$$E\left[H \mid \overline{L}_m, \overline{A}_m\right] = E\left[H \mid \overline{L}_m, \overline{A}_{m-1}\right] \tag{8.4}$$

since, from its definition, H is a deterministic function of $\left(H_m, \overline{L}_m, \overline{A}_{m-1}\right)$.

We say the data follow a SNMM if $\gamma^{\dagger}\left(\overline{\ell}_m, \overline{a}_m\right) = \gamma^{*}\left(\overline{\ell}_m, \overline{a}_m, \psi_0\right)$ where $\gamma\left(\overline{\ell}_m, \overline{a}_m, \psi\right)$ is a known function depending on a finite dimensional parameter ψ satisfying $\gamma^{*}\left(\overline{\ell}_m, \overline{a}_m, \psi\right) = 0$ if $a_m = 0$ or $\psi = 0$, so $\psi_0 = 0$ represents the null hypothesis of no effect of a final blip of treatment of size a_m on the mean of Y for subjects with an observed history $\left(\overline{\ell}_m, \overline{a}_m\right)$. An example of such a function is

$$\gamma^{*}\left(\overline{\ell}_m, \overline{a}_m, \psi\right) = \psi_1 a_m + \psi_2 a_m a_{m-1} + \psi_3 a_m w_m$$

where w_m is a function of $\overline{\ell}_m$. Pseudo-SNMMs are considered in Appendix 2.

Define $H(\psi) \equiv Y - \sum_{k=0}^{K} \gamma^{*}\left(\overline{L}_k, \overline{A}_k, \psi\right)$. It follows from Eq. (8.4) and Theorem (3.2) that, given (3.1), (3.3), and the sequential randomization assumption (3.10),

$$E\left[H(\psi_0) \mid \overline{L}_m, \overline{A}_m\right] = E\left[H(\psi_0) \mid \overline{L}_m, \overline{A}_{m-1}\right] . \tag{8.5}$$

Robins (1994) showed that (8.5) implies that, under regularity conditions, the g-estimate $\widetilde{\psi}$ of ψ_0 of Sec. 8.1 is a CAN estimator of the parameter ψ_0 of the SNMM $\gamma\left(\overline{\ell}_m, \overline{a}_m, \psi\right)$ provided that the covariate $Q_m(\psi) = q^{*}\left\{H(\psi), \overline{L}_m, \overline{A}_{m-1}\right\}$ added to Eq. (6.4) is linear in $H(\psi)$, i.e., $q^{*}\left(H(\psi), \overline{L}_m, \overline{A}_{m-1}\right) = q_1^{*}\left(\overline{L}_m, \overline{A}_{m-1}\right) H(\psi) + q_2^{*}\left(\overline{L}_m, \overline{A}_{m-1}\right)$. It follows that, under sequential randomization, a $n^{\frac{1}{2}}$-consistent estimator of $E\left(Y_{g=(0)}\right)$ is given by $n^{-1} \sum_{i=1}^{n} H_i\left(\widetilde{\psi}\right)$. Robins (1994) shows that the g-estimate $\widetilde{\psi}$ will attain the semiparametric efficiency bound for the model when the choices of $q_1^{*}()$ and $q_2^{*}()$ are optimal. [However, a large sample 95 percent confidence set for ψ_0 can no longer be obtained as the set of ψ for which the score test of the hypothesis $\theta = 0$ rejects. Appropriate confidence procedures are discussed in Robins (1994).]

Further, under sequential randomization assumption (3.10), Robins (1994) shows that the mean effect of regime g on the mean of Y, $E[Y_g] - E\left[Y_{g=(0)}\right]$, is the limit as $V \to \infty$ of the following monte carlo procedure.

MC Procedure: For $v = (1, \ldots, V)$,
Draw for $m = (0, \ldots, K)$, $\ell_{m,v}$ recursively from $f\left[\ell_m \mid \overline{\ell}_{m-1,v}, g\left(\overline{\ell}_{m-1,v}\right)\right]$ and then compute $V^{-1} \sum_{v=1}^{V} \sum_{m=0}^{K} \gamma^{\dagger}\left\{\overline{\ell}_{m,v}, g\left(\overline{\ell}_{m,v}\right)\right\}$.

Thus, under sequential randomization, the g-null mean hypothesis

$$E\left(Y_g\right)=E\left(Y\right), g\in\mathbf{g} \tag{8.6}$$

holds if and only if $\gamma^{\dagger}\left(\overline{\ell}_m,\overline{a}_m\right)=0$ for all m. Further, if, for each m, $\gamma^{\dagger}\left(\overline{\ell}_m,\overline{a}_m\right)\equiv\gamma^{\dagger}\left(\overline{a}_m\right)$ does not depend on $\overline{\ell}_m$, then, for a non-dynamic regime $g=(\overline{a})$, $H=Y-\sum_{m=0}^{K}\gamma^{\dagger}\left(\overline{A}_m\right)$. Thus, given a SNMM $\gamma^*\left(\overline{a}_m,\psi\right)$ for $\gamma^{\dagger}\left(\overline{a}_m\right)$, $E\left[Y_{g=(\overline{a})}-Y_{g=(0)}\right]$ can be consistently estimated by $\sum_{m=0}^{K}\gamma^*\left(\overline{a}_m,\widetilde{\psi}\right)$.

On the other hand, if $\widetilde{\psi}\neq 0$ and either g is dynamic or $\gamma^{\dagger}\left(\overline{\ell}_m,\overline{a}_m\right)$ depends on $\overline{\ell}_m$, one can *(i)* specify a parametric model $f\left[\ell_m\mid\overline{\ell}_{m-1},\overline{a}_{m-1};\eta\right]$; *(ii)* find $\widehat{\eta}$ that maximizes $\prod_{i=1}^{n}\prod_{m=0}^{K}f\left[L_{mi}\mid\overline{L}_{(m-1)i},\overline{A}_{(m-1)i};\eta\right]$; *(iii)* and estimate $E\left[Y_g-Y_{g=(0)}\right]$ by $V^{-1}\sum_{v=1}^{V}\sum_{m=0}^{K}\gamma^*\left(\overline{\ell}_{m,v},g\left(\overline{\ell}_{m,v}\right),\widetilde{\psi}\right)$ where $\ell_{m,v}$ is drawn recursively from $f\left[\ell_m\mid\overline{\ell}_{(m-1),v},g\left(\overline{\ell}_{(m-1),v}\right);\widehat{\eta}\right]$. Appendix 2 provides a more complex "monte carlo procedure" that allows one to estimate the survivor function $S^g\left(y\right)$ under a SNMM.

We have not discussed likelihood-based methods for estimating the parameter ψ_0 of a SNMM. Rather, we have only considered g-estimation. This reflects the fact that likelihood-based inference for the parameter ψ_0 of a SNMM is somewhat complex. It is discussed in Appendix 2.

Multiplicative SNMMs: A SNMM fails to automatically impose the restriction $E\left[Y_g\right]\geq 0$ when Y_g is a non-negative random variable. However, we can use multiplicative SNMMs to impose this restriction. Specifically, redefine $\gamma^{\dagger}\left(\overline{\ell}_m,\overline{a}_m\right)=\ell n\left\{E\left[Y_{g=(\overline{a}_m,0)}\mid\overline{\ell}_m,\overline{a}_m\right]/E\left[Y_{g=(\overline{a}_{m-1},0)}\mid\overline{\ell}_m,\overline{a}_m\right]\right\}$. We then say the data follow a multiplicative SNMM if $\gamma^{\dagger}\left(\overline{\ell}_m,\overline{a}_m\right)=\gamma^*\left(\overline{\ell}_m,\overline{a}_m,\psi_0\right)$ with $\gamma^*\left(\overline{\ell}_m,\overline{a}_m,\psi\right)$ as defined above. Redefine $H=Y\exp\left\{-\sum_{m=0}^{K}\gamma^{\dagger}\left(\overline{L}_m,\overline{A}_m\right)\right\}$ and $H\left(\psi\right)=Y\exp\left\{-\sum_{m=0}^{K}\gamma^*\left(\overline{L}_m,\overline{A}_m,\psi\right)\right\}$. It can be proven that $E\left(H\right)=E\left(Y_{g=(0)}\right)$ and further, when (3.1), (3.3), and the sequential randomization assumption (3.10) hold, the g-null mean hypothesis (8.6) holds if and only if $\gamma^{\dagger}\left(\overline{\ell}_m,\overline{a}_m\right)=0$ for all $\overline{\ell}_m,\overline{a}_m$. Furthermore, under regularity conditions the G-estimate $\widetilde{\psi}$ of ψ_0 and the estimate $n^{-1}\sum_i H_i\left(\widetilde{\psi}\right)$ of $E\left[Y_{g=(0)}\right]$ are CAN. Additionally, if, for each m, $\gamma^{\dagger}\left(\overline{\ell}_m,\overline{a}_m\right)\equiv\gamma^{\dagger}\left(\overline{a}_m\right)$ does not depend on $\overline{\ell}_m$ then $E\left[Y_{g=(\overline{a})}\right]=E\left[Y_{g=(0)}\right]\exp\left[\sum_{m=0}^{K}\gamma^{\dagger}\left(\overline{a}_m\right)\right]$. Thus $\ell n\left\{E\left[Y_{g=(\overline{a})}\right]/E\left[Y_{g=(0)}\right]\right\}$ can be consistently estimated by $\sum_{m=0}^{K}\gamma^*\left(\overline{a}_m,\widetilde{\psi}\right)$. However, if $\widetilde{\psi}\neq 0$ and either g is non-dynamic or $\gamma^{\dagger}\left(\overline{\ell}_m,\overline{a}_m\right)$ depends on $\overline{\ell}_m$, no simple analog of the above SNMM monte carlo procedure exists for estimating $\ell n\left[E\left(Y_g\right)/E\left(Y_{g=(0)}\right)\right]$. Rather, the more complex monte carlo methods described in Appendix 2 must be used.

Neither a SNMM or a multiplicative SNMM automatically imposes the true restriction $0\leq E\left[Y_g\right]\leq 1$ when Y and Y_g are Bernoulli random variables. Unfortunately, there is no simple method to estimate the parameter ψ_0 of a "logistic" SNMM $\gamma^*\left(\overline{\ell}_m,\overline{a}_m,\psi\right)$ that

imposes the restriction that $\gamma^*\left(\overline{\ell}_m, \overline{a}_m, \psi_0\right) =$
$logit\left\{E\left[Y_{g=(\overline{a}_m,0)} \mid \overline{a}_m, \overline{\ell}_m\right]\right\} - logit\left\{E\left[Y_{g=(\overline{a}_{m-1},0)} \mid \overline{a}_m, \overline{\ell}_m\right]\right\}$.

9. g-Estimation without Sequential Randomization

If our fundamental assumption that sequential randomization (3.10) holds for $g \in \mathbf{g}$ is false, the structural blip function $\gamma^\dagger$ will differ from the pseudo-structural blip function γ. Thus if $\gamma^*\left(y, \overline{\ell}_m, \overline{a}_m, \psi\right)$ is a correctly specified SNDM, then it will not represent a correctly specified pseudo-SNDM. In this section, we assume that $\gamma^*\left(y, \overline{\ell}_m, \overline{a}_m, \psi\right)$ is a correctly specified SNDM so that $\gamma^*\left(y, \overline{\ell}_m, \overline{a}_m, \psi_0\right)$ still represents the causal effect of a final brief bit of treatment among subjects with observed history $\overline{\ell}_m, \overline{a}_m$. Now if (3.10) is false, then, according to Eq. (7.6), A_m will no longer be independent of $H(\psi_0)$ given $\overline{L}_m, \overline{A}_{m-1}$, since $H(\psi_0) = H\left(\gamma^\dagger\right)$ but $H(\psi_0) \neq H(\gamma)$. However, even when (3.10) is false, Robins (1993, App. 1) shows that the consistency assumption (3.1) implies $H\left(\gamma^\dagger\right)$ has the same distribution as $Y_{g=(0)}$. More generally, given $\left(\overline{L}_m, \overline{A}_m\right)$, $H_m\left(\gamma^\dagger\right)$ has the same conditional distribution as $Y_{g=\overline{A}_{(m-1,0)}}$. Thus it remains of interest to estimate the parameter ψ_0 of our SNDM. To do so, suppose A_m is dichotomous and we can correctly specify a logistic model

$$f\left(A_m = 1 \mid \overline{L}_m, \overline{A}_{m-1}, H(\psi_0)\right) = \left\{1 + \exp\left[-\left\{\alpha_{10}' W_m + \alpha_{20}' W_m^*(\psi_0)\right\}\right]\right\}^{-1} \tag{9.1}$$

where $W_m^*(\psi) = w_m^*\left(H(\psi), \overline{L}_m, \overline{A}_{m-1}\right)$ is a function of $H(\psi), \overline{L}_m, \overline{A}_{m-1}$ and $W_m = w_m\left(\overline{L}_m, \overline{A}_{m-1}\right)$. Note the hypothesis $\alpha_{20} = 0$ is equivalent to the hypothesis that $H(\psi_0) \amalg A_m \mid \overline{L}_m, \overline{A}_{m-1}$ and thus the hypothesis that our SNDM is also a pseudo-SNDM. The functional form of $w_m^*(\cdot,\cdot,\cdot)$ will be hard to accurately specify. Thus in practice, we suggest, as a sensitivity analysis, repeating the following analysis a number of times with different choices for $w_m^*(\cdot,\cdot,\cdot)$ and $w_m(\cdot,\cdot,\cdot)$.

Remark: Given our SNDM is correctly specified, the ability to specify model (9.1) correctly is equivalent to the ability to specify a model $f\left(A_m = 1 \mid \overline{L}_m, \overline{A}_{m-1}, Y_{g=\overline{A}_{(m-1,0)}}\right)$ since $H\left(\gamma^\dagger\right)$ is a deterministic function of $H_m\left(\gamma^\dagger\right)$ and $\overline{L}_m, \overline{A}_{m-1}$.

Given model (9.1), a G-estimate $\widetilde{\psi}$ of ψ_0 of our SNDM is obtained as in Sec. 8 except now $\theta' Q_m(\psi)$ is added to model (9.1) (with ψ_0 replaced by ψ) rather than to (6.4). $\widetilde{\psi}$ is now the value of ψ for which the score test of $\theta = 0$ is exactly 0 and a 95% confidence set for ψ_0 is still the set of ψ for which the score test of the hypothesis $\theta = 0$ fails to reject.

Under our model (9.1), the true value of ψ_0 may not be identified, the asymptotic variance of $\widetilde{\psi}$ may be infinite, the score test of the hypothesis $\theta = 0$ may not reject for any value of ψ, and there may not be a unique estimate $\widetilde{\psi}$. We now give a sufficient *theoretical* condition that guarantees the estimate $\widetilde{\psi}$ of ψ_0 in a SNDM will indeed have infinite asymptotic variance. We then describe how this theoretical result might be used in practice.

Let $S_\psi(\psi, \alpha) = \partial \log \mathcal{L}(\psi, \alpha) / \partial \psi$ be the score for ψ for a single subject. Based on the likelihood $\mathcal{L}(\psi, \alpha) = \left\{\partial H(\psi) / \partial \psi\right\} f\left\{H(\psi)\right\} \prod_{m=0}^{K} f\left(L_m \mid \overline{L}_{m-1}, \overline{A}_{m-1}, H(\psi)\right)$

$\prod_{m=0}^{K} f\left(A_m \mid \overline{A}_{m-1}, \overline{L}_m, H(\psi); \alpha\right)$. Note in practice, $S_\psi(\psi, \alpha)$ is a theoretical object when we have not specified models for $f\{H(\psi_0)\}$ or $f\left(L_m \mid \overline{L}_{m-1}, \overline{A}_{m-1}, H(\psi_0)\right)$. Let $Q_{opt,m}(\psi_0)$ be the unknown function $E\left[S_\psi(\psi_0, \alpha_0) \mid \overline{L}_m, \overline{A}_{m-1}, H(\psi_0), A_m = 1\right] -$
$E\left[S_\psi(\psi_0, \alpha_0) \mid \overline{L}_m, \overline{A}_{m-1}, H(\psi_0), A_m = 0\right]$. Then $\tilde{\psi}$ that uses $Q_{opt,m}(\psi)$ in place of $Q_m(\psi)$ is the most efficient g-estimator. In fact, it attains the semiparametric variance bound for the semiparametric model characterized by the SNDM $H(\psi_0)$ and the model (9.1)that leaves the law of $H(\psi_0)$ and the law of L_m given $\left\{\overline{L}_{m-1}, \overline{A}_{m-1}, H(\psi_0)\right\}$ completely unspecified. Hence, if the asymptotic variance of $\tilde{\psi}$ that uses $Q_{opt,m}(\psi)$ is not finite, then no g-estimator has a finite asymptotic variance. A necessary and sufficient condition for $\tilde{\psi}$ that uses $Q_{opt,m}(\psi)$ to have infinite asymptotic variance is that, when $\psi = \psi_0$, for each m and each subject, $Q_{opt,m}(\psi)$ is a linear combination of the regressors $\left(W_m', W_m^*(\psi)'\right)'$ in model (9.1). In particular, if $W_m^*(\psi)$ equals $Q_{opt,m}(\psi)$, then all possible g-estimators $\tilde{\psi}$ will have infinite asymptotic variance. $Q_{opt,m}(\psi)$ is also optimal when doing g-estimation under model (6.4) [i.e., when it is known that $\dot{\alpha}_{20} = 0$ a priori]. In that case, $\tilde{\psi}$ that uses $Q_{opt,m}(\psi)$ will have finite asymptotic variance.

We now consider how an analyst might use this theoretical result in practice. Suppose we obtain a confidence interval based on some initial chosen function $Q_m(\psi)$. If this interval is reasonably narrow, then we have carried out a successful g-analysis. However, if our 95 percent confidence interval for ψ is too wide to be substantively useful, then either *(i)* our choice of the function Q_m was quite inefficient or *(ii)* no choice of Q_m, including the optimal choice $Q_{opt,m}$, would give usefully narrow intervals. The best approach to try to discriminate between explanations *(i)* and *(ii)* is as follows. Again specify parametric models, depending on parameter $\rho = (\eta', \phi')'$ for the unparameterized densities in $\mathcal{L}(\psi, \alpha)$ and rewrite $\mathcal{L}(\psi, \alpha)$ as $\mathcal{L}(\psi, \alpha, \rho)$ to reflect this dependence, find the maximum likelihood estimators $\left(\widehat{\psi}, \widehat{\rho}, \widehat{\alpha}\right)$ based on maximizing the product over the n subjects of the $\mathcal{L}(\psi, \alpha, \rho)$, and finally construct an estimate $\widehat{Q}_{opt,m}(\cdot)$ of $Q_{opt,m}(\cdot)$ based on the distribution implied by $\left(\widehat{\psi}, \widehat{\rho}, \widehat{\alpha}\right)$. Now construct a new confidence interval based on g-estimation using the function $\widehat{Q}_{opt,m}(\psi)$ rather than the original $Q_m(\psi)$. The resulting estimator and 95% confidence interval is said to be a locally efficient g-estimate and interval (at the parametric submodel indexed by ρ). It is a valid confidence interval for ψ_0 even when the models parameterized by ρ are misspecified. If this 95 percent confidence interval for ψ_0 is reasonably narrow, we report this interval and conclude that option *(i)* above was true.

If the resulting confidence interval is still uselessly wide, we conclude it is likely that there is insufficient information about ψ_0 in our semiparametric model which leaves the marginal density of $H(\psi_0)$ and the conditional law of L_m given $\overline{L}_{m-1}, \overline{A}_{m-1}$ and $H(\psi_0)$ unspecified. In that case we might use the maximum likelihood estimator $\widehat{\psi}$ described above. $\widehat{\psi}$ will be consistent for ψ_0 provided model (9.1) and the parametric models $f\{H(\psi_0); \eta\}$ and $f\left\{L_m \mid \overline{L}_{m-1}, \overline{A}_{m-1}, H(\psi_0); \phi\right\}$ are also correctly specified. Results based on $\widehat{\psi}$ should be viewed with caution because of the difficulty in correctly specifying the parametric model for L_m given $\overline{L}_{m-1}, \overline{A}_{m-1}, H(\psi_0)$.

Remark: The assumptions that our SNDM is correctly specified and that (9.1) identifies the parameter ψ_0 are not sufficient to identify the law of Y_g for any $g \in \mathbf{g}$ other than

$g = (0)$. However, Theorem A1.2 of Robins (1993 Appendix 1) implies that the law of Y_g for any $g \in \mathbf{g}$ is identified provided our SNDM is correctly specified, Eq. (9.1) identifies the parameter ψ_0 of the SNDM, and, in addition, for $g \in \mathbf{g}$ either *(i)* $Y_g \amalg A_m \mid \overline{L}_m, \overline{A}_{m-1} = g\left(\overline{L}_{m-1}\right), H\left(\gamma^{\dagger}\right)$ [i.e., we have sequential randomization for Y_g given $\overline{L}_m$ and $H\left(\gamma^{\dagger}\right)$] or *(ii)* there is no current treatment interaction with respect to L.

Definition: We say there is no current treatment interaction with respect to L if the treatment effect transformation function $\lambda\left(y, \overline{l}_m, \overline{a}_m, g\right)$ does not depend on a_m for $g \in \mathbf{g}$ where, by definition, $\lambda\left(y, \overline{l}_m, \overline{a}_m, g\right)$ satisfies $pr\left[Y_g > \lambda\left(y, \overline{l}_m, \overline{a}_m, g\right) \mid \overline{l}_m, \overline{a}_m\right] =$ $pr\left[Y_{g=(\overline{a}_{m-1},0)} > y \mid \overline{l}_m, \overline{a}_m\right]$.

Robins (1993, p. 258-260) discusses the substantive plausibility of both assumptions *(i)* and *(ii)*.

10. Continuous Time SNMs

There are two difficulties with the SNDMs of Sec. 7, both of which can be solved by defining SNDMs in continuous time. First, the meaning of the parameter ψ depends on the time between measurements. For example, the meaning of ψ depends on whether time m is a day versus a month later than time $m-1$. Thus it would be advantageous to have the parameter ψ defined in terms of the instantaneous effect of a particular treatment rate. Second, if the covariate process L and/or the treatment process A can (randomly) jump in continuous time rather than just at the pre-specified times $1, 2, \ldots, K$, the SNDMs of Sec. 7 cannot be used.

To extend SNDMs to continuous time, we assume that Y is measured at time t_{K+1}, but now a subject's covariate process $\overline{L}(t) = \{L(u); 0 \leq u \leq t\}$ and treatment process $\overline{A}(t) = \{A(u); 0 \leq u \leq t\}$ are generated by a marked point process where, for example, $A(u)$ is recorded treatment at time u. That is, *(i)* $L(t)$ and $A(t)$ have sample paths that are step functions that are right-continuous with left-hand limits; *(ii)* the $L(t)$ and $A(t)$ process do not jump simultaneously; and *(iii)* the total number of jumps K^* of the joint $\left(\overline{A}(t), \overline{L}(t)\right)$ processes in $[0, t_{K+1}]$ is random and finite, occurring at random times $T_1, \ldots, T_{K^*}$. We choose this restricted class of sample paths because their statistical properties are well understood (Arjas, 1989). Now, given treatment history $\overline{a}(t_{K+1}) = \{a(u); 0 \leq u \leq t_{K+1}\}$ on $[0, t_{K+1}]$, let $Y_{(\overline{a}(t),0)}$ be the counterfactual value of Y under the treatment history $\overline{a}^*(t_{K+1})$ where $a^*(u) = a(u)$ for $u \leq t$ and $a^*(u) = 0$ otherwise. Similarly, let $Y_{(\overline{a}(t^-),0)}$ be the counterfactual value of Y under the treatment history $\overline{a}^*(t_{K+1})$ with $a^*(u) = a(u)$ for $u < t$, $a^*(u) = 0$ otherwise. For convenience, we have suppressed the "$g =$ " in our counterfactual notation. We shall make two additional assumptions, which attempt to capture the fact that we assume that an instantaneously brief bit of treatment has a negligible effect on Y.

Assumption 1: If Y is a continuous outcome, given any history $\overline{a}(t_{K+1})$, $Y_{(\overline{a}(t),0)} = Y_{(\overline{a}(t^-),0)}$.

Assumption 2: $pr\left[Y_{(\overline{A}(u^-),0)} > y \mid \overline{L}(t), \overline{A}(t)\right]$ is continuous as a function of u.

Remark: More elegantly, we could capture the fact that a brief bit of treatment has a negligible effect on Y by the following assumption which implies Assumption 2. Given the square integrable function $a(t), t \in [0, t_{K+1}]$, let $S[a(\cdot), y, t] \equiv$ $pr\left[Y_{(\overline{a}(t_{K+1}))} > y \mid \overline{L}(t), \overline{A}(t)\right]$ so $S[\cdot, \cdot, \cdot]$ maps $L_2[0, t_{K+1}] \times R^1 \times [0, t_{K+1}]$ into R^1 where $L_2[0, t_{K+1}]$ are the set of square integrable functions on $[0, t_{K+1}]$. Our assumption is that $S[a(\cdot), y, t]$ is L_2-continuous in $a(\cdot)$. That is, for all ϵ there exists a σ such that if, for a given $a_1(\cdot)$ and $a_2(\cdot)$, the L_2 distance between $a_1(x)$ and $a_2(x)$, $\left[\int_0^{t_{K+1}} [a_1(x) - a_2(x)]^2 dx\right]^{1/2}$, is less than σ, then the absolute value of the difference between the conditional survival curves at y of $Y_{(\overline{a}_1(t_{K+1}))}$ and $Y_{(\overline{a}_2(t_{K+1}))}$, $\mid S(a_1(\cdot), y, t) - S(a_2(\cdot), y, t) \mid$, is less than ϵ. Here $\sigma = \sigma(y, t)$ may depend on (y, t).

We first shall study the simpler continuous-time structural nested mean models (SNMMs).

Continuous-Time SNMMs:

Let $V(t, h) = E\left[Y_{(\overline{A}(t+h^-),0)} - Y_{(\overline{A}(t^-),0)} \mid \overline{L}(t), \overline{A}(t)\right]$ be the mean causal effect on subjects with observed history $\left(\overline{L}(t), \overline{A}(t)\right)$ of a final blip of observed treatment $\{A(u); t \le u < t+h\}$ in the interval $[t, t+h)$. Note $V(t, 0) = 0$. Assumption 2 implies $V(t, h)$ is continuous in h. To be able to define the effect of an instantaneous treatment rate, we need to assume $V(t, h)$ is differentiable with respect to h.

Assumption 3: We assume *(i)* $D(t) \equiv \lim_{h \downarrow 0} V(t, h)/h$ exists for all $t \in [0, t_{K+1}]$ and *(ii)* $D(t) = \partial V(t, 0)/\partial h$ is continuous on $[T_m, T_{m+1}), m = 0, \ldots, K^* + 1$ where $T_0 \equiv 0, T_{K^*+1} \equiv t_{K+1}$.

$V(t, h)$ and $D(t)$ may be discontinuous in t at the jump times T_m because of the abrupt change in the conditioning event defining $V(t, h)$ at $t = T_m$. $D(t)\, dt$ is the effect of a last blip of observed treatment $A(t)$ sustained for "instantaneous" time dt on the mean of Y.

Define $H(t) = Y - \int_t^{t_{K+1}} D(t)\, dt$ and define H to be $H(0)$. In Appendix 0 we prove

Theorem 10.1: $E\left[H(t) \mid \overline{L}(t), \overline{A}(t)\right] = E\left[Y_{(\overline{A}(t^-),0)} \mid \overline{L}(t), \overline{A}(t)\right]$. In particular, $E(H) = E\left[Y_{(0)}\right]$.

We say the data follow a continuous-time SNMM $D(t, \psi)$ if $D(t) \equiv d\left(t, \overline{L}(t), \overline{A}(t)\right)$ equals $D(t, \psi_0) \equiv d\left(t, \overline{L}(t), \overline{A}(t), \psi_0\right)$ where ψ_0 is an unknown parameter to be estimated and $D(t, \psi)$ is a known function continuous in t on $[T_m, T_{m+1})$ satisfying $D(t, \psi) = 0$ if $\psi = 0$ or $A(t) = 0$. Define $H(t, \psi) \equiv Y - \int_t^{t_{K+1}} D(t, \psi)\, dt$. For simplicity, suppose $A(t)$ takes only the values 0 and 1 and let $\lambda\left(t \mid \overline{L}(t^-), \overline{A}(t^-)\right) dt$ be the probability that the $A(t)$ process will jump to a new state in the infinitesimal interval $[t, t+dt)$ given $\overline{A}(t^-)$ and $\overline{L}(t^-)$. We make the sequential randomization assumption that

$$\lambda\left(t \mid \overline{L}\left(t^-\right), \overline{A}\left(t^-\right), Y_{(0)}\right) = \lambda\left(t \mid \overline{L}\left(t^-\right), \overline{A}\left(t^-\right)\right). \tag{10.1}$$

Given a correctly specified Cox model, i.e.,

$$\lambda\left(t \mid \overline{L}\left(t^-\right), \overline{A}\left(t^-\right)\right) = \lambda_0(t) \exp\left[\alpha' W(t)\right] \tag{10.2}$$

where $W(t)$ is a vector function of $\left\{\overline{L}(t^-), \overline{A}(t^-)\right\}$, α is an unknown vector parameter, and $\lambda_0(t)$ is an unrestricted baseline hazard function, we obtain a G-estimate of the parameter ψ of the continuous-time SNMM $D(t, \psi)$ by adding the term $\theta' g\left(H(\psi), \overline{L}(t^-), \overline{A}(t^-)\right)$

to model (10.2) where $H(\psi) = H(0, \psi)$, $g(\cdot, \cdot)$ is a known function chosen by the investigator. Specifically, the G-estimate $\widehat{\psi}_{ge}$ is the value of ψ for which the Cox partial likelihood estimator of θ in the expanded model is zero. Then, $\widehat{\psi}_{ge}$ and $n^{-1}\sum_i H_i\left(\widehat{\psi}\right)$ will be $n^{\frac{1}{2}}$-consistent for ψ_0 and $E\left[Y_{(0)}\right]$ provided (10.1) is true, Cox model (10.2) is correctly specified, and $g\left\{H(\psi), \overline{L}(t^-), \overline{A}(t^-)\right\}$ is linear in $H(\psi)$, i.e., $g\left\{H(\psi), \overline{L}(t^-), \overline{A}(t^-)\right\} = g_1\left\{\overline{L}(t^-), \overline{A}(t^-)\right\} H(\psi) + g_2\left\{\overline{L}(t^-), \overline{A}(t^-)\right\}$. A consistent estimator of the asymptotic variance of $\widehat{\psi}_{ge}$ can be derived using methods in Robins (1994).

10.1. Continuous SNDM:

Suppose again that Y is a continuous random variable with a continuous distribution function. To describe a continuous-time SNDM, let $Q(y,t,h) \equiv q\left(y,t,h,\overline{L}(t), \overline{A}(t)\right)$ be the unique function such that $Y_{(\overline{A}(t+h^-),0)}$ and $Q\left(Y_{(A(t^-),0)}, t, h\right)$ have the same conditional distribution given $\overline{L}(t), \overline{A}(t)$. This is equivalent to

$$pr\left[Y_{(\overline{A}(t+h^-),0)} > Q(y,t,h) \mid \overline{L}(t), \overline{A}(t)\right] = pr\left[Y_{(\overline{A}(t^-),0)} > y \mid \overline{L}(t), \overline{A}(t)\right] \tag{10.3}$$

for $y \in R^1, t \in [0, t_{K+1}], h \in [0, t_{K+1} - t]$ so $Q(y,t,h)$ represents the causal effect of a final blip of the observed treatment $\{A(u); t \le u < t+h\}$ on quantiles of Y. Note $Q(y,t,0) = y$. We now make a smoothness (differentiability) assumption.

Assumption 4: We assume that *(i)* $D(y,t) \equiv \lim_{h \downarrow 0} \{Q(y,t,h) - Q(y,t,0)\}/h$ exists and is bounded for all $(y,t) \in R^1 \times [0, t_{K+1}]$; *(ii)* further, for $(y,t) \in R^1 \times [T_m, T_{m+1}), m = 0, \ldots, K^*, D(y,t) = \partial Q(y,t,0)/\partial h$ and the matrix $\partial D(y,t)/\partial(y,t)$ has bounded and uniformly continuous entries.

Note $D(y,t)\,dt$ is the effect of a last blip of observed treatment $A(t)$ at t sustained for an instantaneous time dt on quantiles of Y. Hence $D(y,t) \equiv d\left(y, \overline{L}(t), \overline{A}(t)\right) = 0$ if $A(t) = 0$. $Q(y,t,h)$ and $D(y,t)$ may be discontinuous in t at the jump times T_m because of the abrupt change in the conditioning event defining $Q(y,t,h)$ when $t = T_m$.

Remark: It is important to note that we do not have to assume $Y_{(\overline{A}(t),0)}$ is differentiable in t for $t \ne T_m$. This is scientifically important since an unmeasured covariate process $U(t)$ may jump at t, and there may be a treatment-covariate process interaction so that the instantaneous effect on $Y_{(\overline{A}(t^-),0)}$ of a given treatment rate $A(t)$ may change at jump time t, in which case $Y_{(\overline{A}(t^-),0)}$ will not be differentiable (although still continuous). However, since $U(t)$ is unmeasured, Assumption 4 can still hold if, as we assume, the jump times for the $U(t)$ process have a continuous distribution, so that at any time the probability that the $U(t)$ process jumps at t is negligibly small.

It then follows from Theorem (2.3) of Sec. 6 of Loomis and Sternberg (1968) that *(i)* there exists a unique continuous solution $H(t) \equiv h\left(Y, t, \overline{L}(t_{K+1}), \overline{A}(t_{K+1})\right)$ to the differential equation $dH(t)/dt = D(H(t), t)$ satisfying $H(t_{K+1}) = Y$; and *(ii)* the solution is a continuous function of (Y,t) on $R^1 \times [0, t_{K+1}]$ and $\partial h\left(t, Y, \overline{L}(t_{K+1}), \overline{A}(t_{K+1})\right)/\partial(Y,t)$ exists and is bounded and continuous on $R^1 \times [T_m, T_{m+1})$. Hence the Jacobian $\partial H(0)/\partial Y$ for the transformation from Y to $H(0)$ exists with probability 1. The strong smoothness conditions in Assumption 4*(ii)* were required to guarantee the existence of this Jacobian. Our main result is Theorem (8.2).

Theorem 10.2: $H(t)$ and $Y_{(\overline{A}(t^-),0)}$ have the same conditional distribution given $\left(\overline{L}(t), \overline{A}(t)\right)$. In particular, $H \equiv H(0)$ and $Y_{(0)}$ have the same marginal distribution.

Theorem (10.2) is proved in Appendix 0 for the special case in which there is local rank preservation, i.e., $Q\left(Y_{(A(t^-),0)}, t, h\right) = Y_{(\overline{A}(t+h^-),0)}$ with probability one. Although I am nearly certain that Theorem (10.2) holds without local rank preservation by a coupling argument, my current proof attempt still suffers from unresolved technical problems.

We say the data follows a continuous-time SNDM $D(y,t,\psi)$ if $D(y,t,\psi) = D(y,t,\psi_0)$ where ψ_0 is an unknown parameter and $D(y,t,\psi) \equiv d\left(y,t,\overline{L}(t), \overline{A}(t), \psi\right)$ is a known function satisfying *(i)* $D(y,t,0) = 0$, *(ii)* $D(y,t,\psi) = 0$ if $A(t) = 0$, and *(iii)* Assumption 4*(ii)* holds for each fixed value of ψ. It then follows if the explainable non-random non-compliance assumption (10.1) holds, the Cox model (10.2), and the continuous-time SNDM model $D(y,t,\psi)$ are correctly specified, then the estimators $\widehat{\psi}_{ge}$ and $n^{-1}\sum_i H_i\left(\widehat{\psi}_{ge}\right)$ will be $n^{\frac{1}{2}}$-consistent for ψ_0 and $E\left[Y_{(0)}\right]$ respectively, even when the function $g\left(H(\psi), \overline{L}(t^-), \overline{A}(t^-)\right)$ to be added to the Cox model (10.2) is not linear in $H(\psi)$.

11. Faithfulness: From Association to Causation by Philosophy?

Except when there has been physical randomization (e.g., a sequential randomized trial was performed), the methods described in this paper do not allow one to claim that associations found in the data are causal without prior untestable assumptions, such as the assumption (3.10) of no unmeasured confounders. Assumption (3.10) will never be exactly true. Whether it may nonetheless serve as a reasonable working hypothesis will depend critically both on the covariates $\overline{L}_K$ available for data analysis and on the substantive issue under investigation. It is this latter reason that epidemiologic studies need to be conducted by subject matter experts.

In stark contrast, Pearl and Verma (PV) (1991) and SGS (1993) argue that one can go from "association" to "causation" without any subject matter knowledge based on a "philosophical" assumption, the assumption of faithfulness or stability. In fact, SGS have written a computer program, Tetrad, that searches epidemiologic data bases for causal relations by applying the faithfulness assumption. I will show that their argument is unconvincing. I will argue that in observational epidemiologic studies, one cannot, even in principle, go from association to causation without strong subject matter knowledge.

To demonstrate why their argument fails, I will consider the following simple example. The causal DAG in Figure 5 is a non-parametric recursive structural equation (SEM) model (Pearl, 1995). The arrows indicate the potential causal associations between variables A, B, C, V, and W. Variables V and W are unobserved, variables A, B, and C are observed. As indicated on the graph, we assume that we know the temporal ordering: A then B then C. Our goal is to determine whether B causes C. To fix ideas, one might think of C as lung cancer, B as alcohol consumption, A as city of birth, V as race, and W as cigarette consumption. We wish to know whether alcohol causes lung cancer.

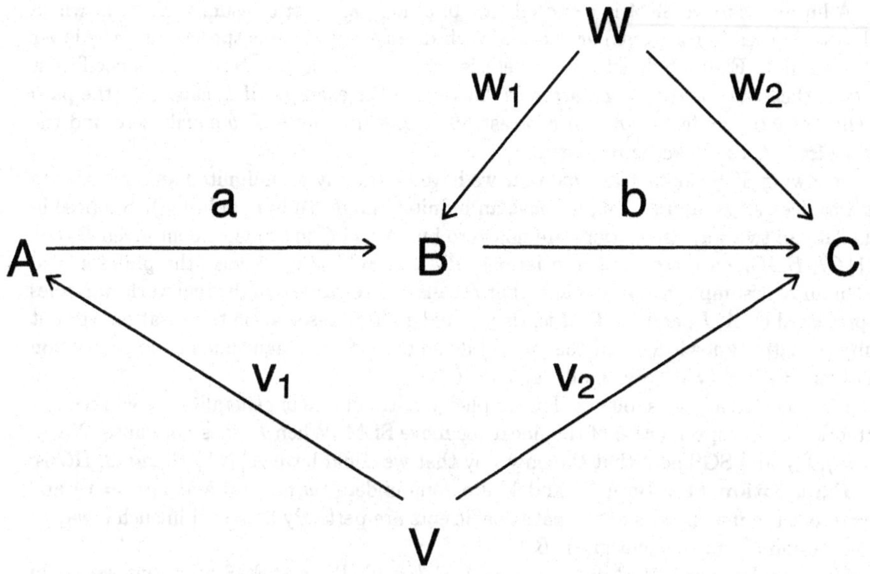

Figure 5: A Causal DAG representing a linear structural equation model with normal errors. Small letters are path coefficients. Variables W and V are unobserved. Variables A, B, and C are observed.

A linear recursive SEM is a special case of a non-parametric recursive SEM in which all disturbances (error terms) are normal with mean zero. For this special case, the lower case letters in Figure 5 denote path coefficients. If, as a fact of nature, a path coefficient is zero, then the corresponding arrow is "missing." For example, if B causes C, the path coefficient b differs from zero. In contrast, if B does not cause C, b equals zero and the arrow from B to C can be removed.

Following SGS, we first assume that we have essentially an unlimited amount of data on variables A, B, and C (i.e., we have an infinitely large study population). Suppose in the data all two-way correlations are non-zero but A and C are independent given B, i.e., $A \amalg C \mid B$. Given these data associations, PV and SGS would invoke the philosophical faithfulness assumption and conclude that B causes C regardless of the real world variables represented by A, B, and C. That is, they would go from association to causation without subject matter knowledge and they would do so even if the magnitude of the correlation between B and C was very small, say $\rho = 10^{-5}$.

PV and SGS argue as follows. For simplicity, without loss of generality, we will restrict attention to the special case of the linear recursive SEM. When B does not cause C (i.e., $b = 0$), PV and SGS note that the only way that we could have $A \amalg C \mid B$ and $B \not\amalg C$ is:

Explanation (1): Both W and V are confounders ($w_1 w_2 \neq 0$ and $v_1 v_2 \neq 0$) and $b = 0$, but the magnitudes of the path coefficients are perfectly balanced in such a way to make A and C independent given B.

On the other hand, if B does cause C (i.e., $b \neq 0$), PV and SGS note that we would have $A \amalg C \mid B$ and $B \not\amalg C$ if the following explanation held.

Explanation (2): Both W and V are non-confounders ($w_1 w_2 = v_1 v_2 = 0$) and $b \neq 0$.

SGS and PV then proceed to rule out explanation (1) (which is always a logical mathematical possibility) by the following "faithfulness" argument. They argue that, since the subset $\{(w_1, w_2, b_1, b_2)\}$ of path coefficients that will make $A \amalg C \mid B$ when $w_1 w_2 \neq 0$ and $v_1 v_2 \neq 0$ has Lebesgue measure zero in R^4, the "prior" probability of explanation (1) occurring is zero (since explanation (1) requires such "fortuitous" values of the path coefficients). I fully agree with SGS and PV that the prior probability of Explanation (1) is zero.

However, SGS and PV then conclude that explanation (2) must be the proper explanation and thus conclude B causes C. It is this last step with which I disagree. This step can only be justified if the prior probability that V and W are both non-confounders ($v_1 v_2 = 0$ and $w_1 w_2 = 0$) is greater than zero. Otherwise, both explanations (1) and (2) are events of probability zero and so their relative likelihood is not defined without further assumptions. Now the Lebesgue measure of the event "$v_1 v_2 = 0$ and $w_1 w_2 = 0$" is also zero, since v_1, v_2,w_1, and w_2 take values in the continuum $[-1, 1]$. It follows that PV and SGS are making the "hidden" assumption that the event that a path coefficient has value zero (i.e., that an arrow is missing on the graph) has a non-zero probability. On this "hidden" assumption, explanation (2) does have a positive probability, and we can conclude that B causes C. Now I do agree with PV and SGS that, for any two given variables, the probability that the path coefficient between them is zero (i.e., neither variable causes the other) is non-zero. But in the epidemiologic example in which we are trying to determine whether drinking causes lung cancer, the unmeasured variables W and V in Fig. 5 do not represent single variables but rather represent all possible unmeasured common causes of B and C and A and C respectively. If there should exist any common cause of B and C or of A and C, explanation (2) is false.

Now I (and every other epidemiologist I know) believe that, given a sufficiently large sample size, a test for independence between any two variables recorded in an observational study will reject due to the existence of unmeasured common causes (confounders). Thus, we teach epidemiology students that, with certainty (i.e., with prior probability 1), explanation (2) will not occur (once we understand that W and V represent the conglomeration of all unmeasured common causes). It follows that we cannot conclude explanation (2) is to be preferred to explanation (1). Thus we cannot conclude that B causes C.

Suppose, however, that, although Explanations (1) and (2) both have zero prior probability, we believe under further reasonable assumptions and with respect to some measure, the relative likelihood of Explanation (1) is infinitely greater than that of Explanation (2). We now argue that even so, nonetheless, in realistic studies with their finite sample size, we should still not invoke the "faithfulness assumption" and conclude that B causes C when, in the data, $A \coprod C \mid B$.

Consider now a realistic study, even a very large one, with its finite sample size. Then, even if, in the data, A is exactly independent of C given B, nonetheless, due to sampling variability, there will exist a confidence interval around the null estimate of the conditional association parameter between A and C given B. Since, as argued above, our prior probability that in truth A and C are independent given B is zero [explanations (1) and (2) both have prior probability zero], our *posterior* odds that $A \coprod C \mid B$ is exactly zero will be zero, compared to the hypothesis that A and C are dependent given B with their conditional association parameter lying in the small confidence interval surrounding zero. Thus, again, we would not necessarily conclude that B causes C since the event that we "almost" have balanced confounding and $b = 0$ does not have a prior probability of 0 and thus its posterior probability need not be small. However, if the BC correlation was large and, based on subject matter knowledge, we believed that the magnitude of the confounding by the conglomerate variables W and V was small, then we would conclude, based on a formal Bayesian analysis, that b was likely non-zero. This analysis would, of course reflect our subject matter knowledge and would not be based on any subject-matter-independent philosophical principle.

To conclude, I would recommend against using the computer program Tetrad as a tool for searching epidemiologic data bases for causal associations, since its search strategy is based solely on the implications of the faithfulness assumption. Finally, I wish to note that my argument against the use of the faithfulness assumption in analyzing epidemiologic data is not an argument against the faithfulness assumption in analyzing data in simple, stereotyped environments where the number of potential unmeasured confounders conglomerated in W and V is known to be small. Thus, in AI programs designed to allow robots to learn from data obtained in a rather stereotyped environment, the faithfulness assumption and thus a Tetrad-like program might be useful.

Appendix 0: Continuous Time SNMs

Proof of Theorem 10.1: Let $M(u,t) = \lim_{\Delta t \downarrow 0} \{Z(u+\Delta t, t) - Z(u,t)\}/\Delta t$ where $Z(u,t) \equiv E\left[Y_{(\overline{A}(u^-),0)} \mid \overline{L}(t), \overline{A}(t)\right]$. Note $M(u,t) = E\left[D(u) \mid \overline{L}(t), \overline{A}(t)\right]$ and thus, by Assumption 3, is continuous in t except at $T_m, m = 1, \ldots, K^*$. Hence $E\left[H(t) \mid \overline{L}(t), \overline{A}(t)\right] =$

$E\left[Y\mid\overline{L}(t),\overline{A}(t)\right]-E\left[\int_t^{t_{K+1}}D(u)\,du\mid\overline{L}(t),\overline{A}(t)\right]=$
$E\left[Y\mid\overline{L}(t),\overline{A}(t)\right]-\left\{\int_t^{t_{K+1}}M(u,t)\,du\right\}=E\left[Y\mid\overline{L}(t),\overline{A}(t)\right]-$
$\left\{E\left[Y\mid\overline{L}(t),\overline{A}(t)\right]+\sum_{\{m;T_m>t\}}\{Z(T_m^-,t)-Z(T_m,t)\}-Z(t,t)\right\}=$
$Z(t,t)=E\left[Y\left(\overline{A}(t^-),0\right)\mid\overline{L}(t),\overline{A}(t)\right]$ where the last equality is definitional, the second to last is by the fact that, by Assumption 2, $Z(u,t)$ is continuous in u, so $Z(T_m^-,t)=Z(T_m,t)$, the third to last is by the facts that $M(u,t)=\partial Z(u,t)/\partial u$ for $u\in[T_m,T_{m+1})$ and $Z(t_{K+1},t)=E\left[Y\mid\overline{L}(t),\overline{A}(t)\right]$ by the consistency assumption.

Proof of Theorem 10.2 under Local Rank Preservation: By Theorem (2.3) of Chpt. 6 of Loomis and Sternberg (1968), there is a unique continuous solution to $dH(t)/dt=D(H(t),t),t\in[T_m,T_{m-1})$ satisfying $H(t_{K+1})=Y$. We now show that under local rank preservation $Y_{(\overline{A}(t^-),0)}$ satisfies these conditions as a function of t. First by Assumption (4) and local rank preservation *(i)* $d\left(Y_{(\overline{A}(t^-),0)}\right)/dt$ exists and equals $D\left(Y_{(\overline{A}(t^-),0)},t\right)$ on $[T_m,T_{m+1})$, *(ii)* $Y_{(A(t^-),0)}$ is continuous in t by Assumption (1) and (2), and *(iii)* $Y_{(\overline{A}(t^-),0)}$ evaluated at $t=t_{K+1}$ is Y by the consistency assumption. Note that, given Assumption (4), the additional assumption of (local) rank preservation implies that $Y_{(\overline{A}(t^-),0)}$ is continuously differentiable for $t\in[T_m,T_{m+1})$. [Note the assumption of local rank preservation is incompatible with the existence of the unmeasured covariate process $U(t)$ that interacts with treatment described in the Remark of Sec. 10.1.]

Appendix 1:

Proof of Theorem 3.2:

It will be sufficient to prove the rightmost equality in (3.8). In the following, set $\overline{A}_{m-1}$ to be $g\left(\overline{L}_{m-1}\right)$. If $f\left(\overline{L}_m,\overline{A}_{m-1}\right)\neq 0$, define $b^*\left(y,\overline{L}_m,\overline{A}_{m-1}\right)=E\left[I(Y_g>y)\mid\overline{L}_m,\overline{A}_{m-1}\right]$ where the dependence on g has been suppressed. Now if $f\left(\overline{L}_m,\overline{A}_{m-1},A_m=g\left(t_m,\overline{L}_m\right)\right)\neq 0$, then $b^*\left(y,\overline{L}_m,\overline{A}_{m-1}\right)=$
$E\left[b^*\left(y,\overline{L}_{m+1},\overline{A}_m\right)\mid\overline{L}_m,\overline{A}_{m-1},A_m=g\left(t_m,\overline{L}_m\right)\right]\equiv$
$\int b^*\left(y,\ell_{m+1},\overline{L}_m,A_m=g\left(t_m,\overline{L}_m\right),\overline{A}_{m-1}\right)f\left[\ell_{m+1}\mid\overline{L}_m,A_m=g\left(t_m,\overline{L}_m\right),\overline{A}_{m-1}\right]d\mu(\ell_{m+1})$
since, by the sequential randomization assumption (3.10),
$b^*\left(y,\overline{L}_m,\overline{A}_{m-1}\right)\equiv E\left[I(Y_g>y)\mid\overline{L}_m,\overline{A}_m\right]=E\left[I(Y_g>y)\mid\overline{L}_m,\overline{A}_{m-1},A_m=g\left(t_m,\overline{L}_m\right)\right]$
$=E\left\{E\left[I(Y_g>y)\mid\overline{L}_{m+1},\overline{A}_m\right]\mid\overline{L}_m,\overline{A}_{m-1},A_m=g\left(t_m,\overline{L}_m\right)\right\}$ by iterated expectations.
Now putting $\overline{A}_{k-1}=g\left(\overline{L}_{k-1}\right)$ invoking (3.3) and arguing recursively, we then obtain
$b^*\left[y,\overline{L}_k,g\left(\overline{L}_{k-1}\right)\right]=\int\cdots\int b^*\left(y,\overline{L}_K,g\left(\overline{L}_{K-1}\right)\right)\prod_{m=k+1}^{K}f\left[\ell_m\mid\overline{\ell}_{m-1},g\left(\overline{\ell}_{m-1}\right)\right]d\mu(\ell_m)$.
But, by (3.10) and (3.3), $b^*\left(y,\overline{\ell}_K,g\left(\overline{\ell}_{K-1}\right)\right)=E\left[I(Y_g>y)\mid\overline{\ell}_K,g\left(\overline{\ell}_K\right)\right]$
$=E\left[I(Y>y)\mid\overline{\ell}_K,g\left(\overline{\ell}_K\right)\right]\equiv S\left(y\mid\overline{\ell}_K,g\left(\overline{\ell}_K\right)\right)$ by the consistency assumption (3.1).

Appendix 2:

Further Results on SNMMs and Pseudo-SNMMs

The semiparametric model induced by (3.1), (3.3), (3.10) and the SNMM $\gamma^*\left(\overline{\ell}_m, \overline{a}_m, \psi\right)$ restricts the joint distribution of the observables $\overline{V} = \left(\overline{L}, \overline{A}\right)$ only through the restriction (8.5). We now construct a pseudo-SNMM that obeys this restriction without reference to counterfactuals. Let $b(g) = \int y dF(y,g)$ and $b\left(\overline{\ell}_m, g\right) = \int y \, dF\left(y, \overline{\ell}_m, g\right)$ where $dF(y,g)$ and $dF\left(y, \overline{\ell}_m, g\right)$ are the measures on Y induced by $b(y,g)$ and $b\left(y, \overline{\ell}_m, g\right)$. For example, if Y has differentiable conditional distribution functions, then $dF(y,g) = \{-\partial b(y,g)/\partial y\} \, dy$ and $dF\left(y, \overline{\ell}_m, g\right) = -\left\{\partial b\left(y, \overline{\ell}_m, g\right)/\partial y\right\} dy$. Define $\gamma\left(\overline{\ell}_m, \overline{a}_m, F\right) = b\left[\overline{\ell}_m, g = (\overline{a}_m, 0)\right] - b\left[\overline{\ell}_m, g = (\overline{a}_{m-1}, 0)\right]$ where F denotes the dependence on the joint distribution F of the observables V. Set $H(\gamma) = Y - \sum_{m=0}^{K} \gamma\left(\overline{\ell}_m, \overline{a}_m, F\right)$. Then it is straightforward to prove both

$$E\left[H(\gamma) \mid \overline{L}_m, \overline{A}_m\right] = E\left[H(\gamma) \mid \overline{L}_m, \overline{A}_{m-1}\right] \tag{A2.1}$$

and the following "g"-null mean theorem.

"g"-null mean theorem: $b(g_1) = b(g_2)$ for all $g_1, g_2 \in \mathbf{g}$ if and only if $\gamma\left(\overline{\ell}_m, \overline{a}_m, F\right) = 0$ for all $\left(\overline{\ell}_m, \overline{a}_m\right)$.

We say the data follow a pseudo-SNMM $\gamma\left(\overline{\ell}_m, \overline{a}_m, \psi\right)$ if the definition of a SNMM is satisfied with $\gamma\left(\overline{\ell}_m, \overline{a}_m, F\right)$ replacing $\gamma^{\dagger}\left(\overline{\ell}_m, \overline{a}_m\right)$. It follows from Theorem (3.2) that when (3.1), (3.3), and (3.10) are true, then $\gamma^*\left(\overline{\ell}_m, \overline{a}_m, \psi\right)$ is a pseudo-SNMM if and only if it is a SNMM.

Our next goal is to reparameterize the likelihood function of the observables V in such a way that the restriction (A2.1) is naturally incorporated. Note that simply writing $f_V\left(y, \overline{\ell}_K, \overline{a}_K\right) = f_{H(\gamma)|\overline{L}_K, \overline{A}_K}\left(h \mid \overline{\ell}_K, \overline{a}_K\right) f\left(\overline{\ell}_K, \overline{a}_K\right)$ fails to impose the restriction (A2.1). To impose this restriction, for notational convenience, define $\overline{X}_m = \left(\overline{L}_m, \overline{A}_m\right)$. Consider the following algebraic decomposition of $H(\gamma)$.

$$\begin{gathered} H(\gamma) = \left\{H(\gamma) - E\left[H(\gamma) \mid \overline{X}_K\right]\right\} + \\ \left\{\sum_{m=0}^{K} E\left(H(\gamma) \mid \overline{X}_m\right) - E\left(H(\gamma) \mid \overline{X}_{m-1}\right)\right\} + E\left(H(\gamma)\right) \equiv \\ \sigma + \sum_{m=0}^{K} \nu\left(\overline{X}_m\right) + \beta_0 \end{gathered} \tag{A2.2}$$

with $\sigma \equiv H(\gamma) - E\left(H(\gamma) \mid \overline{X}_K\right), \nu\left(\overline{X}_m\right) = E\left[H(\gamma) \mid \overline{X}_m\right] - E\left[H(\gamma) \mid \overline{X}_{m-1}\right], \beta_0 = E\left(H(\gamma)\right)$. The restriction (A2.1) implies that $\nu\left(\overline{X}_m\right) \equiv \nu\left(L_m, \overline{X}_{m-1}\right)$ does not depend on A_m. Thus, we can write,

$$\sigma \equiv H(\gamma) - \sum_{m=0}^{K} \nu\left(L_m, \overline{X}_{m-1}\right) - \beta_0 \tag{A2.3}$$

with

$$E\left[\nu\left(L_m,\overline{X}_{m-1}\right)\mid\overline{X}_{m-1}\right]=0 \tag{A2.4}$$

and

$$E\left[\sigma\mid\overline{X}_K\right]=0\,. \tag{A2.5}$$

Conversely, given a distribution function for the random variables $\overline{X}_K\equiv\left(\overline{A}_K,\overline{L}_K\right)$, any constant β_0, any random variable σ satisfying (A2.5), and any function $\nu(\cdot,\cdot)$ satisfying (A2.4), then

$$H\equiv\sigma+\sum_{m=0}^{K}\nu\left(L_m,\overline{X}_{m-1}\right)+\beta_0 \tag{A2.6}$$

is mean independent of A_m given $\left(\overline{L}_m,\overline{A}_{m-1}\right)$. Hence, given a function $\gamma\left(\overline{\ell}_m,\overline{a}_m\right)$ satisfying $\gamma\left(\overline{\ell}_m,\overline{a}_m\right)=0$ when $a_m=0$, if we set

$$Y=H+\sum_{m=0}^{K}\gamma\left(\overline{L}_m,\overline{A}_m\right)$$

with H defined by (A2.6), then the law of $V=\left(Y,\overline{L}_K,\overline{A}_K\right)$ will have a joint distribution with $\gamma\left(\overline{\ell}_m,\overline{a}_m,F\right)$ equal to the chosen function $\gamma\left(\overline{\ell}_m,\overline{a}_m\right)$. Hence the data follow a pseudo-SNMM $\gamma^*\left(\overline{\ell}_m,\overline{a}_m,\psi\right)$ [i.e., $\gamma^*\left(\overline{\ell}_m,\overline{a}_m,\psi_0\right)=\gamma\left(\overline{\ell}_m,\overline{a}_m,F\right)$] if and only if there exists a value ψ_0 of the parameter ψ, a constant β_0, a random variable σ, and a function $\nu(\cdot,\cdot)$ [that satisfy (A2.5) and (A2.4) respectively] such that $H(\psi_0)\equiv Y-\sum_{m=0}^{K}\gamma^*\left(\overline{\ell}_m,\overline{a}_m,\psi_0\right)$ is equal to the RHS of (A2.6).

Likelihood Function: Since (A2.4) implies there exists a unique function $\nu^*\left(L_m,\overline{X}_{m-1}\right)$ such that $\nu\left(L_m,\overline{X}_{m-1}\right)=\nu^*\left(L_m,\overline{X}_{m-1}\right)-E\left[\nu^*\left(L_m,\overline{X}_{m-1}\mid\overline{X}_{m-1}\right)\right]$ and $\nu^*\left(0,\overline{X}_{m-1}\right)=0$, the subject-specific likelihood function of a pseudo-SNMM (and thus for a SNMM under sequential randomization) can be written

$$\mathcal{L}(\psi,\eta)\equiv f_V\left(Y,\overline{X}_K;\psi,\eta\right)=$$

$$\left\{\frac{\partial\sigma(\psi_0,\eta_0)}{\partial H(\psi_0)}\,\frac{\partial H(\psi_0)}{\partial Y}\right\}f\left(\sigma(\psi,\eta)\mid\overline{X}_K;\eta_1\right)\prod_{m=0}^{K}f\left(A_m\mid L_m,\overline{X}_{m-1};\eta_4\right)f\left(L_m\mid\overline{X}_{m-1};\eta_3\right),$$

with

$$\sigma(\psi,\eta)\equiv H(\psi)-\eta_5-\sum_{m=0}^{K}\left\{\nu^*\left(L_m,\overline{X}_{m-1};\eta_2\right)-\int\nu^*\left(\ell_m,\overline{X}_{m-1};\eta_2\right)dF\left(\ell_m\mid\overline{X}_{m-1};\eta_3\right)\right\} \tag{A2.7}$$

subject to the restrictions that

$$\int dF\left(t\mid\overline{X}_K;\eta_1\right)=0 \tag{A2.8}$$

and

$$\nu^*\left(0,\overline{X}_{m-1};\eta_2\right)=0 \tag{A2.9}$$

Here $\eta = (\eta_1, \ldots, \eta_5)'$ are unknown parameters; η_5 is an unknown constant (that plays the role that β_0 did in A2.2); η_4 indexes all conditional laws of A_m given $\left(L_m, \overline{X}_{m-1}\right)$; η_3 indexes all possible laws of L_m given $\overline{X}_{m-1}$; η_1 indexes all laws of $\sigma \equiv \sigma(\psi_0, \eta_0)$ given $\overline{X}_K$ such that $E\left[\sigma \mid \overline{X}_K\right] = 0$ (which implies (A2.8)); and η_2 indexes all possible functions $\nu^*(\cdot, \cdot; \eta_2)$ of $\left(L_m, \overline{X}_{m-1}\right)$ satisfying (A2.9). The Jacobian $\left\{\frac{\partial \sigma(\psi_0, \eta_0)}{\partial H(\psi_0)} \frac{\partial H(\psi_0)}{\partial Y}\right\}$ for the transformation from Y to σ is equal to 1.

It follows that fully parametric likelihood-based inference for the parameter ψ of a pseudo-SNMM is performed by *(i)* choosing parametric models for $f(\cdot \mid \cdot; \eta_1)$, $f\left(\overline{L}_m \mid \overline{X}_{m-1}; \eta_3\right)$ and for $\nu^*(\cdot, \cdot; \eta_2)$ indexed by finite dimensional parameters ω_1, ω_3, and ω_2, *(ii)* maximizing $\prod_i \mathcal{L}_i(\psi, \omega, \eta_4, \eta_5)$ with respect to (ψ, ω, η_5) to obtain $\left(\widehat{\psi}, \widehat{\omega}, \widehat{\eta}_5\right)$. [Note that the MLE $\left(\widehat{\psi}, \widehat{\omega}, \widehat{\eta}_5\right)$ does not depend on whether $f\left(A_m \mid \overline{L}_m, \overline{X}_{m-1}; \eta_4\right)$ is completely unknown or follows a parametric submodel such as $f\left(A_m \mid \overline{L}_m, \overline{X}_{m-1}; \alpha\right)$ of Eq. (6.4).] An alternative to fully parametric likelihood-based inference is to specify a model [such as (6.4)] for $f\left(A_m \mid \overline{L}_m, \overline{X}_{m-1}\right)$ and obtain a g-estimate $\widetilde{\psi}$ of ψ_0. Then one can obtain estimates of $(\widetilde{\omega}, \widetilde{\eta}_5)$ by maximizing $\prod_i \mathcal{L}_i\left(\widetilde{\psi}, \omega, \eta_4, \eta_5\right)$ with respect to ω and η_5.

Drawing From the Law of Y_g Under a SNMM $\gamma^*\left(\overline{\ell}_m, \overline{a}_m, \psi\right)$:

Given the true values (ψ_0, η_0) of (ψ, η) to obtain V-independent draws $y_{g,v}$ from the distribution $b(y, g)$ [and, thus, under (3.1), (3.3) and (3.10) from the law of Y_g], we proceed as follows.

Monte Carlo Procedure:

For $v = (1, \ldots, V)$

1). Recursively for $m = (0, \ldots, K)$, draw ℓ_{mv} from $f\left(\ell_m \mid \overline{\ell}_{(m-1)v}, g\left(\overline{\ell}_{(m-1)v}\right); \eta_{0,3}\right)$;

2). Draw σ_v from $f\left(\sigma \mid \overline{\ell}_{Kv}, g\left(\overline{\ell}_{Kv}\right); \eta_{0,1}\right)$;

3). Compute $h_v = \sigma_v + \eta_{0,5} + \sum_{m=0}^{K} \{\nu^*\left(\ell_{mv}, \overline{\ell}_{(m-1)v}, g\left(\overline{\ell}_{(m-1)v}\right); \eta_{0,2}\right) - \int \nu^*\left(\ell_m, \overline{\ell}_{(m-1)v}, g\left(\overline{\ell}_{(m-1)v}\right); \eta_{0,2}\right) f\left(\ell_m \mid \overline{\ell}_{(m-1)v}, g\left(\overline{\ell}_{(m-1)v}\right); \eta_{0,3}\right) d\mu(\ell_m)\}$;

4). Compute $y_{g,v} = h_v + \sum_{m=0}^{K} \gamma\left(\overline{\ell}_{mv}, g\left(\overline{\ell}_{mv}\right), \psi_0\right)$.

In practice, we replace (ψ_0, η_0) with suitable estimates.

Multiplicative SNMMs:

Results of this Appendix go over straightforwardly to multiplicative SNMMs. In particular, when (3.1), (3.3), and the sequential randomization assumption (3.10) hold, the likelihood function for a multiplicative SNMM is as above. However, the Jacobian $\{\partial \sigma(\psi_0, \eta_0) / \partial H(\psi_0)\} \{\partial H(\psi_0 / \partial Y)\}$ now equals $\exp\left\{-\sum_{m=0}^{K} \gamma^*\left(\overline{L}_m, \overline{A}_m, \psi_0\right)\right\}$. Furthermore, the monte carlo procedure to draw from the law of Y_g is as above, except Step 4 is replaced by the following:

4'). Compute $y_{g,v} = h_v \exp\left\{\sum_{m=0}^{K} \gamma\left(\overline{\ell}_{mv}, g\left(\overline{\ell}_{mv}\right), \psi_0\right)\right\}$.

Appendix 3

Suppose, as in Sec. 8.2, $Y \equiv L_{K+1}$ and $A_m = (A_{Pm}, A_{Zm})$ is comprised of two distinct treatments, A_{Pm} and A_{Zm}. In this Appendix, we provide a new extension of SNMs, the direct-effect SNMs, for which, under sequential randomization (3.10), there exists a parameter ψ_P that takes the value zero if and only if the *direct-effect g-null hypothesis* of no direct effect of $\overline{a}_P$ on Y controlling for any $\overline{a}_Z$ treatment holds. To formalize this null hypothesis, let

$$\mathbf{g}_P = \left\{g \in \mathbf{g}; g\left(t_k, \overline{\ell}_k\right) \equiv \left(g_P\left(t_k, \overline{\ell}_k\right), a_{Zk}\right) \text{ has the same component } a_{Zk} \text{ for all } \overline{\ell}_k\right\} .$$

We shall write Y_g, $b\left(y, \overline{\ell}_m, g\right)$, and $b\left(\overline{\ell}_m, g\right)$ for $g \in \mathbf{g}_P$ as $Y_{(g_P, \overline{a}_Z)}$, $b\left(y, \overline{\ell}_m, \{g_P, \overline{a}_Z\}\right)$, and $b\left(\overline{\ell}_m, \{g_P, \overline{a}_Z\}\right)$ in obvious notation where $b\left(\overline{\ell}_m, g\right)$ is defined in Appendix 2. Thus $Y_{(g_P, \overline{a}_Z)}$ is the outcome Y in a hypothetical study in which the treatment $\overline{a}_Z \equiv \overline{a}_{ZK}$ is assigned non-dynamically, but $\overline{a}_P$ may be assigned dynamically. Note that the function $g_P\left(\overline{\ell}_k\right) \equiv \left\{g_P\left(t_m, \overline{\ell}_m\right); m = 0, \ldots, k\right\}$ are functions taking values in the support $\overline{\mathbf{a}}_{Pk}$ of $\overline{a}_{Pk}$.

The *direct-effect g-null and g-null mean hypotheses* of no direct effect of $\overline{a}_P$ controlling for $\overline{a}_Z$ are, respectively,

$$pr\left[Y_{(g_{P1}, \overline{a}_Z)} > y\right] = pr\left[Y_{(g_{P2}, \overline{a}_Z)} > y\right] \tag{A3.1}$$

and

$$E\left[Y_{(g_{P1}, \overline{a}_Z)} - Y_{(g_{P2}, \overline{a}_Z)}\right] = 0 \tag{A3.2}$$

for all $\overline{a}_Z \equiv \overline{a}_{ZK}, g_{P1}, g_{P2}$. Under Assumptions (3.1), (3.3), and (3.10), (A3.1) and (A3.2) are, respectively, equivalent to the direct effect "g"-null and "g"-null mean hypotheses

$$b\left(y, \{g_{P1}, \overline{a}_Z\}\right) = b\left(y, \{g_{P2}, \overline{a}_Z\}\right) \tag{A3.3}$$

and

$$b\left(\{g_{P1}, \overline{a}_Z\}\right) = b\left(\{g_{P2}, \overline{a}_Z\}\right) . \tag{A3.4}$$

Direct Effect SNMMs:

We shall first study direct-effect SNMMs and direct-effect pseudo-SNMMs. To describe these models, set $t_k\left(\overline{\ell}_k, \overline{a}_k\right) = \gamma\left(\overline{\ell}_k, \overline{a}_k, F\right)$ with the right hand side as defined in Appendix 2. For $m = k, \ldots, 1$, define $t_k\left(\overline{\ell}_{m-1}, \overline{a}_k\right) =$
$E\left[t_k\left(\left\{\overline{\ell}_{m-1}, L_m\right\}, \overline{a}_k\right) \mid \overline{L}_{m-1} = \overline{\ell}_{m-1}, \overline{A}_{m-1} = \overline{a}_{m-1}\right]$. Let $t_k\left(\overline{a}_k\right) \equiv t_k\left(\overline{\ell}_{-1}, \overline{a}_k\right) \equiv E\left[t_k\left(L_0, \overline{a}_k\right)\right]$. Set $r_k\left(\overline{\ell}_m, \overline{a}_k\right) = t_k\left(\overline{\ell}_m, \overline{a}_k\right) - t_k\left(\overline{\ell}_{m-1}, \overline{a}_k\right)$. Hence,

$$\gamma\left(\overline{\ell}_k, \overline{a}_k, F\right) = \sum_{m=0}^{k} r_k\left(\overline{\ell}_m, \overline{a}_k\right) + t_k\left(\overline{a}_k\right) \tag{A3.5}$$

where, by construction,

$$E\left[r_k\left(\overline{L}_m, \overline{a}_k\right) \mid \overline{L}_{m-1}, \overline{A}_{m-1} = \overline{a}_{m-1}\right] = 0$$

and, by $\gamma\left(\overline{\ell}_k, \overline{a}_k, F\right) = 0$ when $a_k = 0$,

$$r_k\left(\overline{\ell}_m, \overline{a}_k\right) = t_k\left(\overline{a}_k\right) = 0 \text{ when } a_k = 0 .$$

Now set $r_k^*\left(\overline{\ell}_m, \overline{a}_k\right) \equiv r_k\left(\overline{\ell}_m, \overline{a}_k\right) - r_k\left(\left\{\overline{\ell}_{m-1}, \ell_m = 0\right\}, \overline{a}_k\right)$, so that

$$r_k\left(\overline{\ell}_m, \overline{a}_k\right) = r_k^*\left(\overline{\ell}_m, \overline{a}_k\right) - E\left[r_k^*\left(\left\{\overline{\ell}_{m-1}, L_m\right\}, \overline{a}_k\right) \mid \overline{L}_{m-1} = \overline{\ell}_{m-1}, \overline{A}_{m-1} = \overline{a}_{m-1}\right] \tag{A3.5a}$$

and

$$r_k^*\left(\overline{\ell}_m, \overline{a}_k\right) = 0 \text{ if } a_k = 0 \text{ or } \ell_m = 0 . \tag{A3.6}$$

Thus

$$\gamma\left(\overline{\ell}_k, \overline{a}_k, F\right) = \tag{A3.7}$$

$$\sum_{m=0}^{k} r_k^*\left(\overline{\ell}_m, \overline{a}_k\right) - E\left[r_k^*\left(\left\{\overline{\ell}_{m-1}, L_m\right\}, \overline{a}_k\right) \mid \overline{L}_{m-1} = \overline{\ell}_{m-1}, \overline{A}_{m-1} = \overline{a}_{m-1}\right] + t_k\left(\overline{a}_k\right)$$

If the mean of Y is unrestricted, (A3.7) is an unrestricted parameterization of $\gamma\left(\overline{\ell}_k, \overline{a}_k, F\right)$ in the sense that given any function $t_k\left(\overline{a}_k\right)$ that is zero when $a_k = 0$, any function $r_k^*\left(\overline{\ell}_m, \overline{a}_k\right)$ satisfying (A3.6), and any law of $\overline{X}_K = \left(\overline{L}_K, \overline{A}_K\right)$, the function $\gamma\left(\overline{\ell}_k, \overline{a}_k\right)$ defined by the R.H.S. of (A3.7) will equal $\gamma\left(\overline{\ell}_k, \overline{a}_k, F\right)$ when the law of $\left(Y, \overline{X}_K\right)$ is generated as described following Eq. (A2.5); furthermore, all possible distributions of $\left(Y, \overline{X}_K\right)$ can be constructed in this manner.

An alternative interesting unrestricted parameterization is as follows. Let $b\left(\overline{\ell}_m, \overline{a}\right)$ be shorthand for $b\left(\overline{\ell}_m, g = (\overline{a})\right)$ as defined in Appendix 2. Now define $v\left(\overline{\ell}_m, \overline{a}\right) = b\left(\overline{\ell}_m, \overline{a}\right) - b\left(\overline{\ell}_{m-1}, \overline{a}\right)$ and $v^*\left(\overline{\ell}_m, \overline{a}\right) = v\left(\overline{\ell}_m, \overline{a}\right) - v\left(\left\{\overline{\ell}_{m-1}, \ell_m = 0\right\}, \overline{a}\right)$ so that

$$v^*\left(\overline{\ell}_{m-1}, \ell_m = 0, \overline{a}\right) = 0 \tag{A3.8}$$

Then, we have

$$b\left(\overline{\ell}_K, \overline{a}\right) = \tag{A3.9}$$

$$\sum_{m=0}^{K} v^*\left(\overline{\ell}_m, \overline{a}\right) - E\left[v^*\left(\left\{\overline{\ell}_{m-1}, L_m\right\}, \overline{a}\right) \mid \overline{L}_{m-1} = \overline{\ell}_{m-1}, \overline{A}_{m-1} = \overline{a}_{m-1}\right] + b\left(\overline{a}\right)$$

(A3.9) is an unrestricted parameterization in the sense that given any law for $\left(\overline{L}_K, \overline{A}_K\right)$, any random variable σ satisfying (A2.5), any function $b\left(\overline{a}\right)$, and any function $v^*\left(\overline{\ell}_m, \overline{a}\right)$ satisfying (A3.8), if we set $Y = b\left(\overline{L}_K, \overline{A}_K\right) + \sigma$ with $b\left(\overline{L}_K, \overline{A}_K\right)$ computed by the RHS of (A3.9), then the joint distribution of $\left(Y, \overline{L}_K, \overline{A}_K\right)$ will be consistent with our choices of $v^*\left(\overline{\ell}_m, \overline{a}\right)$ and $b\left(\overline{a}\right)$. Further, all possible distributions of $\left(Y, \overline{L}_K, \overline{A}_K\right)$ can be constructed in this manner.

The relationship between the two parameterizations is given by

$$t_k\left(\overline{a}_k\right) = b\left(\left(\overline{a}_k, 0\right)\right) - b\left(\left(\overline{a}_{k-1}, 0\right)\right) \text{ and } r_k^*\left(\overline{\ell}_m, \overline{a}_k\right) = v^*\left(\overline{\ell}_m, \left(\overline{a}_k, 0\right)\right) - v^*\left(\overline{\ell}_m, \left(\overline{a}_{k-1}, 0\right)\right) .$$

If Y is a positive random variable, an analogous multiplicative version of both parameterizations can be obtained.

Our main theorem is the following.

Theorem A3.1: The following are equivalent: *(i)* The direct effect "g"-null mean hypothesis (A3.4) holds;

(ii) $t_k\left(\overline{\ell}_m, \overline{a}_k\right)$ does not depend on $(a_{Pm}, \ldots, a_{Pk})$;

(iii) $t_k(\overline{a}_k) = t_k(\overline{a}_{Zk})$ does not depend on $\overline{a}_{Pk}$ and $r_k^*\left(\overline{\ell}_m, \overline{a}_k\right) = r_k^*\left(\overline{\ell}_m, \overline{a}_{m-1}, a_{Zm}, \ldots, a_{Zk}\right)$ does not depend on $(a_{Pm}, \ldots, a_{Pk})$;

(iv) $v^*\left(\overline{\ell}_m, \overline{a}\right)$ does not depend on $(a_{Pm}, \ldots, a_{PK})$ and $b(\overline{a})$ does not depend on $(\overline{a}_{PK})$;

(v) $b\left(\overline{\ell}_m, \overline{a}\right)$ does not depend on $a_{Pm}, \ldots, a_{PK}$.

The proof, which we do not include, follows easily by examining the limiting version of the monte carlo procedure of Sec. 8.3. The following is an easy corollary of Theorem (A3.1) and Theorem (3.2).

Corollary: If (3.1), (3.3), and (3.10) hold, then the direct effect g-null mean hypothesis (A3.2) holds if and only if for $0 \le m \le k, 0 \le k \le K$

$$E\left[Y_{g=(\overline{a}_k,0)} \mid \overline{L}_m, \overline{A}_{m-1} = \overline{a}_{m-1}\right] \text{ does not depend on } (a_{Pm}, \ldots, a_{Pk}).$$

We could create models based on either of the unrestricted parameterizations (A3.7) or (A3.9). We choose here (A3.7) because of the connection with SNMMs. We say the data follow a direct effect pseudo-SNMM for the direct effect of treatment $\overline{a}_P$ controlling for treatment $\overline{a}_Z$ if $t_k(\overline{a}_k) = t_k(\overline{a}_k, \psi_{Pt0}, \psi_{Zt0})$ and $r_k^*\left(\overline{\ell}_m, \overline{a}_k\right) = r_k^*\left(\overline{\ell}_m, \overline{a}_k, \psi_{Pr0}, \psi_{Zr0}\right)$ where *(i)* $t_k(\overline{a}_k, \psi_{Pt}, \psi_{Zt})$ and $r_k^*\left(\overline{\ell}_m, \overline{a}_k, \psi_{Pr}, \psi_{Zr}\right)$ are known functions that take the value zero if $a_k = 0$ or $\psi = 0$ with $\psi \equiv (\psi_P, \psi_Z)$, $\psi_P = (\psi_{Pt}, \psi_{Pr})$, $\psi_Z = (\psi_{Zt}, \psi_{Zr})$, *(ii)* $r_k^*\left(\overline{\ell}_m, \overline{a}_k, \psi_{Pr}, \psi_{Zr}\right) = 0$ if $\ell_m = 0$, *(iii)* $t_k(\overline{a}_k, \psi_{Pt} = 0, \psi_{Zt})$ depends on $\overline{a}_k$ only through $\overline{a}_{Zk}$, *(iv)* and $r_k^*\left(\overline{\ell}_m, \overline{a}_k, \psi_{Pr} = 0, \psi_{Zr}\right)$ depends on $\overline{a}_k$ only through $\overline{a}_{m-1}, a_{Zm}, \ldots, a_{Zk}$.

Remark: There are no restrictions on the functions $r_k^*\left(\overline{\ell}_m, \overline{a}_k, \psi_r\right)$ and $t_k(\overline{a}_k, \psi_t)$ except for (i)-(iv) above. Here $\psi_r \equiv (\psi_{Pr}, \psi_{Zr})$.

Example: With ℓ_m univariate,

$$r_k^*\left(\overline{\ell}_m, \overline{a}_k, \psi_r\right) = \psi_{Pr} a_{Pk} \ell_m \left[\sum_{j=0}^{k-1} (a_{Pj} + a_{Zj})\right] + \psi_{Zr} a_{Zk} \ell_m \left[\sum_{j=0}^{m-1} a_{Pj} + \sum_{j=0}^{k-1} a_{Zj}\right]$$

$$t(\overline{a}_k, \psi_t) = \psi_{Pt1} a_{Pk} \left(\sum_{m=0}^{k-1} a_{Pm} + a_{Zm}\right) + \psi_{Pt2} a_{Zk} \left(\sum_{m=0}^{k-1} a_{Pm}\right) + \psi_{Zt} a_{Zk} \left(\sum_{m=0}^{k-1} a_{Zm}\right).$$

It follows from the "g"-null mean theorem of Appendix 2 and Theorem (A3.1) that *(i)* $\psi_0 = 0$ if and only if the "g"-null mean hypothesis that $b(g_1) = b(g_2)$ for all g_1, g_2 holds, and *(ii)* $\psi_{P0} = 0$ if and only if the direct effects "g"-null mean hypothesis (A3.4) holds.

If (3.1), (3.3), and (3.10) hold so that $\gamma\left(\overline{\ell}_k, \overline{a}_k, F\right) = \gamma^{\dagger}\left(\overline{\ell}_k, \overline{a}_k\right)$, we also call our direct-effect pseudo-SNMM a direct-effect SNMM. In this case, *(i)* $\psi_0 = 0$ if and only if the g-null mean hypothesis (5.1) holds and *(ii)* $\psi_{P0} = 0$ if and only if the direct effect g-null mean hypothesis (A3.2) holds.

To estimate the parameter ψ_0 of a direct-effect pseudo-SNMM by "generalized" g-estimation, we proceed as follows. Write $\psi_r = (\psi_{Pr}, \psi_{Zr})$ and $\psi_t = (\psi_{Pt}, \psi_{Zt})$. First we estimate $f\left[\ell_m \mid \overline{\ell}_{m-1}, \overline{a}_{m-1}\right]$ by specifying a parametric model $f\left[\ell_m \mid \overline{\ell}_{m-1}, \overline{a}_{m-1}; \omega_3\right]$ depending on an unknown parameter ω_3 and find $\widehat{\omega}_3$ that maximizes $\prod_{i=1}^{n} f\left[\ell_{mi} \mid \overline{\ell}_{(m-1)i}, \overline{a}_{(m-1)i}; \omega_3\right]$. We then construct $r_k\left(\overline{\ell}_m, \overline{a}_k; \psi_r, \widehat{\omega}_3\right)$ by evaluating Eq. (A3.5a) at $\widehat{\omega}_3$ and $r_k^*\left(\overline{\ell}_m, \overline{a}_k; \psi_r\right)$. Next we use $r_k\left(\overline{\ell}_m, \overline{a}_k; \psi_r, \widehat{\omega}_3\right)$ and $t_k\left(\overline{a}_k; \psi_t\right)$ to obtain $\gamma\left(\overline{\ell}_k, \overline{a}_k; \psi, \widehat{\omega}_3\right)$ by Eq. (A3.5). We next compute $H\left(\psi, \widehat{\omega}_3\right) = Y - \sum_{m=0}^{K} \gamma\left(\overline{L}_m, \overline{A}_m; \psi, \widehat{\omega}_3\right)$. Finally we estimate the unknown parameter ψ by g-estimation as in Sec. 8.3 with $H\left(\psi, \widehat{\omega}_3\right)$ replacing $H(\psi)$.

To estimate ψ by maximum likelihood, we would proceed as in Appendix 2 in the paragraph following Eq. (A2.9), except that, in computing the residual $\sigma(\psi, \omega)$, *(i)* $H(\psi)$ is replaced by $H(\psi, \omega_3)$ and *(ii)* all η's are replaced by ω's in (A2.7).

Direct-Effect SNDMs:

We now describe direct-effect SNDMs and pseudo-SNDMs. We first provide the notation necessary to describe our models. Write $b\left(y, \overline{\ell}_k, \overline{a}\right)$ and $b(y, \overline{a})$ as shorthand for $b\left(y, \overline{\ell}_k, g = (\overline{a})\right)$ and $b(y, g = (\overline{a}))$. Note $b\left(y, \overline{\ell}_{m-1}, \overline{a}\right) =$
$E\left[b\left(y, \left(\overline{\ell}_{m-1}, L_m\right), \overline{a}\right) \mid \overline{L}_{m-1} = \overline{\ell}_{m-1}, \overline{A}_{m-1} = \overline{a}_{m-1}\right]$. Under (3.1), (3.3), and the sequential randomization assumption (3.10), Theorem (3.2) implies
$b\left(y, \overline{\ell}_m, (\overline{a}_k, 0)\right) = pr\left[Y_{g=(\overline{a}_k,0)} > y \mid \overline{L}_m = \overline{\ell}_m, \overline{A}_{m-1} = \overline{a}_{m-1}\right]$. Now define $t_k\left(y, \overline{\ell}_m, \overline{a}_k\right)$ for $m = -1, 0, \ldots, k$ by

$$b\left[y, \overline{\ell}_m, (\overline{a}_k, 0)\right] = b\left[t_k\left(y, \overline{\ell}_m, \overline{a}_k\right), \overline{\ell}_m, (\overline{a}_{k-1}, 0)\right]$$

so, in particular, $t_k\left(y, \overline{\ell}_k, \overline{a}_k\right) = \gamma\left(y, \overline{\ell}_k, \overline{a}_k, F\right)$. Hence, under (3.1), (3.3), and (3.10), $t_k\left(y, \overline{\ell}_m, \overline{a}_k\right)$ satisfies

$$\begin{aligned} &pr\left[Y_{g=(\overline{a}_k,0)} > y \mid \overline{L}_m = \overline{\ell}_m, \overline{A}_{m-1} = \overline{a}_{m-1}\right] = \\ &pr\left[Y_{g=(\overline{a}_{k-1},0)} > t_k\left(y, \overline{\ell}_m, \overline{a}_k\right) \mid \overline{L}_m = \overline{\ell}_m, \overline{A}_{m-1} = \overline{a}_{m-1}\right]. \end{aligned}$$

In particular, $t_k\left(y, \overline{\ell}_k, \overline{a}_k\right) = \gamma^{\dagger}\left(y, \overline{\ell}_k, \overline{a}_k\right)$. Our main results are the following theorem and corollary.

Theorem (A3.2): The following are equivalent: *(i)* The direct effect "g"-null hypothesis (A3.3) holds; *(ii)* for $0 \le m \le k$, $0 \le k \le K$,

$$t_k\left(y, \overline{\ell}_m, \overline{a}_k\right) \text{ does not depend on } \left(a_{Pm}, \ldots, a_{Pk}\right); \qquad \text{(A3.10)}$$

$$\textit{(iii)}\ b\left(y, \overline{\ell}_k, \overline{a}\right) \text{ does not depend on } \left(a_{Pk}, \ldots, a_{PK}\right), 0 \le k \le K \qquad \text{(A3.11)}$$

Corollary (A3.1): If (3.1), (3.3), and (3.10) hold, the following are equivalent: the direct-effect g-null hypothesis (A3.1) holds; *(ii)* (A3.10) is true; *(iii)* (A3.11) is true.

The proof of Theorem A3.2 which we do not give follows easily from the limiting form of the monte carlo algorithm of Sec. 7.5. Now define $v\left(y,\overline{\ell}_m,\overline{a}\right)$ by

$$b\left(y,\overline{\ell}_m,\overline{a}\right)=b\left\{v\left(y,\overline{\ell}_m,\overline{a}\right),\overline{\ell}_{m-1},\overline{a}\right\}\ . \tag{A3.12}$$

Now let

$$v^*\left(y,\overline{\ell}_m,\overline{a}\right)\equiv v\left(y,\overline{\ell}_m,\overline{a}\right)-v\left(y,\left\{\overline{\ell}_{m-1},\ell_m=0\right\},\overline{a}\right)\ , \tag{A3.12a}$$

so

$$v^*\left(y,\overline{\ell}_m,\overline{a}\right)=0 \text{ if } \ell_m=0\ . \tag{A3.13}$$

We have from these definitions that

$$\begin{aligned} &E[b\{\left[v^*\left(y,\overline{L}_m,\overline{a}\right)+v\left(y,\left\{\overline{\ell}_{m-1},\ell_m=0\right\},\overline{a}\right)\right],\overline{\ell}_{m-1},\overline{a}\}\mid \\ \overline{L}_{m-1} \;=\; &\overline{\ell}_{m-1},\overline{A}_{m-1}=\overline{a}_{m-1}]=b\left(y,\overline{\ell}_{m-1},\overline{a}\right)\ . \end{aligned} \tag{A3.14}$$

Now $b\left(y,\overline{a}\right)$ and $v^*\left(y,\overline{\ell}_m,\overline{a}\right)$ constitute an unrestricted parameterization in the sense that any survivor function $b\left(y,\overline{a}\right)$, function $v^*\left(y,\overline{\ell}_m,\overline{a}\right)$ satisfying (A3.13), and law for $\left(\overline{L}_K,\overline{A}_K\right)$ determine a unique law F for $\left(Y,\overline{L}_K,\overline{A}_K\right)$ by $pr\left[Y>y\mid\overline{L}_K,\overline{A}\right]=b\left(y,\overline{L}_K,\overline{A}\right)$ with $\overline{A}\equiv\overline{A}_K$ and $b\left(y,\overline{\ell}_K,\overline{a}\right)$ determined recursively from $b\left(y,\overline{a}\right)\equiv b\left(y,\overline{\ell}_{-1},\overline{a}\right)$, $v^*\left(y,\overline{\ell}_m,\overline{a}\right)$, (A3.12) and (A3.14). Specifically, given $b\left(y,\overline{\ell}_{m-1},\overline{a}\right)$ and $v^*\left(y,\overline{\ell}_m,\overline{a}\right)$, we solve (A3.14) for $v\left(y,\left\{\overline{\ell}_{m-1},\ell_m=0\right\},\overline{a}\right)$, and then compute $v\left(y,\overline{\ell}_m,\overline{a}\right)$ by (A3.12a). Finally this allows us to compute $b\left(y,\overline{\ell}_m,\overline{a}\right)$ from (A3.12). Furthermore, any law F for $\left(Y,\overline{L}_K,\overline{A}_k\right)$ can be obtained by this method.

Further, the "g"-null hypothesis (5.2) holds if and only if both $b\left(y,\overline{a}\right)$ does not depend on $\overline{a}$ and $v^*\left(y,\overline{\ell}_m,\overline{a}\right)$ does not depend on $a_m,\ldots,a_K$. Further, the direct-effect "g"-null hypothesis holds if and only if *(i)* $b\left(y,\overline{a}\right)$ does not depend on $\overline{a}_{PK}$ and *(ii)* $v^*\left(y,\overline{\ell}_m,\overline{a}\right)$ does not depend on $a_{Pm},\ldots,a_{PK}$. Thus, likelihood-based inference can be carried out by specifying parametric models for $b\left(y,\overline{a}\right)$ and $v^*\left(y,\overline{\ell}_m,\overline{a}\right)$ (satisfying (A3.13)) with the parameters ψ_P and ψ_Z such that $\psi_P=0$ implies *(i)* and *(ii)* just above. This parameterization is not closely related to SNDM models. We now describe another unrestricted parameterization that allows a connection to SNDM models.

Now for $m=0,\ldots,k$, define $r_k\left(y,\overline{\ell}_m,\overline{a}_k\right)$ by $r_k\left[t_k\left(y,\overline{\ell}_{m-1},\overline{a}_k\right),\overline{\ell}_m,\overline{a}_k\right]=t_k\left(y,\overline{\ell}_m,\overline{a}_k\right)$. Then the following lemma is easy to prove.

Lemma A3.1: $r_k\left(y,\overline{\ell}_m,\overline{a}_k\right)$ satisfies

$$E[b\left\{r_k\left(y,\overline{L}_m,\overline{a}_k\right),\overline{L}_m,\left(\overline{a}_{k-1},0\right)\right\}\mid\overline{L}_{m-1}=\overline{\ell}_{m-1},\overline{A}_{m-1}=\overline{a}_{m-1}]=b\left(y,\overline{\ell}_{m-1},\left(\overline{a}_{k-1},0\right)\right).$$

Lemma (A3.1) has the following Corollary.

Corollary (A3.2): If (3.1), (3.3), and (3.10) hold, $r_k\left(y,\overline{\ell}_m,\overline{a}_k\right)$ satisfies

$$\begin{aligned} &E\{pr\left[Y_{g=(\overline{a}_{k-1},0)}>r_k\left(y,\overline{L}_m,\overline{a}_k\right)\mid\overline{L}_m,\overline{A}_{m-1}=\overline{a}_{m-1}\right]\mid \\ &\overline{L}_{m-1}=\overline{\ell}_{m-1},\overline{A}_{m-1}=\overline{a}_{m-1}\}=pr\left[Y_{g=(\overline{a}_{k-1},0)}>y\mid\overline{L}_{m-1}=\overline{\ell}_{m-1},\overline{A}_{m-1}=\overline{a}_{m-1}\right]. \end{aligned}$$

Now define $r_k^*\left(y,\overline{\ell}_m,\overline{a}_k\right) \equiv r_k\left(y,\overline{\ell}_m,\overline{a}_k\right) - r_k\left(y,\left\{\overline{\ell}_{m-1},\ell_m=0\right\},\overline{a}_k\right)$ so that $r_k^*\left(y,\overline{\ell}_m,\overline{a}_k\right) = 0$ if $\ell_m = 0$ or $a_k = 0$. We now have the following easy corollary to Theorem (A3.2). Set $t_k\left(y,\overline{a}_k\right) \equiv t_k\left(y,\overline{\ell}_{-1},\overline{a}_k\right)$.

Corollary (A3.3): The direct-effect "g"-null hypothesis (A3.3) holds if and only if for all k and $m \leq k$, $r_k^*\left(y,\overline{\ell}_m,\overline{a}_k\right) = r_k^*\left(y,\overline{\ell}_m,\overline{a}_{m-1},a_{Zm},\ldots,a_{Zk}\right)$ does not depend on $\left(a_{Pm},\ldots,a_{Pk}\right)$ and $t_k\left(y,\overline{a}_k\right) = t_k\left(y,\overline{a}_{Zk}\right)$ does not depend on $\overline{a}_{Pk}$.

We say the data follow a direct effect pseudo-SNDM for the direct effect of treatment $\overline{a}_P$ controlling for treatment $\overline{a}_Z$ if $t_k\left(y,\overline{a}_k\right) = t_k\left(y,\overline{a}_k,\psi_{Pt0},\psi_{Zt0}\right)$ and $r_k^*\left(y,\overline{\ell}_m,\overline{a}_k\right) = r_k^*\left(y,\overline{\ell}_m,\overline{a}_k,\psi_{Pr0},\psi_{Zr0}\right)$ where *(i)* $t_k\left(y,\overline{a}_k,\psi_{Pt},\psi_{Zt}\right)$ and $r_k^*\left(y,\overline{\ell}_m,\overline{a}_k,\psi_{Pr},\psi_{Zr}\right)$ are known functions that take the values y or zero respectively if $a_k = 0$ or $\psi = 0$ with $\psi \equiv \left(\psi_P,\psi_Z\right)$; $\psi_P = \left(\psi_{Pt},\psi_{Pr}\right)$, $\psi_Z = \left(\psi_{Zt},\psi_{Zr}\right)$, *(ii)* $r_k^*\left(y,\overline{\ell}_m,\overline{a}_k,\psi_{Pr},\psi_{Zr}\right) = 0$ if $\ell_m = 0$, *(iii)* $t_k\left(y,\overline{a}_k,\psi_{Pt}=0,\psi_{Zt}\right)$ depends on $\overline{a}_k$ only through $\overline{a}_{Zk}$, *(iv)* and $r_k^*\left(\gamma,\overline{\ell}_m,\overline{a}_k,\psi_{Pr}=0,\psi_{Zr}\right)$ depends on $\overline{a}_k$ only through $\overline{a}_{m-1},a_{Zm},\ldots,a_{Zk}$.

It follows from the "g"-null theorem and Corollary (A3.3) that *(i)* $\psi_0 = 0$ if and only if the "g"-null hypothesis (5.2) holds, and *(ii)* $\psi_{P0} = 0$ if and only if the direct effect "g"-null hypothesis (A3.3) holds.

If (3.1), (3.3), and (3.10) hold so that $\gamma\left(y,\overline{\ell}_k,\overline{a}_k,F\right) = \gamma^\dagger\left(y,\overline{\ell}_k,\overline{a}_k\right)$, we also call our direct-effect pseudo-SNDM a direct-effect SNDM. In this case, *(i)* $\psi_0 = 0$ if and only if the g-null hypothesis (5.1) holds and *(ii)* $\psi_{P0} = 0$ if and only if the direct effect g-null hypothesis (A3.1) holds.

We now consider estimation of the parameter ψ_0 for a direct-effect pseudo-SNDM with maximum likelihood. We do not have a simple generalization of g-estimation. As in Sec. 7, let $f\left(\ell_m \mid \overline{\ell}_{m-1},\overline{a}_{m-1},h;\phi\right)$ be a model indexed by ϕ for $L_m \mid \overline{L}_{m-1},\overline{A}_{m-1},H(\gamma)$ and $f(h;\eta)$ be a model indexed by η for the marginal law of $H(\gamma)$. Now let $\theta = (\psi,\phi,\eta)$ where ψ is the parameter vector for the direct-effect pseudo-SNMM and let F_θ be a law of the observed data generated under parameter θ. Since our parameterization can be shown to be unrestricted, the above models need not be restricted. In addition, our pseudo-SNDM model is unrestricted except as in *(i)*-*(iv)* of its definition.

Now using the fact that $H(\gamma) \amalg A_m \mid \overline{L}_m,\overline{A}_{m-1}$, Lemma A3.1 can be rewritten as follows after some algebraic manipulation.

$$\Gamma_k\left(\overline{\ell}_{m-1},\overline{a}_{k-1},y;F_\theta\right) = \tag{A3.15}$$
$$\Gamma_k\left[\overline{\ell}_{m-1},\overline{a}_{k-1},r_k^*\left(y,\overline{\ell}_m,\overline{a}_k;\psi\right) + r_k\left(y,\left\{\overline{\ell}_{m-1},\ell_m=0\right\},\overline{a}_k;F_\theta\right)\right]$$

where, for any function $c\left(y,\overline{\ell}_m,\overline{a}_k\right)$, $\Gamma_k\left(\overline{\ell}_{m-1},\overline{a}_{k-1},c\left(y,\overline{\ell}_m,\overline{a}_k\right);F_\theta\right) \equiv$ $\int\cdots\int I\left\{q\left[h,\overline{\ell}_{k-1},\overline{a}_{k-1};F_\theta\right] > c\left(y,\overline{\ell}_m,\overline{a}_k\right)\right\}\prod_{j=0}^{k-1} f\left[\ell_j \mid \overline{\ell}_{j-1},\overline{a}_{j-1},h;\phi\right]\prod_{j=m}^{k-1} d\mu\left(\ell_j\right)dF(h;\eta)\Big/$ $\int\prod_{j=0}^{m-1} f\left[\ell_j \mid \overline{\ell}_{j-1},\overline{a}_{j-1},h;\phi\right]dF(h;\eta)$. The key identity in deriving (A3.15) is the fact that $H(\gamma) \amalg A_m \mid \overline{L}_m,\overline{A}_{m-1}$ implies $f\left(h \mid \overline{\ell}_k,\overline{a}_{k-1}\right) = \prod_{j=0}^{k} f\left(\ell_j \mid \overline{\ell}_{j-1},\overline{a}_{j-1},h\right)f(h)\Big/$ $\int\prod_{j=0}^{k} f\left(\ell_j \mid \overline{\ell}_{j-1},\overline{a}_{j-1},h\right)f(h)\,dh$ where h represents realizations of $H(\gamma)$.

Now, as in Sec. 7.5, we note $q\left(h,\overline{\ell}_k,\overline{a}_k;F_\theta\right)$ depends on F_θ through $\gamma^{-1}\left(y,\overline{\ell}_m,\overline{a}_m,F_\theta\right)$ for $m\le k$, and, in our model, $\gamma\left(y,\overline{\ell}_k,\overline{a}_k;F_\theta\right)\equiv t_k\left(y,\overline{\ell}_k,\overline{a}_k;F_\theta\right)$ is obtained recursively by $t_k\left(y,\overline{\ell}_m,\overline{a}_k;F_\theta\right)=r_k\left[t_k\left(y,\overline{\ell}_{m-1},\overline{a}_k;F_\theta\right);\overline{\ell}_m,\overline{a}_k;F_\theta\right]$ with $t_k\left(y,\overline{\ell}_{-1},\overline{a}_k;F_\theta\right)\equiv t_k\left(y,\overline{a}_k;\psi\right)$ and $r_k\left(y,\overline{\ell}_m,\overline{a}_k;F_\theta\right)=r_k^*\left(y,\overline{\ell}_m,\overline{a}_k;\psi\right)+r_k\left(y,\left\{\overline{\ell}_{m-1},\ell_m=0\right\},\overline{a}_k;F_\theta\right)$. Thus, given $\theta=(\psi,\phi,\eta)$ and $r_j\left[y,\left\{\overline{\ell}_{m-1},\ell_m=0\right\},\overline{a}_j;F_\theta\right],0\le j\le k-1$ [and thus $q\left(h,\overline{\ell}_{k-1},\overline{a}_{k-1};F_\theta\right)$] are known, it follows that all terms in (A3.15) [including $q\left(h,\overline{\ell}_{k-1},\overline{a}_{k-1};F_\theta\right)$] are known except for $r_k\left[y,\left\{\overline{\ell}_{m-1},\ell_m=0\right\},\overline{a}_k;F_\theta\right]$ which can therefore be solved for recursively. In general, $r_k\left[y,\left\{\overline{\ell}_{m-1},\ell_m=0\right\},\overline{a}_k;F_\theta\right]$ and thus $\gamma\left(y,\overline{\ell}_k,\overline{a}_k;F_\theta\right)$ and $H(\gamma)$ will be functions of all the parameters ψ, ϕ, and η of θ. Thus write $r_k\left(y,\left\{\overline{\ell}_{m-1},\ell_m=0\right\},\overline{a}_k;\theta\right)$ and $H(\theta)=H(\psi,\phi,\eta)$ to emphasize the functional dependence on all of θ. The maximum likelihood estimator of $\theta=(\psi,\phi,\eta)$ is obtained by maximizing Eq. (7.8) with $h_i(\psi)$ replaced by the realization $h_i(\theta)$ of $H(\theta)$. Additional study of the computation problems involved will be useful.

Finally, Robins and Wasserman (1996) provide another unrestricted parameterization for the law of $\left(Y,\overline{L}_K,\overline{A}_K\right)$ based on multiplicative models for the hazard functions of the survivor functions $b(y,\overline{a})$ and $b\left(y,\overline{\ell}_k,\overline{a}\right)$. The models contain a parameter ψ_P that is zero under the direct-effect "g"-null hypothesis.

REFERENCES

Arjas, E. (1989). "Survival models and martingale dynamics (with discussion)." *Scandinavian Journal of Statistics*, 15, 177-225.

Gill, R. and Robins, J. (1996). "Some measure theoretic aspects of causal models." (In preparation.)

Heckerman, D. and Shachter, R. (1995). "Decision-theoretic foundations for causal inference. *Journal of Artificial Intelligence Research*, 3, 405-430.

Holland, P. (1989), "Reader reaction: Confounding in epidemiologic studies," *Biometrics*, 45, 1310-1316.

Lewis, D.K. (1973), **Counterfactuals**. Cambridge: Harvard University Press.

Loomis, B. and Sternberg, S. (1968). **Advanced Calculus**. Addison Wesley.

Pearl, J. (1995), "Causal diagrams for empirical research," *Biometrika*, 82, 669-690..

Pearl, J. and Robins, J.M. (1995). "Probabilistic evaluation of sequential plans from causal models with hidden variables," From: **Uncertainty in Artificial Intelligence:** Proceedings of the Eleventh Conference on Artificial Intelligence, August 18-20, 1995, McGill University, Montreal, Quebec, Canada. San Francisco, CA: Morgan Kaufmann. pp. 444-453.

Pearl, J. and Verma, T. (1991). "A Theory of Inferred Causation." In: **Principles of Knowledge, Representation and Reasoning: Proceedings of the Second International Conference.** (Eds. J.A. Allen, R. Fikes, and E. Sandewall). 441-452.

Robins, J.M. (1986), "A new approach to causal inference in mortality studies with sustained exposure periods – application to control of the healthy worker survivor effect," *Mathematical Modelling*, 7, 1393-1512.

Robins, J.M. (1987), "Addendum to 'A new approach to causal inference in mortality studies with sustained exposure periods – application to control of the healthy worker survivor effect'," *Computers and Mathematics with Applications*, 14, 923-945.

Robins, J.M. (1989), "The analysis of randomized and non-randomized AIDS treatment trials using a new approach to causal inference in longitudinal studies," In: **Health Service Research Methodology: A Focus on AIDS**, eds. Sechrest, L., Freeman, H., Mulley, A., NCHSR, U.S. Public Health Service, 113-159.

Robins, J.M. (1992), "Estimation of the time-dependent accelerated failure time model in the presence of confounding factors," *Biometrika*, 79, 321-334.

Robins, J.M. (1993), "Analytic methods for estimating HIV-treatment and cofactor effects," In: **Methodological Issues in AIDS Mental Health Research**, eds. Ostrow, D.G., and Kessler, R.C., NY: Plenum Press, 213-290.

Robins, J.M. (1994), "Correcting for non-compliance in randomized trials using structural nested mean models," *Communications in Statistics*, 23, 2379-2412.

Robins, J.M. (1995a). "Estimating the Causal Effect of a Time-varying Treatment on Survival using Structural Nested Failure Time Models," (To appear, *Statistica Neederlandica*).

Robins, J.M. (1995b). "Discussion of 'Causal Diagrams for empirical research' by J. Pearl," *Biometrika*, 82, 695-698.

Robins, J.M. (1996). "Correction for non-compliance in bioequivalence trials," *Statistics in Medicine* (To appear).

Robins, J.M., Blevins, D., Ritter, G. and Wulfsohn, M. (1992), "G-estimation of the effect of prophylaxis therapy for pneumocystis carinii pneumonia on the survival of AIDS patients," *Epidemiology*, 3, 319-336.

Robins, J.M. and Pearl, J. (1996). "Causal effects of dynamic policies." In preparation.

Robins, J.M. and Wasserman, L. (1996). "Parameterizations of directed acyclic graphs for the estimation of overall and direct effects of multiple treatments." Technical Report, Department of Epidemiology, Harvard School of Public Health.

Rosenbaum, P.R. (1984), "Conditional permutation tests and the propensity score in observational studies," *Journal of the American Statistical Association*, 79, 565-574.

Rosenbaum, P.R. (1984), "The consequences of adjustment for a concomitant variable that has been adversely affected by treatment," *Journal of the Royal Statistical Society A*, 147, 656-666.

Rosenbaum, P.R., and Rubin, D.B. (1983), "The central role of the propensity score in observational studies for causal effects," *Biometrika*, 70, 41-55.

Rubin, D.B. (1978), "Bayesian inference for causal effects: The role of randomization," *The Annals of Statistics*, 6, 34-58.

Spirtes, P., Glymour, C., and Scheines, R. (1993). **Causation, Prediction, and Search**. New York: Springer Verlag.

Acknowledgements: Financial support for this research was provided by NIH Grant AI32475. Richard Gill, Larry Wasserman, and Sander Greenland provided valuable comments.

MODELS AS INSTRUMENTS, WITH APPLICATIONS TO MOMENT STRUCTURE ANALYSIS

JAN DE LEEUW

ABSTRACT. The paper discusses some very general model to evaluate the quality of estimates, and of models that these estimates are based on. Methods are based on a simple geometrical argument, and on expansions of the loss functions around the estimate, the target, and the replication. We give both delta-method and Jackknife computational procedures to estimate the relevant quantities.

1. INTRODUCTION

In De Leeuw [7] several techniques for comparison of estimates and models in multinomial experiments are discussed. These techniques differ from the usual ones, because they do not assume that a particular unsaturated model is true (even approximately). They also are not based on any preferred loss function (such as the log-likelihood) or any preferred estimation principle (such as maximum likelihood). Moreover the methods have computer-intensive versions which do not require evaluating complicated expansion formulas, but use the Jackknife instead. In this paper I continue the work of [7]. I use a geometric framework, in keeping with the recent work on the use of differential geometry in statistics, and with the recent emphasis on minimum distance estimation. See, for instance, the impressive book by Bickel et al. [2], or the recent paper by Lindsay [11]. My approach is much more pedestrian, however, and it uses models (if at all) in a thoroughly instrumentalist way, as devices to improve estimates.

As mentioned above, the results in De Leeuw [7] are limited to multinomial experiments. Moreover, they concentrate on minimum distance estimates, which means that the notion of a restrictive model for the porportions continues to play an important role. In this paper I generalize the more interesting results to a larger class of estimators, based on fairly arbitrary sequences of statistics (which could be proportions, means, covariances, correlations, and so on). These sequences, which are called *data*, converge in probability to a limit value, the *target*. The target is sometimes also called the *truth*, but as you perhaps already have guessed, we do not like that term.

In this paper, I distinguish various types of errors an estimate can have, and I estimate these errors using delta method and Jackknife techniques. The notion of a distance-type loss function still plays a role. I evaluate the distance between the target and the sequence of estimates. This is called the *overall error*. I also compute the distance between the estimates and a sequence of variables with the same distribution as the target sequence, but independent of the target sequence. This is called the *prediction error*. Finally I compute estimates of the distance between the target and the limit of the estimation sequence, which is the *specification error*. Finally, in the computations I need the distance between the target sequence and the estimation sequence, which I call the *deviance*. Just words, of course, but chosen to suggest the various types of errors that are involved.

Date: June 11, 1996.
Written while visiting the National Institute of Statistical Sciences, Research Triangle Park, NC.

The paper is closely related to the work of Linhart and Zucchini [12], which has been used by Cudeck and Brown [5] and Browne and Cudeck [3, 4]. A related Bayesian approach, also based on predictive distribution, has been discussed by Gelfand, Dey, and Chang [8]. Related research in covariance structure modeling has been published by McDonald [13], McDonald and Marsh [14], and Bentler [1].

2. Estimation Problem

We start, as in De Leeuw [7], with simple experiments in which we estimate a vector of proportions (this could also be a multidimensional cross table, or a discreticized multivariate distribution). This will be generalized slightly in a later section.

Thus we start by considering an *estimation problem* with the following ingredients.

Definition 2.1. The *data* is sequence of random vectors of *proportions* $\underline{p}_n$, taking values in the unit simplex $\mathbf{S}^m$. The *target* is a fixed element π of $\mathbf{S}^m$.

Of course this is highly suggestive, and to some extent misleading, terminology. Actual data are not a random variable, and certainly not a sequence of random variables. To arrive at this representation we must first embed the actual data in a *framework of replication*, which is modeled by a single random variable, and we then have to embed that random variable in a sequence to get our asymptotics under way. Thus the distance from what we call "data" to what an ordinary person would call "data" is quite large. In the same way calling a particular point in parameter space "the target" must be seen as have some heuristic value, at the most.

We still have to connect the data and the target in some convenient way. The three assumptions below are enough to justify our expansions and expectations, and they also serve to simply the notation.

Assumption 2.2. For all $n \in \mathbf{N}$ we have $\mathbf{E}(\underline{p}_n) = \pi$.

Assumption 2.3. There is a positive semi-definite $V(\pi)$ such that for all $n \in \mathbf{N}$ we have $n\mathbf{V}(\underline{p}_n) = V(\pi)$. $V(\pi)$ is of rank $m-1$, and only has the constant vectors in its null-space.

Assumption 2.4. For all $j = 1, \cdots, m$ we have $\mathbf{E} \mid \underline{p}_{jn} - \pi_j \mid^6 = \mathcal{O}(n^{-3})$

By Liaponoff's Inequality on absolute moments, Assumption 2.4 implies that $\mathbf{E} \mid \underline{p}_{jn} - \pi_j \mid^s = \mathcal{O}(n^{-\frac{s}{2}})$ for all $1 \leq s \leq 6$. Our assumptions do *not* imply asymptotic normality, and they do *not* assert that the $\underline{p}_n$ are averages of a sequence of independent indicators. In the multinomial case, of course, the assumptions are true, and we have $V(\pi) = \Pi - \pi\pi'$, where Π is a diagonal matrix with the elements of π along the diagonal.

Definition 2.5. An *estimator* Φ is a function from $\mathbf{S}^m$ to $\mathbf{S}^m$.

Definition 2.5 is somewhat non-standard, because often estimators map data into lower-dimensional (parameter) spaces. Thus $\Phi : \mathbf{S}^m \Rightarrow \mathbb{R}^r$, with $r \leq m$. But in fact this amounts to much the same thing, it merely means that the range of our estimator Φ is a manifold of dimension r. We assume

Assumption 2.6. Φ is three times totally differentiable in a neighborhood of π, and the derivatives are all bounded in that neighborhood.

To measure quality of estimators, we basically look at the distance to the target. Thus we need another definition.

Definition 2.7. A *loss function* Δ is a real-valued function on $\mathbf{S}^m \times \mathbf{S}^m$ satisfying

$$\Delta(x,y) = \begin{cases} \geq 0 & \text{for all } x, y \in \mathbf{S}^m, \\ = 0 & \text{for all } x, y \in \mathbf{S}^m \text{ with } x = y. \end{cases}$$

The function Δ is distance-like, but it need not be symmetric, nor need it satisfy the triangular inequality. For technical purposes we assume

Assumption 2.8. Δ is three times totally differentiable in a neighborhood of $(\pi, \Phi(\pi))$, and the derivatives are bounded in that neighborhood.

In De Leeuw [7] it is assumed that the estimate Φ is a *minimum distance estimate*, i.e. it is computed by minimizing $\Delta(\underline{p}_n, p)$ over p in some sort of *model* Ω, where a model is just a subset of $\mathbf{S}^m$. In the work of Brown and Cudeck, it is assumed, in addition, that the discrepancy function is *appropriate* for the multinomial experiment, in the sense that it gives efficient estimates if the model is true. We do not make any of these assumptions in this paper, in fact we do not even assume that we actually deal with a multinomial or even an asymptotically normal experiment. In our setup, there can be dependence or overdispersion, and some of the higher order moments need not even exist. The only requirement is that our three assumptions 2.2, 2.3, 2.4 are true.

3. Quality of Estimators

We measure the quality of the estimate by using the loss function 2.7 to measure the distance between the target and the estimate.

Definition 3.1. The *overall error* is

$$\underline{\beta}_n \triangleq \Delta(\pi, \Phi(\underline{p}_n)).$$

We also define $\beta_n \triangleq \mathbf{E}(\underline{\beta}_n)$, the *expected overall error* or EOE.

Remark 3.2. Obviously the errors depend on π, but we surpress that dependence in our notation. Linhart and Zucchini calls the EOE the *overall discrepancy*, De Leeuw simply calls it the *bias*.

Now suppose $\underline{q}_n$ is a second sequence of random proportions with the same asymptotic distribution as $\underline{p}_n$, but independent of $\underline{p}_n$. We call $\underline{q}_n$ the *replication*, i.e. it is the outcome of an independent replication of our experiment. Actually, we merely need to assume that Assumptions 2.2, 2.3, and 2.4 are true for $\underline{q}_n$ as well, i.e. the $\underline{q}_n$ are an independent sequence of random variables, converging to the same target as $\underline{p}_n$, with the same speed.

Definition 3.3. The *prediction error* is defined as

$$\underline{\mu}_n \triangleq \Delta(\underline{q}_n, \Phi(\underline{p}_n)).$$

The *expected prediction error* or EPE is, obviously, $\mu_n \triangleq \mathbf{E}(\underline{\mu}_n)$, where expectation is both over $\underline{q}_n$ and $\underline{p}_n$.

Remark 3.4. The EPE is called the *distortion* by De Leeuw. Linhart and Zucchinido not use prediction errors in their work. Cudeck and Browne [5] call the prediction error the *cross validation index*, and give ways to approximate it in [3].

Definition 3.5. The *specification error* is defined as

$$\delta \triangleq \Delta(\pi, \Phi(\pi)).$$

Remark 3.6. Linhart and Zucchini [12] calls this the *discrepancy due to approximation.* The specification error is a non-random quantity, so there is no need for expectations. It also does not vary with n.

Definition 3.7. The *deviance* is

$$\underline{\lambda}_n \triangleq \Delta(\underline{p}_n, \Phi(\underline{p}_n)).$$

And of course we also have the *expected deviance* or EDE $\lambda_n \triangleq \mathbf{E}(\underline{\lambda}_n)$.

Remark 3.8. The deviance does not seem to be a very interesting quantity in itself, but in our interpretation of the estimation situation we only observe $\underline{p}_n$, or at least a realization of $\underline{p}_n$. We do not observe $\underline{q}_n$ or π, and thus the prediction error and specification error cannot be observed directly. There is, of course, a familiar trick to emulate an observable $\underline{q}_n$. If we split our data in two halves, we can use the first half as a realization of $\underline{p}_{n/2}$ and the second half as a realization of $\underline{q}_{n/2}$. We shall use other, Jackknife-based, techniques below, because they seem to be somewhat more systematic.

Definition 3.9. The *estimation error* is defined as

$$\underline{\epsilon}_n \triangleq \Delta(\Phi(\pi), \Phi(\underline{p}_n)).$$

It also has an *expected estimation error* or EEE, which is $\epsilon_n \triangleq \mathbf{E}(\underline{\epsilon}_n)$.

Remark 3.10. Linhart and Zucchini call this the *discrepancy due to estimation.*

Of course generally the various error measures we have defined above do not only have expectations, but also variances, and if suitably normed perhaps even asymptotic distributions. In this paper, as in [7], we concentrate on the expectations.

4. The Distance Geometry of Estimation

A study of the definitions in Section 3 shows that we have six basic quantities between which we measure discrepancies. There is the target π and its image $\Phi(\pi)$, the data $\underline{p}_n$ and its image $\Phi(\underline{p}_n)$, and the replication $\underline{q}_n$ and its image $\Phi(\underline{q}_n)$. These six basic quantities have 30 nontrivial discrepancies between them. If the loss function is symmetric in its two arguments, then there are still 15 discrepancies. The situation is graphically shown in Figure 4.1.

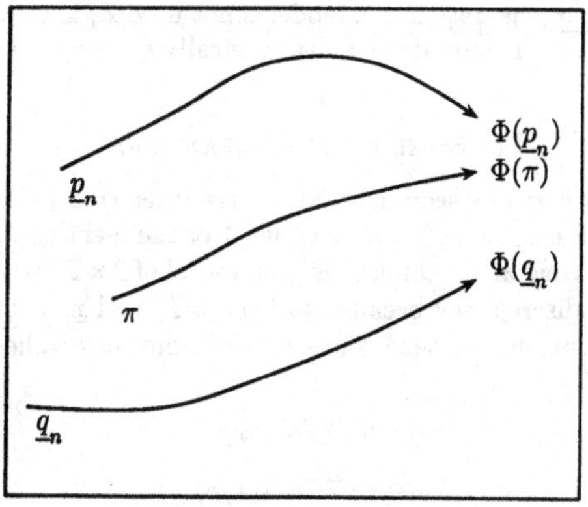

FIGURE 4.1. Geometry of Estimation

Some of these 30 discrepancies, and their expected values, were actually named in Section 3, but many remained anonymous. Nevertheless, even the anonymous ones sometimes are of interest. For instance, $\Delta(\underline{p}_n, \underline{q}_n)$ could be used as a measure of dispersion, and it could be compared with $\Delta(\Phi(\underline{p}_n), \Phi(\underline{q}_n))$ to show that estimation enhances precision. Also, as we shall see further on, we can also put Jackknifed or Bootstrapped versions of $\underline{p}_n$ in this same space.

The *idea* behind out setup is that is that $\hat{\pi}_n \stackrel{\Delta}{=} \Phi(\underline{p}_n)$ estimates π, but this notion is not really defined anywhere. In fact, we could say that the *estimation sequence* $\Phi(\underline{p}_n)$ really estimates $\Phi(\pi)$, because obviously

(4.1) $$\Phi(\underline{p}_n) \stackrel{\mathcal{P}}{\Longrightarrow} \Phi(\pi).$$

This is why $\epsilon_n(\pi)$ is called the estimation error. But to a certain extent, this is cheating. Although trivially we are estimating $\Phi(\pi)$, we want to know how good our estimate of $\Phi(\pi)$ is as an estimator of π. To assert that we are estimating what we are estimating, and then by definition consistently, has been referred to as "debilitating relativism".

We can make things a little bit more intuitive by defining the *model* Ω associated with an estimator Φ as the set of fixed points of Φ. Thus

(4.2) $$\Omega = \{p \in \mathbb{S}^m \mid \Phi(p) = p\}.$$

Since $\Phi(\underline{p}_n) \stackrel{\mathcal{P}}{\Longrightarrow} \Phi(\pi)$, we know that if $\pi \in \Omega$, then $\Phi(\underline{p}_n) \stackrel{\mathcal{P}}{\Longrightarrow} \pi$, i.e. we have consistency if π is on the model. This is sometimes expressed as saying that *the model is true.*

On the other hand we can also start with the model Ω, i.e. add it as an ingredient to our estimation problem, and then define Φ to be *F-consistent* for Ω if $\Phi(p) = p$ for all $p \in \Omega$. There is a standard way of guaranteeing F-consistency, which we already indicated in Section 2. We define Φ by the minimum distance or minimum discrepancy principle, which means that we choose $\Phi(\underline{p}_n)$ as the argmin of $\Delta(\underline{p}_n, p)$ over $p \in \Omega$. Or, to put it more geometrically, $\Phi(\underline{p}_n)$ is the *projection* of $\underline{p}_n$ on the model, in the metric defined by the loss function. This is illustrated in Figure 4.2.

For most of this paper, the notion of a model is not needed, however. It siffices to consider the situation where an estimate of π has magically appeared out of the statisticians hat.

5. Some Simple Expansions

It is clear from the previous sections that we are interested in statistics of the form $\Delta(F_1(\underline{x}_n), F_2(\underline{y}_n))$, where F_1 and F_2 can be either Φ or the identity, and $\underline{x}_n$ and $\underline{y}_n$ can be either $\underline{p}_n, \underline{q}_n$ or π. Choosing all combinations give a total of $2 \times 2 \times 3 \times 3 = 36$ possibilities, of which 6 give a zero discrepancy because both $F_1 = F_2$ and $\underline{x}_n = \underline{y}_n$.

To derive our basic result, we need some simple definitions. The first partials of the loss function are a

$$t(x,y) \triangleq \mathcal{D}_1\Delta(x,y), \tag{5.1a}$$

$$u(x,y) \triangleq \mathcal{D}_2\Delta(x,y), \tag{5.1b}$$

and second partials are a

$$A(x,y) \triangleq \mathcal{D}_{11}\Delta(x,y), \tag{5.2a}$$

$$B(x,y) \triangleq \mathcal{D}_{12}\Delta(x,y), \tag{5.2b}$$

$$C(x,y) \triangleq \mathcal{D}_{22}\Delta(x,y). \tag{5.2c}$$

We also need a

$$G_1(p) \triangleq \mathcal{D}F_1(p), \tag{5.3a}$$

$$G_2(p) \triangleq \mathcal{D}F_2(p), \tag{5.3b}$$

and a

$$H_1(p) \triangleq \mathcal{D}G_1(p), \tag{5.4a}$$

$$H_2(p) \triangleq \mathcal{D}G_2(p). \tag{5.4b}$$

Observe that for $k = 1, 2$ the function $H_k(\bullet)$ maps $\mathbb{R}^m \otimes \mathbb{R}^m$ into $\mathbb{R}^m$, which means that $H_k(\bullet)(z,z)$ can be written as

$$H_k(\bullet)(z,z) = \begin{bmatrix} z'H_{k1}(\bullet)z \\ \vdots \\ z'H_{km}(\bullet)z \end{bmatrix} \tag{5.5}$$

with each of the $H_{kj}(\bullet)$ square symmetric matrices. The fact that the H_k are not matrices is actually only a minor nuisance, because the matrices we really need are a

$$\Gamma_1(x,y) \triangleq \sum_{j=1}^{m} t_j(x,y)H_{1j}(\pi), \tag{5.6a}$$

$$\Gamma_2(x,y) \triangleq \sum_{j=1}^{m} u_j(x,y)H_{2j}(\pi). \tag{5.6b}$$

We continue the orgy of definitions with a

$$V_{11}(\pi) \triangleq n\mathbf{C}\,(\underline{x}_n, \underline{x}_n), \tag{5.7a}$$

$$V_{12}(\pi) \triangleq n\mathbf{C}\,(\underline{x}_n, \underline{y}_n), \tag{5.7b}$$

$$V_{22}(\pi) \triangleq n\mathbf{C}\,(\underline{y}_n, \underline{y}_n), \tag{5.7c}$$

and a

$$W_{11}(x,y) \triangleq \Gamma_1(x,y) + G_1(\pi)A(x,y)G_1'(\pi), \tag{5.8a}$$

$$W_{12}(x,y) \triangleq G_1(\pi)B(x,y)G_2'(\pi), \tag{5.8b}$$

$$W_{22}(x,y) \triangleq \Gamma_2(x,y) + G_2(\pi)C(x,y)G_2'(\pi). \tag{5.8c}$$

Finally, $V_{21}(\pi)$ and $W_{21}(x,y)$ are defined by symmetry, and the four submatrices from Equations 5.7a-5.7c and 5.8a-5.8c are collected in matrices $\overline{V}(\pi)$ and $\overline{W}(x,y)$. We are now finally read to state the main result of this section.

Theorem 5.1.

$$2n\mathbf{E}\,(\Delta(F_1(\underline{x}_n), F_2(\underline{y}_n)) - \Delta(F_1(\pi), F_2(\pi))) \Rightarrow$$
$$tr\ \overline{W}(F_1(\pi), F_2(\pi))\overline{V}(\pi).$$

Proof. Apply the first half of Hurt's Theorem, from the Appendix, with $q = 2$. □

6. Estimating the Various Errors

We now apply Theorem 5.1 to some of the expected error measures we have defined earlier. Unfortunately, we need more definitions. Let

$$G(r) \triangleq \left.\frac{\partial\Phi}{\partial p}\right|_{p=r}, \tag{6.1}$$

Corollary 6.1. *For the expected overall error*

$$2n(\beta_n - \delta) \Rightarrow\ tr\ [\Gamma(\pi, \Phi(\pi)) + G(\pi)C(\pi, \Phi(\pi))G'(\pi)]V(\pi).$$

Proof. Here F_1 is the identity and $F_2 = \Phi$. Also $\underline{x}_n = \pi$, and $\underline{y}_n = \underline{p}_n$. □

Corollary 6.2. *For the expected prediction error*

$$2n(\mu_n - \delta) \Rightarrow$$
$$tr\ [A(\pi, \Phi(\pi)) + \Gamma(\pi, \Phi(\pi)) + G(\pi)C(\pi, \Phi(\pi))G'(\pi)]V(\pi)$$

Proof. Here F_1 is the identity, $F_2 = \Phi$, $\underline{x}_n = \underline{q}_n$, and $\underline{y}_n = \underline{p}_n$. □

Corollary 6.3. *For the expected deviance*

$$2n(\lambda_n - \delta) \Rightarrow$$
$$tr\ [A(\pi, \Phi(\pi)) + B(\pi, \Phi(\pi))G'(\pi) + G(\pi)B'(\pi, \Phi(\pi)) +$$
$$\Gamma(\pi, \Phi(\pi)) + G(\pi)C(\pi, \Phi(\pi))G'(\pi)]V(\pi)$$

Proof. Here F_1 is the identity, $F_2 = \Phi$, and $\underline{x}_n = \underline{y}_n = \underline{p}_n$. □

Corollary 6.4. *For the expected estimation error*

$$2n\epsilon_n \Rightarrow tr\,[G(\pi)C(\Phi(\pi),\Phi(\pi))G'(\pi)]V(\pi).$$

Proof. Here $F_1 = F_2 = \Phi$, $\underline{x}_n = \pi$, and $\underline{y}_n = \underline{p}_n$. Moreover we use $\Gamma_2(\Phi(\pi),\Phi(\pi)) = 0$, because $u_j(\Phi(\pi),\Phi(\pi)) = 0$ for all $j = 1,\cdots,m$. □

The formula's in the corrollaries are a bit hard to use, partly because they involve the complicated matrix Γ of second derivatives of the estimator. As an example, we give the following result.

Theorem 6.5. *Define*

$$2n\underline{D}_n \triangleq 2n\underline{\lambda}_n - tr\,[A(\underline{p},\Phi(\underline{p})) + B(\underline{p},\Phi(\underline{p}))G'(\underline{p}) + G(\underline{p})B'(\underline{p},\Phi(\underline{p})) + \Gamma(\underline{p},\Phi(\underline{p})) + G(\underline{p})C(\underline{p},\Phi(\underline{p}))G'(\underline{p})]V(\underline{p}).$$

Then

$$n\mathbf{E}\,(\underline{D}_n) \Rightarrow \delta.$$

Proof. Direct from Corollary 6.3. □

Another problem is that the formulas require values for some of the unobserved quantities. But, as explained in De Leeuw [7], there is a simple and fairly straightforward way around this problem.

Theorem 6.6. *Define*

$$2n\underline{P}_n \triangleq 2n\underline{\lambda}_n - tr\,[B(\underline{p}_n,\Phi(\underline{p}_n))G'(\underline{p}_n) + G(\underline{p}_n)B'(\underline{p}_n,\Phi(\underline{p}_n))]V(\underline{p}_n),$$

$$2n\underline{B}_n \triangleq 2n\underline{\lambda}_n - tr\,[A(\underline{p}_n,\Phi(\underline{p}_n)) + B(\underline{p}_n,\Phi(\underline{p}_n))G'(\underline{p}_n) + G(\underline{p}_n)B'(\underline{p}_n,\Phi(\underline{p}_n))]V(\underline{p}_n).$$

Then

$$\mathbf{E}\,(\underline{P}_n) \Rightarrow \mu_n,$$
$$\mathbf{E}\,(\underline{B}_n) \Rightarrow \beta_n.$$

Proof. From Corollaries 6.2 and 6.1. □

7. Moment Structure Analysis

We now generalize our results to statistics of the form

(7.1) $$\underline{x}_n = \Psi(L\underline{p}_n).$$

Here $\underline{p}_n$ are $m-$vectors of proportions, as before, L is a known $p \times m$ matrix, and Ψ maps $\mathbb{R}^p$ into $\mathbb{R}^s$. We estimate $\underline{x}_n$ by functions of the form $\Phi(\underline{x}_n)$, which map $\mathbb{R}^s$ into $\mathbb{R}^s$. Correspondingly, we define a

(7.2a) $$\mu \triangleq \Psi(L\pi),$$

(7.2b) $$\underline{y}_n \triangleq \Psi(L\underline{q}_n).$$

With these definitions, our previous results and definitions remain valid, with $\underline{x}_n$ substituted for $\underline{p}_n$, $\underline{y}_n$ substituted for $\underline{q}_n$, and μ substituted for π.

To indicate the generality of this situation, think of a p-dimensional random variable $\underline{\ell}$ which takes the values $\ell_1, \cdots \ell_m$ with probabilities $\pi_1, \cdots, \pi_m$. The ℓ_j are the columns of L. Then $\mu = L\pi$ is the expectation of $\underline{\ell}$, and n replications of $\underline{\ell}$ bring us to the $\underline{x}_n$ of 7.1. But this is not all.

If $\underline{z}$ takes the scalar values $z_1, \cdots, z_m$, with probabilities $\pi_1, \cdots, \pi_m$, then $\underline{\ell} = (\underline{z}, \underline{z}^2, \cdots, \underline{z}^p)$ takes on values $\ell_j = (z_j, z_j^2, \cdots, z_j^p)$ with the same probabilities, and the $\underline{x}_n$ are the first p sample moments. Thus the theory covers functions of the sample moments.

A slightly more general setup extends it to functions of sample moments and product moments, and functions of suitably discreticized versions of the empirical distribution function or the empirical characteristic function. The usual analysis of covariance structures is covered painlessly, because it concerns itself with functions of the second order product moments (and, in the case of asymptotically distribution-free methods, functions of the product moments up to order four, cf [6]).

8. Use of the Jackknife

Although Theorems 6.5 and 6.6 can be applied in considerable generality, they do require quite heavy derivative-type calculations. This may not seem to be a large disadvantage, since differentiation is usually straightforward, although tedious. The major problem, however, is that automating differentiation is not simple. Unless sophisticated symbolic mathematics packages are used, differentiation has to be done by hand, and then programmed into subroutines which are problem-specific.

Alternatively, we can do numerical differentiation. This is automatic, and comparatively easy to program in full generality, but we still have to make some strategic choices on how to implement it precisely. In this paper we use the Jackknife for our numerical differentiation, because it is specifically designed for statistical problems, and in particular for statistical problems dealing with proportions.

Jackknife-based techniques are somewhat less general than the similar delta-method expansions, because they actually suppose a strictly multinomial model. I am sure they can be adapted quite easily, but here we simply assume that $V(\pi) = \Pi - \pi\pi'$. If we want to use the Jackknife, we use perturbations in our differentiation of the form

$$p_{n:j} \triangleq p_n + \frac{1}{n-1}(p_n - e_j), \tag{8.1}$$

where the e_j are the unit-vectors in $\mathbb{R}^m$. Thus

$$p_n = \sum_{j=1}^{m} p_{nj} e_j. \tag{8.2}$$

We also set

$$V(p_n) \triangleq \sum_{j=1}^{m} p_{nj}(p_n - e_j)(p_n - e_j)' = P_n - p_n p_n', \tag{8.3}$$

with P_n the diagonal matrix with the p_n. Define the following functions. They are defined to mimic the expansions of the error measures we have derived in the previous sections.

Definition 8.1.

$$\mathcal{J}_n(p_n) \triangleq 2(n-1)^2\{\sum_{j=1}^{m} p_{nj}\Delta(p_{n:j}, \Phi(p_n)) - \Delta(p_n, \Phi(p_n))\},$$

$$\mathcal{K}_n(p_n) \triangleq 2(n-1)^2\{\sum_{j=1}^{m} p_{nj}\Delta(p_n, \Phi(p_{n:j})) - \Delta(p_n, \Phi(p_n))\},$$

$$\mathcal{L}_n(p_n) \triangleq 2(n-1)^2\{\sum_{j=1}^{m} p_{nj}\Delta(p_{n:j}, \Phi(p_{n:j})) - \Delta(p_n, \Phi(p_n))\}.$$

Lemma 8.2.

$$\mathcal{J}_n(p_n) \Rightarrow tr\ A(p_n, \Phi(p_n))V(p_n),$$
$$\mathcal{K}_n(p_n) \Rightarrow tr\ [\Gamma(p_n, \Phi(p_n)) + G(p_n)C(p_n, \Phi(p_n))G'(p_n)]V(p_n),$$

and

$$\mathcal{L}_n(p_n) \Rightarrow tr\ [A(p_n, \Phi(p_n)) + \Gamma(p_n, \Phi(p_n)) + G(p_n)B(p_n, \Phi(p_n)) + B(p_n, \Phi(p_n))G'(p_n) + G(p_n)C(p_n, \Phi(p_n))G'(p_n)]V(p_n).$$

Proof. These are just simple expansions. □

Theorem 8.3.

$$\mathbf{E}\,(2n\underline{\lambda}_n - \mathcal{L}_n(\underline{p}_n)) \Rightarrow \delta$$

Proof. From Lemma 8.2 and Theorem 6.5. □

Theorem 8.4.

$$\mathbf{E}\,(2n\underline{\lambda}_n - \{\mathcal{L}_n(\underline{p}_n) - \mathcal{K}_n(\underline{p}_n)\}) \Rightarrow \beta_n,$$
$$\mathbf{E}\,(2n\underline{\lambda}_n - \{\mathcal{J}_n(\underline{p}_n) + \mathcal{K}_n(\underline{p}_n) - \mathcal{L}_n(\underline{p}_n)\}) \Rightarrow \mu_n.$$

Proof. From Lemma 8.2 and Theorem 6.6. □

It is clear from the form of Theorems 8.3 and 8.4, and the type of reasoning that is used, that they generalize without modification to the more general situation explained in Section 7.

9. Discussion

It is quite obvious what the computational costs are of the methods in Theorems 8.3 and 8.4. In stead of computing $\Phi(p_n)$ once, we have to compute it the m values $\Phi(p_{n:j})$. If m is larger than n, we can also develop a version which recomputes the estimate n times (leave out one observation at a time). For complicated estimates, such as those generated by LISREL or EQS runs, and for large samples this may be a problem (although we can start the iterative process to compute the estimates in $\Phi(p_n)$, which will in general be quite close to the pertubed estimate). It is also clear that for large samples, we need pretty high precision in our computations, otherwise the effect of the perturbations will get lost in the rounding errors.

If the Jackknife based computations are inpractical, there are some alternatives. We can use a random version of the Jackknife, which leaves out cells or observations at random.

Or we can use the expansions in Theorems 6.5 and 6.6. Or we can use the Bootstrap (as tried in De Leeuw [7]).

I prefer these computational techniques to techniques that make very specific assumptions. Some of these assumptions have been mentioned above. We can assume that the model is true, i.e. that $\Phi(\pi) = \pi$, or that the estimates are minimum distance, or that the loss functions are "appropriate", or even that we use likelihood methods throughout. I see no compelling reason for making any of these assumptions in many cases, and I can think of many reasons for not making them.

REFERENCES

1. P. M. Bentler, *Comparative Fit Indices in Structural Models*, Psychological Bulletin **107** (1990), 238–246.
2. P. J. Bickel, C. A. J. Klaassen, Y. Ritov, and J. A. Wellner, *Efficient and Adaptive Estimation for Semiparametric Models*, John Hopkins University Press, Baltimore, MD, 1993.
3. M. W. Browne and R. Cudeck, *Single Sample Cross-Validation Inidices for Covariance Structures*, Multivariate Behavioral Research **24** (1989), 445–455.
4. ———, *Alternative Ways of Assessing Model Fit*, Sociological Methods and Research **21** (1992), 230–258.
5. R. Cudeck and M. W. Browne, *Cross-Validation of Covariance Structures*, Multivariate Behavioral Research **18** (1983), 147–167.
6. J. de Leeuw, *Models and Methods for the Analysis of Correlation Coefficients*, Journal of Econometrics **22** (1983), 113–137.
7. ———, *Model Selection in Multinomial Experiments*, On Model Uncertainty and its Statistical Implications (T. J. Dijkstra, ed.), Springer Verlag, New York, NY, 1988.
8. A.E. Gelfand, D.K. Dey, and H. Chang, *Model Distributions using Predictive Distributions with Implementation via Sampling-Based Methods*, Bayesian Statistics 4, Oxford University Press, New York, NY, 1992.
9. J. Hurt, *Asymptotic Expansions of Functions of Statistics*, Aplikace Matematiky **21** (1976), 444–456.
10. ———, *Asymptotic Expansions for Moments of Functions of Statistics*, Second Prague Symposium on Asymptotic Statistics, Akademie, Prague, Czechoslovakia, 1978.
11. B. Lindsay, *Efficiency versus Robustness: The Case for Minimum Hellinger Distance and Related methods*, Annals of Statistics **22** (1994), 1081–1114.
12. H. Linhart and W. Zucchini, *Model Selection*, Wiley, New York, NY, 1986.
13. R. P. McDonald, *An Index of Goodness-of-Fit Based on Noncentrality*, Journal of Classification **6** (1989), 97–103.
14. R. P. McDonald and H. W. Marsh, *Choosing a Multivariate Model: Noncentrality and Goodness of Fit*, Psychological Bulletin **107** (1990), 247–255.

Appendix A. Hurt's Theorem

In this Appendix we give a relevant theorem that can be used to justify the approximations to the expected values used in the paper. It is, to some extent, classical, but a particulary clear and useful statement appears in two publications by Jan Hurt [9, 10].

Theorem A.1. *Suppose*

1. $\{\phi_n(x)\}$ *is a sequence of real-valued functions on* $\mathbb{R}^m$,
2. $\{\underline{x}_n\}=\{(\underline{x}_{1n},\cdots \underline{x}_{mn})\}$ *is a sequence of* $m-$*dimensional statistics.*

Assume that

1. *For all* n, ϕ_n *is* $(q+1)$*-times totally differentiable with respect to the* x_j *in the interval* $\mathbb{K}=\prod_{j=1}^m[\mu_j-\delta_j,\mu_j+\delta_j], \delta_j>0, \delta_j$ *independent of* n.
2. *For all* n, ϕ_n *is bounded on* $\mathbb{R}^m$.
3. *For all* n, *the derivatives* $\phi_n^{(1)},\cdots,\phi_n^{(q+1)}$ *are bounded on* $\mathbb{K}$.
4. *For all* j *and all* n, $\underline{x}_{jn}$ *has finite absolute moments up to the order* $2(q+1)$.
5. *For all* $j=1,\cdots,m$

$$\mathbf{E}\mid \underline{x}_{jn}-\mu_j\mid^{2(q+1)}=\mathcal{O}(n^{-(q+1)}).$$

Then

$$\mathbf{E}\left(\phi_n(\underline{x}_n)-\phi_n(\mu)\right)=\sum_{j=1}^{q}\frac{1}{q!}\sum_{\substack{i_1+\cdots+i_m=j\\ i_1,\cdots,i_m\geq 0}}\!\!\cdots\sum\left(\frac{\partial^j\phi_n}{\partial x_1^{i_1}\cdots\partial x_m^{i_m}}\right)_{x=\mu}\times$$
$$\times\,\mathbf{E}\left((\underline{x}_{1n}-\mu_1)^{i_1}\cdots(\underline{x}_{mn}-\mu_m)^{i_m}\right)+\mathcal{O}(n^{-\frac{(q+1)}{2}}).$$

and

$$\mathbf{V}\left(\phi_n(\underline{x}_n)-\phi_n(\mu)\right)=\sum_{\substack{j=1\\ j+k\leq q+1}}^{q}\sum_{k=1}^{q}\frac{1}{j!}\frac{1}{k!}\sum_{\substack{i_1+\cdots+i_m=j\\ i_1,\cdots,i_m\geq 0}}\!\!\cdots\sum\ \sum_{\substack{\ell_1+\cdots+\ell_m=k\\ \ell_1,\cdots,\ell_m\geq 0}}\!\!\cdots\sum\times$$
$$\times\left(\frac{\partial^j\phi_n}{\partial x_1^{i_1}\cdots\partial x_m^{i_m}}\right)_{x=\mu}\left(\frac{\partial^k\phi_n}{\partial x_1^{\ell_1}\cdots\partial x_m^{\ell_m}}\right)_{x=\mu}\times$$
$$\times\,\mathbf{C}\left((\underline{x}_{1n}-\mu_1)^{i_1}\cdots(\underline{x}_{mn}-\mu_m)^{i_m},(\underline{x}_{1n}-\mu_1)^{i_1}\cdots(\underline{x}_{mn}-\mu_m)^{i_m}\right)+$$
$$+\,\mathcal{O}(n^{-\frac{(q+2)}{2}}).$$

Proof. See Hurt [9]. □

Remark A.2. In [10] the same result is proved, but the residuals are not expressed in terms of powers of n, but in terms of the absolute moments. This extends the result to a more general class of statistics.

UCLA Statistics Program, 8118 Mathematical Sciences Building, University of California at Los Angeles

E-mail address: deleeuw@@stat.ucla.edu

Bias and mean square error of the maximum likelihood estimators of the parameters of the intraclass correlation model

Maia Berkane and Peter Bentler
University of California Los Angeles

Abstract

The differential geometry of the exponential family of distributions is applied to derive the bias and the mean square error of the maximum likelihood estimator of the parameters of the intraclass correlation model.

1 Introduction

The last two decades have seen an increased trend in the use of differential geometry in statistics. Although the first work in the area was done by Mahalanobis who gave a measure of distance between multivariate normal distributions with equal scatter matrices, followed by Rao in 1945 who introduced the Fisher information matrix as a Riemannian metric on the space of distributions, research in the area slowed down until 1978 when Efron defined the curvature of a one-parameter exponential family. Efron's work opened the ways to more thorough investigations of the differential geometry approach in statistics. Amari (1980, 1982) generalized the work of Efron to the multiparameter exponential family and looked at conditioning and asymptotic ancillarity. Researchers were trying to provide a new understanding of the metricized space of distributions using the known concepts of differential geometry, on the one hand, and hoping to find elegant results in statistical inference either by using the differential geometry or by directly utilizing results already known in differential geometry on the other hand. In this paper, we use known results of differential geometry to give an expression for the bias and the mean square error of the parameters of the intraclass correlation model. This model is commonly used in genetics and in psychometrics. Let $y_1, y_2, ...y_n$ be a sample of independent and identically distributed p-dimensional vectors of observations. Assume the parent distribution is the multivariate normal with mean 0 and covariance matrix $\boldsymbol{\Sigma}$. The intraclass correlation model is a covariance structure model where the covariance matrix $\boldsymbol{\Sigma}$ has the form

$$\boldsymbol{\Sigma} = \boldsymbol{\lambda}[(1-\rho)I + \rho\underline{1}\underline{1}']$$

where $\boldsymbol{\lambda}(> 0)$ and $\boldsymbol{\rho}(-1 \leq \rho \leq 1)$ are the parameters to be estimated by use of the maximum likelihood method, I is the identity matrix and $\underline{1}$ is the p-dimensional vector of ones. The m.l.e. estimators $\hat{\boldsymbol{\lambda}}$ and $\hat{\boldsymbol{\rho}}$ have been obtained in the literature but they are known to be biased. Olkin and Pratt (1957) computed unbiased estimator of $\boldsymbol{\rho}$ by projecting the slope of the regression line of y_1 on y_2 onto the complete sufficient statistic (τ_1, τ_2) with

$$\tau_1 = \boldsymbol{\lambda}[1 + (p-1)\boldsymbol{\rho}]$$

being the largest eigenvalue of $\boldsymbol{\Sigma}$ and

$$\tau_2 = \boldsymbol{\lambda}(1 - \boldsymbol{\rho})$$

being the second eigenvalue of $\boldsymbol{\Sigma}$ with multiplicity $p-1$. However, this method led to a cumbersome series form of the estimator and its use has been quite limited as was observed in Donoghue and Collins (1990). Berkane an Bentler (1992) used the differential geometry method to derive the bias, to the order n^{-1}, of $\hat{\boldsymbol{\lambda}}$ and $\hat{\boldsymbol{\rho}}$ in an easy closed form, as well as the mean square error of the bias corrected estimators for the case $p = 2$. This article generalizes their result.

2 Review and notation of differential geometry

Skovgaard (1984) considered the family of distributions

$$M = \{N_p(\boldsymbol{\mu}, \boldsymbol{\Sigma}) | \boldsymbol{\Sigma} \in A\},$$

where A is the the set of positive definite matrix of order p, $N_p(\boldsymbol{\mu}, \boldsymbol{\Sigma})$ is the mutivariate normal density with mean $\boldsymbol{\mu}$ and covariance matrix $\boldsymbol{\Sigma}$. Let $\boldsymbol{\mu} = 0$ and $p* = p(p+1)/2$, using the isomorphism

$$f : M \to U \subset R^{p*}$$

where U is open, with

$$\phi(N_p(0, \boldsymbol{\Sigma})) = \boldsymbol{\sigma}$$

with $\boldsymbol{\sigma}$ being the column vector formed by stacking the elements of lower diagonal elements of $\boldsymbol{\Sigma}$. M is then viewed as a differentiable manifold of dimension $p*$ with (U, ϕ) as a global coordinate system. A basis of the set of vector fields is then defined on M by the holomorphic set

$$\frac{\partial}{\partial \boldsymbol{\sigma}_{ij}}, i, j = 1, ...p, i \leq j.$$

Subsequently, these basis vector fields are identified with the matrices

$$E_{ij} = \left\{ \begin{array}{ll} \underline{1}_{i,i} & \text{if } i = j \\ \underline{1}_{i,j} + \underline{1}_{j,i} & \text{if } i \neq j \end{array} \right\},$$

where $\underline{1}_{kl}$ is a $p * \times p*$ matrix with (k, l)th element equal 1 and all the others are 0. Each vector field B can be expressed as

$$B = \sum_{i,j=1}^{p} B_{i,j} E_{i,j}, i \leq j$$

where the $B_{i,j}$ are smooth real functions on M. Let $\partial_{i,j} = \frac{\partial}{\partial \sigma_{ij}}, i,j = 1,...p, i \leq j$. By introducing the inner product

$$\begin{aligned} < \partial_{i,j}, \partial_{k,l} > &= \frac{1}{2} tr(\boldsymbol{\Sigma}^{-1} E_{i,j} \boldsymbol{\Sigma}^{-1} E_{kl}) \\ &= \boldsymbol{\sigma}^{ij} \boldsymbol{\sigma}^{kl} + \boldsymbol{\sigma}^{ik} \boldsymbol{\sigma}^{jl} \end{aligned}$$

M becomes a Riemannian manifold. Let $G = (\boldsymbol{\sigma}^{ij}\boldsymbol{\sigma}^{kl} + \boldsymbol{\sigma}^{ik}\boldsymbol{\sigma}^{jl})_{ij,kl}$. G is a $p* \times p*$ positive definite and is called the Riemannian metric matrix or the information metric matrix. As the point $\boldsymbol{\sigma}$ moves to $\boldsymbol{\sigma} + d\boldsymbol{\sigma}$, the tangent spaces $T_{\boldsymbol{\sigma}}$ and $T_{\boldsymbol{\sigma}+d\boldsymbol{\sigma}}$ are different, therefore it is not possible to compare two vectors $A_1 \in T_{\boldsymbol{\sigma}}$ and $A_2 \in T_{\boldsymbol{\sigma}+d\boldsymbol{\sigma}}$. It is then necessary to establish a one-to- one correspondence between the vector spaces so that one vector space is mapped to another. This mapping is called affine connection and is determined by $(p*)^3$ functions called coefficients of the affine connection. The Riemannian metric induces the Riemannian affine connection or covariant derivative denoted by ∇. Here and in the rest of the paper, we adopt the Einstein notation, i.e. whenever an index appears twice, once as a subscript and once as a superscript, a summation over that index is meant, for example $a^i x_i = \sum a^i x_i$. The vector field $\boldsymbol{\nabla}_{\partial_{ij}} \partial_{kl}$ denotes the rate of intrinsic change of ∂_{kl} in the direction of ∂_{ij} as $\boldsymbol{\sigma}$ changes. It is called the covariant derivative of vector field ∂_{kl} along ∂_{ij} and is in $T_{\boldsymbol{\sigma}}$. Hence it is expressed as

$$\boldsymbol{\nabla}_{\partial_{ij}} \partial_{kl} = \boldsymbol{\Gamma}_{ij,kl}^{tu} \partial_{tu}$$

for some smooth functions $\boldsymbol{\Gamma}_{ij,kl}^{tu}$ on M. The coefficient of the affine connection are given by

$$\boldsymbol{\Gamma}_{ijkltu}(\boldsymbol{\sigma}) = < \boldsymbol{\nabla}_{\partial_{ij}} \partial_{kl}, \partial_{rs} > = \boldsymbol{\Gamma}_{ij,kl}^{tu}(\boldsymbol{\sigma}) g_{tu,rs}.$$

An infinite number of statistical affine connections were introduced by Amari (1982) in the space of statistical model in such a way that each one will represent the intrinsic properties of the family of distributions. First, define the dual of the tangent space $T_{\boldsymbol{\sigma}}$ as the space of 1-representation $T_{\boldsymbol{\sigma}}^{(1)}$,

$$T_{\boldsymbol{\sigma}}^{(1)} = \{A(x) | A(x) = A^{ij} \partial_{ij} l(x, \boldsymbol{\sigma})\}.$$

The corresponding affine connection is defined as

$$\boldsymbol{\Gamma}_{ijklrs}(\boldsymbol{\sigma}) = E[\partial_{ij} \partial_{kl} l(x, \boldsymbol{\sigma}) \partial_{rs} l(x, \boldsymbol{\sigma})].$$

The $\boldsymbol{\alpha}$-connection is defined as

$$\boldsymbol{\Gamma}_{ijklrs}^{(\alpha)}(\boldsymbol{\sigma}) = E[\partial_{ij} \partial_{kl} l(x, \boldsymbol{\sigma}) + \frac{1-\alpha}{2} \partial_{ij} l(x, \boldsymbol{\sigma}) \partial_{kl} l(x, \boldsymbol{\sigma}) \partial_{rs} l(x, \boldsymbol{\sigma})].$$

The term $E[\partial_{ij} l(x, \boldsymbol{\sigma}) \partial_{kl} l(x, \boldsymbol{\sigma}) \partial_{rs} l(x, \boldsymbol{\sigma})]$ is called the skewness tensor and is denoted by T_{ijklrs}. The term $E[\partial_{ij} \partial_{kl} l(x, \boldsymbol{\sigma}) \partial_{rs} l(x, \boldsymbol{\sigma})]$ is called the exponential connection mainly because the affine connection reduces to that term for the exponential family of distributions, which leads to $\alpha = 1$. The α- connection can then be expressed as

$$\boldsymbol{\Gamma}_{ijklrs}(\boldsymbol{\sigma}) = \boldsymbol{\Gamma}_{ijklrs}^{(1)}(\boldsymbol{\sigma}) + \frac{1-\alpha}{2} T_{ijklrs}.$$

The above expression is also given as

$$\boldsymbol{\Gamma}_{ijklrs}(\boldsymbol{\sigma}) = \boldsymbol{\Gamma}^{(0)}_{ijklrs}(\boldsymbol{\sigma}) + \frac{\alpha}{2} T_{ijklrs}.$$

where the first term on the right hand side of the equation is known as the information connection or the Christofel symbol of the first kind. For $\boldsymbol{\alpha} = -1, \boldsymbol{\Gamma}^{(-1)}_{ijklrs}$ is called the mixture connection.

Mitchell (1989) derived the expression of the α-connections for general elliptical distributions. In the particular case of the normal distribution, we obtain for the skewness tensor

$$T_{ijklrs} = tr(\boldsymbol{\Sigma}^{-1} E_{ij} \boldsymbol{\Sigma}^{-1} E_{kl} \boldsymbol{\Sigma}^{-1} E_{rs}),$$

and for the α-connections

$$\boldsymbol{\Gamma}^{(\alpha)}_{ijklrs} = -\frac{1}{2} tr(\boldsymbol{\Sigma}^{-1} E_{ij} \boldsymbol{\Sigma}^{-1} E_{kl} \boldsymbol{\Sigma}^{-1} E_{rs}).$$

M can also be studied in the context of the exponential family of distributions. The canonical parameters are

$$\boldsymbol{\beta}_{ij} = \boldsymbol{\sigma}^{ij}, \boldsymbol{\beta}_{ii} = -\frac{1}{2}\boldsymbol{\sigma}^{ii}, i, j = 1, ...p; j > i,$$

where $\boldsymbol{\sigma}^{ij}$ are the entries of the inverse of $\boldsymbol{\Sigma}$. The $p* = p \times (p+1)/2$ nonduplicated elements of $\boldsymbol{\Sigma}$ are the dual parameters usually denoted by η_{ij}. Let $\beta = (\beta_{ij})$ the vector with elements β_{ij}. The potential function is

$$\phi(\beta) = -\frac{1}{2} \ln |\boldsymbol{\Sigma}| + \frac{p-1}{2} \ln 2 + \frac{p}{2} \ln \pi.$$

With respect to the dual coordinate system, the metric tensor is, in matrix form

$$G = \frac{1}{2} K_p' (\boldsymbol{\Sigma}^{-1} \otimes \boldsymbol{\Sigma}^{-1}) K_p,$$

where K_p is a $p^2 \times p*$ commutation matrix. The matrix of α-connection is

$$\frac{1+\alpha}{2} (\boldsymbol{\Sigma}^{-1} \otimes \boldsymbol{\Sigma}^{-1} \otimes \boldsymbol{\Sigma}^{-1}).$$

Let

$$M_0 = \{N_p(0, \boldsymbol{\Sigma}(\theta)), \boldsymbol{\Sigma} \in A, \theta \in V \subset R^q, q < p*\}$$

M_0 has the structure of a q-dimensional manifold, where now, each element of M_0 is identified with an element $\theta = (\theta_1, ...\theta_q) \in V$. The equations

$$\beta_{ij} = \beta_{ij}(\theta), i, j = 1, ...p; i \leq j,$$

where β_{ij} are the canonical parameters, give a parametric representation of M_0 in M. If we assume the jacobian matrix $\Delta = \frac{\partial \beta}{\partial \theta'}$ to be of full rank, then M_0 is said to be imbedded in M or a submanifold of M. Denote by $\partial_a, a = 1, ...q$ the basis of the vector fields on M_0, then $\partial_a = d_a^{ij}\partial_{ij}$, where $d_a^{ij} = \frac{\partial \beta_{ij}}{\partial \theta_a}$. The induced Riemannian metric in M_0 is given by

$$G* = \Delta' G \Delta,$$

and the induced α-covariant derivative is

$$\nabla^{\alpha}_{\partial_a}\partial_b = (d_a^{kl}\nabla_{\partial_{ij}}(d_b^{kl}\partial_{kl}) = \partial_a d_b^{kl} + d_a^{ij}d_b^{rs}\Gamma^{(\alpha)kl}_{ijrs}\partial_{kl}.$$

This α-covariant derivative is decomposed into a tangential component parallel to the tangent plane $T_\theta(M_0)$ and a normal component perpendicular to $T_\theta(M_0)$. The normal component measures the α-curvature of M_0 in M, this curvature is intrinsic to M_0. In order to simplify matters, Amari (1985) defines $p * -q$ vectors $\partial_\kappa, \kappa = q+1, ...p*$, in $T_{\sigma(\theta)}(M)$ such that $\{\partial_a, \partial_\kappa\}\, a = 1, ..q; \kappa = q+1, ...p*$ form a basis of $T_{\beta(\theta)}(M)$. Additionally, ∂_κ's are orthogonal to ∂_a's.
In our situation, the ∂_κ's will form a basis of the orthogonal complement of $T_\theta(M_0)$ and in terms of the ∂_{ij}, the basis of the complement is given by

$$\partial_\kappa = c_\kappa^{ij}\partial_{ij},$$

where the c_κ^{ij} are the entries of the jacobian matrix of the dual parameter vector η with respect to $(v^\kappa)\kappa = q+1, ...p*$, which is a coordinate system of the ancillary submanifold $A = A(\theta)$. A is a $p * -q$-dimensional submanifold attached to each point $\eta = \eta(\theta)$ and transversing M_0 at $\sigma(\theta)$. With this definition, the normal component of $\nabla^{\alpha}_{\partial_a}\partial_b$ is

$$H^{\alpha}_{ab\kappa} =< \nabla^{\alpha}_{\partial_a}\partial_b, \partial_\kappa >= (\partial_a d_a^{kl} + d_b^{ij}d_a^{rs}\Gamma^{(\alpha)kl}_{ijrs})c_\kappa^{tv}g_{tv,kl}$$

and is called the α-curvature of M_0 in M.
Remark
When the covariance structure is linear, d_a^{kl} is constant and its derivative with respect to θ^a is 0. The above expression becomes

$$H^{\alpha}_{ab\kappa} = d_b^{ij}d_a^{rs}\Gamma^{(\alpha)kl}_{ijrs}c^{tv}_{\alpha}g_{tv,kl}.$$

The coefficients of the α-connection in M are

$$\Gamma^{\alpha}_{abc} =< \nabla^{\alpha}_{\partial_a}\partial_b, \partial_c >= d_a^{ij} + d_b^{kl}d_c^{tu}\Gamma^{\alpha}_{ijkltu} + (\partial_a d_b^{rs})d_c^{tu}g_{rs,tu}$$

This shows that only linear covariance structures are -1-flat (intrinsically) even though M is -1-flat. The α-curvature of M_0 in M depends on the ancillary submanifold.

3 Intraclass correlation model: the two-parameters model

Let M_0 be the family of multivariate normal distributions with mean 0 and covariance matrix

$$\Sigma = \Sigma(\lambda, \rho) = \lambda((1-\rho)I + \rho \underline{1}\underline{1}'),$$

where $\underline{1}$ is the p-dimensional vector with all elements equal to 1. $11'$ is usually denoted by J and I is the p-identity matrix. Let $\theta = (\boldsymbol{\lambda}, \boldsymbol{\rho})$, we have the following proposition
Proposition 1
Denote by R^+ the positive half of the real line, then

$$M_0 = \left\{q(x, \boldsymbol{\lambda}\boldsymbol{\rho}) = f_{\boldsymbol{\Sigma}(\boldsymbol{\theta})}(x) = 2\pi^{-p/2}|\boldsymbol{\Sigma}(\boldsymbol{\theta})|^{-1/2} \exp \mathrm{x}'\boldsymbol{\Sigma}^{-1}(\boldsymbol{\theta})\mathrm{x}\,\boldsymbol{\theta} \in \mathrm{R}^+ \times [-1, 1]\right\}$$

is a $(p*, 2)$-curved exponential family.
proof: straightforward since $\frac{\partial vec(\boldsymbol{\Sigma})}{\partial \theta}$ is of rank 2.
The inverse is known to be (Graybill, 1969)

$$\boldsymbol{\Sigma}^{-1} = \frac{1}{a}(I_p - \frac{b}{a + pb}),$$

where $a = \boldsymbol{\lambda}(1 - \boldsymbol{\rho})$ and $b = \boldsymbol{\lambda}\boldsymbol{\rho}$. Hence

$$\begin{aligned}
\boldsymbol{\beta}_{ij} &= \frac{b}{a(a + pb)}\; i, j = 1, ...p;\; j > i \\
\boldsymbol{\beta}_{ii} &= -\frac{1}{2a}(1 - \frac{b}{a + pb})\; i = 1, ...p.
\end{aligned}$$

It is useful to note that the transformation $\boldsymbol{\beta}_{ii} = \boldsymbol{\beta}_{11}, \boldsymbol{\beta}_{ij} = \boldsymbol{\beta}_{12}, i, j = 1, ...p, j > i$ is linear therefore it induces a $(p*, 2)$ submanifold M_1 in M and is also an exponential family of distributions. The family M_0 is just a $(2, 2)$ reparametrization of M_1. We will then work with M_1.
Each element of M_1 has the parametric representation

$$p(z, \boldsymbol{\beta}) = \exp\{\boldsymbol{\beta}_1 z_1 + \boldsymbol{\beta}_2 z_2 - \phi(\boldsymbol{\beta})\}$$

with

$$\begin{aligned}
\boldsymbol{\beta}_1 &= \frac{-1}{2}\frac{1 + (p - 2)\boldsymbol{\rho}}{\boldsymbol{\lambda}(1 - \boldsymbol{\rho})(1 + (p - 1)\boldsymbol{\rho})} \\
\boldsymbol{\beta}_2 &= \frac{\boldsymbol{\rho}}{\boldsymbol{\lambda}(1 - \boldsymbol{\rho})(1 + (p - 1)\boldsymbol{\rho})} \\
z_1 &= \sum_{i=1,...p} x_i^2 \\
z_2 &= \sum_{j>i} x_i x_j,
\end{aligned}$$

and the potential function is

$$\psi(\boldsymbol{\beta}) = -(1/2)\ln(-2\boldsymbol{\beta}_1 - \boldsymbol{\beta}_2)^{\mathrm{p}-1}(-2\boldsymbol{\beta}_1 + (\mathrm{p} - 1)\boldsymbol{\beta}_2).$$

The dual parameters are then

$$\begin{aligned}
\eta_1 &= \frac{p(-2\boldsymbol{\beta}_1 + (p - 2)\boldsymbol{\beta}_2)}{(2\boldsymbol{\beta}_1 + \boldsymbol{\beta}_2)(2\boldsymbol{\beta}_1 + (p - 1)\boldsymbol{\beta}_2)} \\
\eta_2 &= \frac{p(p - 1)\boldsymbol{\beta}_2}{2(2\boldsymbol{\beta}_1 + \boldsymbol{\beta}_2)(2\boldsymbol{\beta}_1 + (p - 1)\boldsymbol{\beta}_2))}
\end{aligned}$$

It is convenient to express the above expressions of the dual parameters in terms of θ for upcoming computations. We find

$$\begin{aligned}\eta_1 &= \frac{\lambda p(-1+\rho)(-1+4\rho-2p\rho)(1-\rho+p\rho)}{(1-3\rho+p\rho)(1-3\rho+2p\rho)}\\ \eta_2 &= \frac{\lambda(1-p)p(-1+\rho)\rho\lambda(1-r+p\rho)}{2(1-3\rho+p\rho)(1-3\rho+2p\rho)}\end{aligned}$$

and the elements of the information metric matrix can be obtained as

$$\begin{aligned}g_{11} &= \frac{2p\lambda^2(\rho-1)^2(1-\rho+p\rho)^2(1+4(p-2)\rho+(15-15p_4p^2)\rho^2}{(1+(p-3)br)^2(1+(2p-3)\rho)^2}\\ g_{12} &= \frac{p\lambda^2\rho(1-\rho)^2(p-1)(1+(p-1)\rho)^2(2+3(p-2)\rho)}{(1+(p-3)br)^2(1+(2p-3)\rho)^2}\\ g_{22} &= \frac{p\lambda^2(1-\rho)^2(p-1)(1+(p-1)\rho)^2(1+2(p-2)\rho+(3-3p+p^2)\rho^2}{(1+(p-3)br)^2(1+(2p-3)\rho)^2}\end{aligned}$$

Now we need to compute the entries of the matrix Δ of derivatives of β with respect to θ. We find

$$\begin{aligned}\Delta^1_\lambda &= \frac{\partial\beta_1}{\partial\lambda} = \frac{-1+(p-2)\rho}{2\lambda^2(1-\rho)(1+(p-1)\rho)^2}\\ \Delta^1_\rho &= \frac{\partial\beta_1}{\partial\rho} = -\frac{(p-1)\rho(2+(p-2)\rho)}{2\lambda^2(1-\rho)^2(1+(p-1)\rho)^2}\\ \Delta^2_\lambda &= \frac{\partial\beta_2}{\partial\lambda} = -\frac{\rho}{\lambda^2(1-\rho)(1+(p-1)\rho)}\\ \Delta^2_\rho &= \frac{\partial\beta_2}{\partial\rho} = -\frac{1+(p-1)\rho^2}{\lambda(1-\rho)^2(1+(p-1)\rho)^2}\end{aligned}$$

The induced Riemannian metric matrix in M_0 is then

$$G_{\lambda\rho} = \Delta' G\Delta$$

and has entries

$$\begin{aligned}g_{\lambda\lambda} &= \frac{p}{2\lambda^2}\\ g_{\lambda\rho} &= \frac{p(p-1)\rho(1+5(p-2)\rho)+(9-9p+2p^2)\rho^2)}{2\lambda(-1+\rho)(1+(p-3)\rho)(1+(p-1)\rho)(1+(2p-3)\rho)}\\ g_{\rho\rho} &= \frac{P_6(\rho)}{Q_8\rho}\end{aligned}$$

where $P_6(\rho)$ and $Q_8(\rho)$ are polynomials of degree 6 and 8 respectively with coefficients dependent on p, in fact

$$\begin{aligned}P_6(\rho) &= p(p-1)((1+2(p-2)\rho+(5-5p+p^2)\rho^2+(16-24p+8p^2)\rho^3\\ &+ (-117+234p-146p^2+29p^3)\rho^4+(180-450p+400p^2-150p^3+20p^4)\rho^5\\ &+ (-81+243p-279p^2+153p^3-40p^4+4p^5)\rho^6)\end{aligned}$$

and

$$Q_8(\rho) = 2(1-\rho)^2(1+(p-3)\rho)^2(1+(p-1)\rho)^2(1+(2p-3)\rho)^2.$$

4 Bias and mean square error

Amari (1985) gave the expressions for the bias and the mean square error of any first order efficient estimator. These expressions involve the mixture connection ($\alpha = -1$) of M_0 which is common to all efficient estimators, and the mixture curvature of the ancillary submanifold which depends on the estimator. Therefore, the estimation procedure is the only way to minimize the bias of a sufficient estimator. In our set up this can be seen from

$$\begin{aligned} E(\hat{\lambda}-\lambda) &= \frac{-1}{2n}(\Gamma^{(-1)\lambda}_{\lambda\lambda} g^{\lambda\lambda} \\ &+ 2\Gamma^{(-1)\lambda}_{\lambda\rho} g^{\lambda\rho} + \Gamma^{(-1)\lambda}_{\rho\rho} g^{\rho\rho} + O(n^{-3/2}) \\ E(\hat{\rho}-\rho) &= \frac{-1}{2n}(\Gamma^{(-1)\rho}_{\lambda\lambda} g^{\lambda\lambda} \\ &+ 2\Gamma^{(-1)\rho}_{\lambda\rho} g^{\lambda\rho} + \Gamma^{(-1)\rho}_{\rho\rho} g^{\rho\rho} + O(n^{-3/2}) \end{aligned}$$

where

$$\Gamma^{(-1)a}_{cd} = g^{ae}\Gamma^{(-1)}_{cde}, a, b, c, d, e = \lambda, \rho,$$

and g^{ae} are the entries of the inverse of G. Let

$$\Delta_{\lambda_i} = \frac{\partial \eta_i}{\partial \lambda}, \Delta_{\rho_i} = \frac{\partial \eta_i}{\partial \rho}.$$

Since the α-connection is not a tensor, a reparemetrization of an α-affine coordinate system is not necessarily α-affine. From the expressions of the α-connections in M_0 derived in section 2, and since the dual coordinate system η is -1-affine in M_1, the mixture connections of M_0 are given by the following equations

$$\Gamma^{-1}_{\lambda\lambda\lambda} = (\partial_\lambda \Delta_{\lambda_i})\Delta^i_\lambda = 0, \Gamma^{(-1)}_{\lambda\lambda\rho} = (\partial_\lambda \Delta_{\lambda_i})\Delta^i_\rho = 0$$

$$\Gamma^{-1}_{\lambda\rho\rho} = (\partial_\lambda \Delta_{\rho_i})\Delta^i_\rho = \frac{P'_6(\rho)}{Q'_8(\rho,\lambda)}$$

$$\Gamma^{-1}_{\lambda\rho\lambda} = (\partial_\lambda \Delta_{\rho_i})\Delta^i_\lambda = \frac{P'_3(\rho)}{Q'_4(\rho,\lambda)}$$

$$\Gamma^{-1}_{\rho\lambda\rho} = (\partial_\rho \Delta_{\lambda_i})\Delta^i_\rho = \Gamma^{-1}_{\rho\lambda\rho} = (\partial_\rho \Delta_{\lambda_i})\Delta^i_\rho = \Gamma^{-1}_{\lambda\rho\rho}$$

$$\Gamma^{-1}_{\rho\rho\lambda} = (\partial_\rho \Delta_{\rho_i})\Delta^i_\lambda = \frac{P_2(\rho)}{Q'_6(\rho,\lambda)}$$

$$\Gamma^{-1}_{\rho\rho\rho} = (\partial_\rho \Delta_{\rho_i})\Delta^i_\rho = \frac{P_5(\rho)}{Q'_{10}(\rho,\lambda)}$$

where the P's and Q's are polynomials of degrees corresponding to the indicated subscript. and will be given in Appendix.
The bias of $\hat{\lambda}$ and $\hat{\rho}$ are then

$$\begin{aligned} b(\hat{\lambda}) &= \frac{1}{np(1-\rho)^2(1+(\mathrm{p}-1)\rho)}\{4\lambda(-2+\mathrm{p})\rho(3-(24\rho-12\mathrm{p})\rho \\ &+(66-66p+16p^2)\rho^2-(72+180\mathrm{p}-54\mathrm{p}^2+19\mathrm{p}^3)\rho^3)\}+O(n^{-3/2}) \end{aligned}$$

$$b(\hat{\rho}) = \frac{1}{np(1-p)}\{-2(-4+2p+(33-33p+8p^2)\rho-(90-135\mathrm{p}+65\mathrm{p}^2-10\mathrm{p}^3)\rho^2$$
$$+(81-162p+117p^2-36p^3+4p^4)\rho^3)\}+\mathrm{O}(\mathrm{n}^{-3/2})$$

For $p = 2$, the bias of $\hat{\lambda}$ is 0 and the bias of $\hat{\rho}$ is

$$b_2(\hat{\rho}) = -\rho(\rho+1)(1-\rho)/\mathrm{n}+\mathrm{O}(\mathrm{n}^{-3/2}).$$

Table 1 gives the first order bias corrected estimator of the intraclass correlation coefficient for various sample sizes, with λ fixed to 1. The table given by Olkin and Pratt (1958) show larger bias for sample sizes less than 18, for all values of the estimated coefficient. Both tables give approximate values of the bias corrected estimator, the true value being unknown. We now obtain the mean square of the bias corrected estimators. Amari (1985) gave the expression of the mean square error of any bias corrected first order efficient estimator $\hat{\theta}$ terms of of the mixture connections, the exponential imbedding curvature of the submanifold, and the mixture curvature of the ancillary submanifold as

$$E(\hat{\theta}^a\hat{\theta}^b) = (1/n)g^{ab} + (1/2n)\{(\boldsymbol{\Gamma}^{\mathrm{m}})^{2\mathrm{ab}} + 2(\mathrm{H}_{\mathrm{M}_0})^{2\mathrm{ab}} + (\mathrm{H}^{\mathrm{m}}_{\mathrm{A}})^{2\mathrm{ab}}\} + 0(\mathrm{n}^{-5/2})$$

Where g^{ab} are the elements of the inverse of the Fisher information matrix of M_0 and

$$\begin{aligned}
(\boldsymbol{\Gamma}^{\mathrm{m}})^{2\mathrm{ab}} &= g^{ac}g^{bd}(\boldsymbol{\Gamma}^{-1})^2_{\mathrm{cd}} = \mathrm{g}^{\mathrm{ac}}\mathrm{g}^{\mathrm{bd}}\boldsymbol{\Gamma}^{(-1)}_{\mathrm{efc}}\boldsymbol{\Gamma}^{(-1)}_{\mathrm{hld}}\mathrm{g}^{\mathrm{ch}}\mathrm{g}^{\mathrm{fl}} \\
(H^e_{M_0})^{2ab} &= g^{ac}g^{bd}H^{(e)}_{ce\kappa}H(e)_{dh\lambda}g^{eh}g^{\kappa\lambda} \\
(H^m_A)^{2ab} &= g^{ac}g^{bd}H^{(m)}_{\kappa\lambda c}H(m)_{\nu\mu d}g^{\kappa\nu}g^{\lambda\mu}.
\end{aligned}$$

Terms of order $n^{-1/2}$ vanish for all first order efficient estimators and a bias corrected first order efficient estimator is automatically second-order efficient. It is third order efficient if the associated ancillary submanifold has zero mixture curvature at $v = 0$ (this translates into $(H^m_A)^{2ab} = 0$). The (bias corrected) m.l.e is always third order efficient since its associated ancillary submanifold is everywhere mixture flat.
Since in our intraclass correlation model, M_0 is only a repramêtrization of M_1 and since M_1 is an exponential, the exponential curvature which is an intrinsic quantity of M_1 does not change with reparemetrization, it follows that the exponential curvature of M_0 is 0. Hence the previous formula for the mean square error of $\hat{\lambda}^* = \hat{\lambda} - b(\hat{\lambda})$ and $\hat{\rho}^* = \hat{\rho} - b(\hat{\rho})$ simplifies to

$$E(\hat{\lambda}^* - \lambda)(\hat{\rho}^* - \rho) = (1/\mathrm{n})\mathrm{g}^{\mathrm{ab}} + (1/2\mathrm{n}^2)(\boldsymbol{\Gamma}^{\mathrm{m}})^{2\mathrm{ab}} + \mathrm{O}(\mathrm{n}^{-5/2}).$$

Hence we only need to compute $(\boldsymbol{\Gamma}^{\mathrm{m}})^{2\mathrm{ab}}$. The formulas are very lenghty and can be requested from the author along with the Mathematica codes used to derive them. for $p = 2$ the terms of order n^{-2} simplify to

$$\frac{\lambda^2\rho^2(1+\rho^2)^2(1+2\rho^2)}{\mathrm{n}^2} \text{ for } \lambda$$

$$\frac{(1-\rho^2)^2(1+\rho^2)^2(1+2\rho^2)}{\mathrm{n}^2} \text{ for } \rho$$

5 Intraclass correlation model: the one-parameter model

If we fix $\boldsymbol{\lambda}$ to 1 in the two-parameters intraclass correlation model, we obtain another known variant of the model. It is interesting to note that the class of multivariate normal distributions with this structure on the covariance matrix is now a $(2,1)$-curved exponential family imbedded in the manifold of the two-parameters exponential family. The elements of $\boldsymbol{\Sigma}$ are

$$\begin{aligned} \sigma_{ij} &= 1 \text{ for } i = j \\ \sigma_{ij} &= \theta \text{ otherwise.} \end{aligned}$$

The elements of M_0 are now

$$q(x,\boldsymbol{\theta}) = f_{\boldsymbol{\Sigma}(\boldsymbol{\theta})}(x) = (2\pi)^{-p/2}|\boldsymbol{\Sigma}|^{-1/2}\exp\{-1/2x'\boldsymbol{\Sigma}^{-1}(\boldsymbol{\theta})x\}, \boldsymbol{\theta} \in [-1,1].$$

The supporting manifold M_1 defined by $\sigma_{11} = \sigma_{22} = \cdots = \sigma_{pp}$ and $\sigma_{12} = \sigma_{13} = \ldots = \sigma_{p-1,p}$ is itself a (p*, 2)-submanifold but since the transformation from M to M_1 is linear, M_1 is also an exponential family. It follows that M_0 is a $(2,1)$-curved exponential family of distributions. This remark will allow an extensive simplification of computations. Each element of M_0 is expressed in terms of $\boldsymbol{\theta}$ as

$$q(x,\boldsymbol{\theta}) = \exp\{-1/2\frac{1}{(1-\boldsymbol{\theta})(1+(p-1)\boldsymbol{\theta}}((1+(p-2)\boldsymbol{\theta}))\sum_{i=1,\cdots,p} x_i^2 - 2\boldsymbol{\theta}\sum_{j>i} x_i x_j) + \ln\{1+(p-1)\boldsymbol{\theta})(1-\boldsymbol{\theta})^{p-}$$

this is because the determinant of $\boldsymbol{\Sigma}$ structured by $\boldsymbol{\theta}$ is $\ln(1+(p-1)\boldsymbol{\theta})(1-\boldsymbol{\theta})^{p-1}$. It is then clear that M_0 is imbedded in $M_1 = \{p(z,\boldsymbol{\beta})\}$ with

$$\begin{aligned} p(z,\boldsymbol{\beta}) &= \exp \boldsymbol{\beta}_1 z_1 + \boldsymbol{\beta}_2 z_2 - \phi(\boldsymbol{\beta}) \\ \boldsymbol{\beta}_1 &= -\frac{1+(p-1)\boldsymbol{\theta}}{2(1-\boldsymbol{\theta})(1+(p-1)\boldsymbol{\theta})} \\ \boldsymbol{\beta}_2 &= \frac{\boldsymbol{\theta}}{(1-\boldsymbol{\theta})(1+(p-1)\boldsymbol{\theta})} \\ z_1 &= \sum_{i,=1,\ldots p} x_i^2 \\ z_2 &= \sum_{j>i} x_i x_j \end{aligned}$$

The parametric equations of the ancillary submanifold are derived as

$$\begin{aligned} \eta_1(\boldsymbol{\theta},v) &= \frac{p(1-\boldsymbol{\theta})(1-\boldsymbol{\theta}+p\boldsymbol{\theta})(1-4\boldsymbol{\theta}+2p\boldsymbol{\theta})}{(1-3\boldsymbol{\theta}+p\boldsymbol{\theta})(1-3\boldsymbol{\theta}+2p\boldsymbol{\theta})} + \frac{1+(p-1)\boldsymbol{\theta}^2}{\boldsymbol{\theta}(2-2\boldsymbol{\theta}-2p+3p\boldsymbol{\theta}-p^2\boldsymbol{\theta})}v \\ \eta_2(\boldsymbol{\theta},v) &= \frac{p\boldsymbol{\theta}(p-1)(1-\boldsymbol{\theta})(1-\boldsymbol{\theta}+p\boldsymbol{\theta})}{2(1-3\boldsymbol{\theta}+p\boldsymbol{\theta})(1-3\boldsymbol{\theta}+2p\boldsymbol{\theta})} + v \end{aligned}$$

The mixture connection coefficient of M_0 is

$$\begin{aligned} \boldsymbol{\Gamma}^{-1}_{\boldsymbol{\theta\theta\theta}} = \; & 2p(2-3p+p^2)(-1+6\boldsymbol{\theta}-3p\boldsymbol{\theta}-14\boldsymbol{\theta}^2 14p\boldsymbol{\theta}^2-3p^2\boldsymbol{\theta}^2 \\ & 36\boldsymbol{\theta}^3-54p\boldsymbol{\theta}^3+20p^2\boldsymbol{\theta}^3-p^3\boldsymbol{\theta}^3-81\boldsymbol{\theta}^4+162p\boldsymbol{\theta}^4 \\ & -101p^2\boldsymbol{\theta}^4+20p^3\boldsymbol{\theta}^4+54\boldsymbol{\theta}^5-135p\boldsymbol{\theta}^5+120p^2\boldsymbol{\theta}^5-45p^3\boldsymbol{\theta}^5+6p^4\boldsymbol{\theta}^5) \end{aligned}$$

and denominator

$$(-1+\theta)^2\,(1-3\,\theta+\mathrm{p}\,\theta)^3\,(1-\theta+\mathrm{p}\,\theta)^2\,(1-3\,\theta+2\,\mathrm{p}\,\theta)^3$$

the bias of $\hat{\theta}$ is

$$\begin{aligned} b(\hat{\theta}) \;&=\; \frac{1}{np(p-1)num^2}\{(4(p-2)(1-\theta)^2(1+(\mathrm{p}-3)\theta)(1+(\mathrm{p}-1)\theta)^2 \\ &+(1+(2p-3)\theta)(1+(3\mathrm{p}-6)\theta+(3\mathrm{p}^2-14\mathrm{p}+14)\theta^2 \\ &+(p^3-20p^2+54p-36)\theta^3+(-20p^3+101p^2-162p+81)\theta^4 \\ &+(-6p^4+45p^3-120p^2+135p-54)\theta^5)\}+O(n^{-3/2}) \end{aligned}$$

where *num* will be given in appendix. We see that for $p=2$ the bias of θ vanishes up to order $1/n$, this is not true for $p>2$.
The quantities needed for evaluating the mean square errors of θ, namely the Fisher information matrix, the mixture connections, the exponential curvature of M_0 and the mixture curvature of the ancillary submanifold will be given in Appendix.
Since M_0 is curved in M_1,its exponential curvature is nonvanishing, hence the mean square error of the bias corrected esti previous section, will involve an extra term. It is in fact

$$E(\theta^*\theta^*) = (1/\mathrm{n})\mathrm{g}^{\theta\theta} + (1/2\mathrm{n}^2)\{(\Gamma^{\mathrm{m}})^{2\theta\theta} + 2(\mathrm{H}^{\mathrm{e}}_{\mathrm{M}_0})^{2\theta\theta}\} + \mathrm{O}(\mathrm{n}^{-5/2}).$$

This expression is a rational function of θ and will be given in appendix. We note that for $p=2$ the expression simplifies to

$$E(\theta^*\theta^*) = \frac{4(1-\theta^2)^4}{\mathrm{n}^2(1+\theta^2)^4}.$$

6 References

Amari, S. (1980). Theory of information spaces, a differential geometric foundation of statistics. Post Raag Report, no. 106.

Amari, S. (1982). Differential geometry of curved exponential families, curvatures and information loss. Ann. Statist., 10, 357-87.

Amari, S. (1985). Differential- Geometrical Methods in Statistics. Springer Verlag.

Bates, D. M. and Watts, D, G. (1980). Relative Curvature measures of non-linearity. J. Roy. Stat. Soc. , B40, 1-25.

Berkane, M. and Bentler, P., M. (1992). The geometry of the mean and covariance structure models in multivariate normal distributions: A unified approach. Multivariate Analysis, Future Directions 2. C. M. Quadras and C. R. Rao, editors. North Holland.

Cook, D. R. and Goldberg, M. L. (1986). Curvatures for parameter subsets in nonlinear regression. Ann. Statist., Vol 14, 1399-1418.

Donoghue, J. R. and Collins, M. (1990). A note on the unbiased estimation of the intraclass correlation. Psychometrika, Vol 55, 159-164.

Efron, B. (1978). The geometry of exponential families. Ann. Statist, Vol 6, 362-376.

Graybill, F.A.(1969). Matrices with Applications in Statistics. Wadsworth.

Kass, R. E.(1984). Canonical parametrizations and zero parameter effects curvature. J. Roy. Statist. Soc. B., Vol 46, 86-92.

Mitchell, A. F. S.(1989). The information matrix, skewness tensor and α-connections for the general multivariate elliptical distribution. Ann. Inst. Statist. Math., Vol 41, 289-304.

Olkin, I. and Pratt, J. W. (1958). Unbiased estimation of certain correlation coefficients. Ann. Math. Statist., 29, 202-211.

Seber, G. A. F. and Wild, C. J. (1989). Nonlinear Regression. Wiley.

Skovgaard, L.,T.(1984). A Riemannian geometry of the multivariate normal model. Scand. J. Statist., 11, 211-223.

APPENDIX

The polynomials involved in the expressions of the coefficients of the mixture connection are

$$
\begin{aligned}
P_6'(\rho) &= p(p-1)\{(1+2(p-2)\rho+(5-5p-p^2)\rho^2+(16-24p+8p^2)\rho^3+ \\
&\quad (-117+234p-146p^2+29p^3)\rho^4+(180-450p+400p^2-150p^3+20p^4)\rho^5+ \\
&\quad (-81+243p-279p^2+153p^3-40p^4+4p^5)\rho^6) \\
Q_8'(\rho) &= 2\lambda(1-\rho)^2(1+(p-3)\rho)^2(1+(p-1)\rho)^2(1+(2p-3)\rho)^2 \\
P_3'(\rho) &= p(p-1)\rho(1+5(p-2)\rho+(9-9p+2p^2)\rho^2) \\
Q_4'(\rho) &= 2\lambda^2(1-\rho)(1+(p-1)\rho)(1+(p-3)\rho)(1+(2p-3)\rho) \\
P_5'(\rho) &= 2p\rho(2-3p+p^2)(2+3(p-2)\rho) \\
Q_6'(\rho) &= \lambda(-1+\rho)^2(1+(p-1)\rho)(1+(p-3)\rho)^2(1+(2p-3)\rho)^2 \\
P_5'(\rho) &= 2p(2-3p+p^2)\{(-1+3(p-2)\rho+(-14+14p-3p^2)\rho^2+ \\
&\quad (36-54p+20p^2-p^3)\rho^3+(-81+162p-101p^2+20p^3)\rho^4+ \\
&\quad (54-135p+120p^2-45p^3+6p^4)\rho^5)\} \\
Q_{10}'(\rho) &= \lambda(1-\rho)^2(1+(p-1)\rho)2(1+(p-3)\rho)^3(1+(2p-3)\rho)^2
\end{aligned}
$$

The induced metric $\tilde{G}$ in M_0 is a rational function with numerator

$$p(p-1)(1+(-4+2p\boldsymbol{\theta}+(5-5\mathrm{p}+\mathrm{p}^2)\boldsymbol{\theta}^2 \quad +(16-24p+8p^2)\boldsymbol{\theta}^3+$$
$$(-117+234p-146p^3)\boldsymbol{\theta}^4+(81+243\mathrm{p}-279\mathrm{p}^2 \quad +153p^3-40p^4+4p^5)\boldsymbol{\theta}^6$$

and denominator

$$D=2(1-\boldsymbol{\theta})^2(1+(\mathrm{p}-3)\boldsymbol{\theta})^2(1+(\mathrm{p}-1)\boldsymbol{\theta})^2(1+(2\mathrm{p}-3)\boldsymbol{\theta})^2.$$

The numerator of $\tilde{G}$ has been denoted by *num* in the text. The numerator of the exponential connection $\boldsymbol{\Gamma}^{(\mathrm{e})}_{\boldsymbol{\theta\theta\theta}}$ of M_0 is

$$2p(p-2)(p-1)^2p\boldsymbol{\theta}^2(3+(6p-12)\boldsymbol{\theta}+(2\mathrm{p}^2-9\mathrm{p}+9)\boldsymbol{\theta}^2)(1+(3\mathrm{p}-3)\boldsymbol{\theta}^2+$$
$$(p^2-3p+2)\boldsymbol{\theta}^3))-\mathrm{p}(\mathrm{p}-1)(\mathrm{p}-2-(3\mathrm{p}-3)\boldsymbol{\theta}-(\mathrm{p}^2-2\mathrm{p}+1)\boldsymbol{\theta}^3)(1+(2\mathrm{p}-4)\boldsymbol{\theta}+$$
$$(p^22-6p+6)\boldsymbol{\theta}^2+(-6\mathrm{p}^22+18\mathrm{p}-12)\boldsymbol{\theta}^3+(-2\mathrm{p}^3+11\mathrm{p}^2-18\mathrm{p}+9)\boldsymbol{\theta}^4))$$

and the denominator is

$$(1-\boldsymbol{\theta})^3(1+(\mathrm{p}-3)\boldsymbol{\theta})^2(1+(\mathrm{p}-3)\boldsymbol{\theta})^2(1+(\mathrm{p}-1)\boldsymbol{\theta})^3(1+(2\mathrm{p}-3)\boldsymbol{\theta})^2$$

The numerator of the mixture connection is

$$2\,p\,(2-3\,p+p^2)(-1+6t-3pt-14\boldsymbol{\theta}^2+14\mathrm{p}\boldsymbol{\theta}^2-3\mathrm{p}^2\boldsymbol{\theta}^2+36\,\boldsymbol{\theta}^3-54\,\mathrm{p}\,\boldsymbol{\theta}^3+20\mathrm{p}^2\boldsymbol{\theta}^3$$
$$-p^3\,\boldsymbol{\theta}^3-81\,\boldsymbol{\theta}^4+162p\boldsymbol{\theta}^4-101p^2\boldsymbol{\theta}^4+20p^3\boldsymbol{\theta}^4+54\boldsymbol{\theta}^5-135p\boldsymbol{\theta}^5+120p^2\boldsymbol{\theta}^5-45p^3\boldsymbol{\theta}^5+6p^4\boldsymbol{\theta}^5)$$

and denominator

$$(-1+\boldsymbol{\theta})^2\,(1-3\,\boldsymbol{\theta}+\mathrm{p}\,\boldsymbol{\theta})^3\,(1-\boldsymbol{\theta}+\mathrm{p}\,\boldsymbol{\theta})^2\,(1-3\,\mathrm{t}+2\,\mathrm{p}\,\boldsymbol{\theta})^3$$

The components of the exponential curvature of M_0 are

$$\begin{aligned}
H^{(e)}_{\boldsymbol{\theta\theta}_1} &= -\frac{1}{D}\{2p(p-1)(1-\boldsymbol{\theta})(1+(\mathrm{p}-1)\boldsymbol{\theta}+(\mathrm{p}-1)\boldsymbol{\theta}^2+(\mathrm{p}-1)^2\boldsymbol{\theta}^3)\}\\
H^{(e)}_{\boldsymbol{\theta\theta}_2} &= -\frac{1}{D}\{p(p-1)^2(1-\boldsymbol{\theta})\boldsymbol{\theta}^2(2+(3\mathrm{p}-4)\boldsymbol{\theta}+(\mathrm{p}-3\mathrm{p}+2)\boldsymbol{\theta}^2)\}
\end{aligned}$$

The square of the exponential curvature is

$$\frac{1}{D^2}\{4(p-1)(1-\boldsymbol{\theta})^2(1+(4\mathrm{p}-7)\boldsymbol{\theta}+(5\mathrm{p}^2-18\mathrm{p}+15)\boldsymbol{\theta}^2+(2\mathrm{p}^3-11\mathrm{p}^2+18\mathrm{p}-9)\boldsymbol{\theta}^3)^2\}$$

and the mean square error of the bias corrected estimator $\boldsymbol{\theta}^*$ of $\boldsymbol{\theta}$ is

$$\begin{aligned} E(\boldsymbol{\theta}^* - \boldsymbol{\theta})^2 &= \frac{1}{n\tilde{G}} + \frac{1}{2n^2p^2(p-1)^2D^4}(\\ &32(p-1)\{(\boldsymbol{\theta}-1)(1+(\mathrm{p}-3)\boldsymbol{\theta})(1+\mathrm{p}-1)\boldsymbol{\theta}(1+(2\mathrm{p}-3)\boldsymbol{\theta})\}^6 \\ &64(p-2)^2(\boldsymbol{\theta}-1)^4(1+(\mathrm{p}-3)\boldsymbol{\theta})^2(1+(\mathrm{p}-1)\boldsymbol{\theta})^4(1+(2\mathrm{p}-3)\boldsymbol{\theta})^2 \\ &\{(-1-(3p-6)\boldsymbol{\theta}-(3\mathrm{p}^2-14\mathrm{p}-14)\boldsymbol{\theta}^2-(\mathrm{p}^3-20\mathrm{p}^2+54\mathrm{p}=36)\boldsymbol{\theta}^3+ \\ &(20p^3-101p^2+162p-81)\boldsymbol{\theta}^3+(6\mathrm{p}^4-45\mathrm{p}^3+120\mathrm{p}^2-135\mathrm{p}+54)\boldsymbol{\theta}^5\}^2) \end{aligned}$$

Latent Variable Growth Modeling with Multilevel Data

Bengt Muthén *

Graduate School of Education & Information Studies
University of California, Los Angeles
Los Angeles, CA 90024-1521

Abstract

Growth modeling of multilevel data is presented within a latent variable framework that allows analysis with conventional structural equation modeling software. Latent variable modeling of growth considers a vector of observations over time for an individual, reducing the two-level problem to a one-level problem. Analogous to this, three-level data on students, time points, and schools can be modeled by a two-level growth model. An interesting feature of this two-level model is that contrary to recent applications of multilevel latent variable modeling, a mean structure is imposed in addition to the covariance structure. An example using educational achievement data illustrates the methodology.

1 Introduction

Longitudinal studies of growth in educational achievement typically use cluster sampling of students within schools. This gives rise to hierarchical data with three levels: student, time point, and school. With large numbers of students per school, ignoring the clustering of students within schools may give strongly distorted inference even with modest intraclass (school) correlations. While three-level modeling is well established with manifest dependent variables (see, e.g., Bock, 1989; Bryk & Raudenbush, 1992; Goldstein, 1987), less work has been done in the area of latent variables.

*This paper was presented at the UCLA conference Latent Variable Modelling with Applications to Causality, March 19-20, 1994. The research was supported by a grant from the Office of Educational Research and Improvement, Department of Education to the National Center for Research on Evaluation, Standards, and Student Testing (CRESST) and grant AA 08651-01 from NIAAA. I thank Siek-Toon Khoo, Guanghan Liu, and Ginger Nelson Goff for helpful assistance.

The general framework of latent variable modeling is particularly well suited to growth studies of educational achievement data. Longitudinal educational data is frequently collected at fixed time points corresponding to grades. The relevant time dimension for achievement is grade because this reflects the amount of learning that has taken place. In this way, all individuals have the same value on the time dimension at all measurement occasions. This special case offers a convenient simplification which puts the problem in a conventional latent variable framework.

Formulated as a latent variable growth model, the time dimension is transformed into a multivariate vector and the three-level data hierarchy is reduced to a two-level hierarchy. Two-level latent variable models have been studied in the context of covariance structure models, but the growth model also imposes a mean structure. Relative to conventional three-level modeling, the two-level latent variable growth model formulation offers considerable flexibility in the modeling. Using the Muthén (1989) approach to multilevel data, maximum-likelihood estimation under normality assumptions is carried out with conventional structural equation modeling software.

The methodology is illustrated using mathematics achievement data on students in grades 7 - 10 from the Longitudinal Study of American Youth. Here, an average of about 50 students per school are observed in 50 schools for mathematics achievement scores having intraclass correlations of 0.15-0.20.

Sections 2 and 3 will review growth modeling using a simple random sample of individuals. Subsequent sections add the multilevel complication of analyzing individuals observed within schools.

2 Random coefficient growth modeling: Two-level hierarchical modeling

Random coefficient growth modeling (see, e.g. Laird & Ware, 1982) goes beyond conventional structural equation modeling of longitudinal data and its focus on auto-regressive models (see, e.g. Jöreskog & Sorbom, 1977; Wheaton, Muthén, Alwin & Summers, 1977) to describe individual differences in growth. The modeling is of the following type. Consider the indices

Individual i : $i = 1, 2, ..., n$
Time point t : $t = 1, 2, ..., T$

and the variables

x_{it} : time-related variable (age, grade)
w_{it} : time-varying covariate
z_i : time-invariant covariate

and the model

$$y_{it} = \alpha_i + \beta_i x_{it} + \gamma_t w_{it} + \zeta_{it} \tag{1}$$

where

$$\begin{cases} \alpha_i = \alpha + \pi_\alpha z_i + \delta_{\alpha i} \\ \beta_i = \beta + \pi_\beta z_i + \delta_{\beta i} \end{cases} \tag{2}$$

Here, ζ_{it} is a residual which is independent of x, $\delta_{\alpha i}$ and $\delta_{\beta i}$, $\delta_{\alpha i}$ and $\delta_{\beta i}$ are correlated, and $\delta_{\alpha i}$ and $\delta_{\beta i}$ are independent of z. Here, the special case of linear growth is described, where α_i is the random intercept or initial status and β_i is the random slope or growth rate. Non-linear growth can also be accomodated.

An important special case that will be the focus of this chapter is where the time-related variable $x_{it} = x_t$. An example of this is educational achievement studies where x_t corresponds to grade. The x_t values are for example 0, 1, 2, ..., $T-1$ for linear growth.

3 Latent variable modeling formulation of growth

As shown in Meredith and Tisak (1984, 1990), the random coefficient model of the previous section can be formulated as a latent variable model (for applications in psychology, see McArdle & Epstein, 1987; for applications in education, see Muthén 1993 and Willett & Sayer, 1993; for applications in mental health, see Muthén, 1983, 1991). The basic idea can be simply described as follows. In equation 1, α_i is a latent variable varying across individuals. Assuming the special case of $x_{it} = x_t$, the x variable becomes a constant which multiplies a second latent variable β_i. The equation 1 outcome variable y_{it} for individual i is now reformulated as a $T \times 1$ vector for individual i.

The model may be shown as in the path diagram of Figure 1. Note for example that the constants of x_t are the coefficients for the influence of the β factor on the y variables. This makes it clear that non-linear growth can be accomodated by estimating the x_t coefficients, e.g. holding the first two values fixed at 0 and 1, respectively, for identification purposes.

Given the path diagram in Figure 1, input specifications can be given for conventional structural equation modeling software.

The general SEM framework may be described as

$$\begin{cases} \mathbf{y}^* = \nu + \Lambda\eta + \epsilon \\ B\eta = \kappa + \zeta \end{cases} \tag{3}$$

where the mean and variance of $\mathbf{y}^*$,

$$\begin{cases} E(\mathbf{y}^*) = \nu + \Lambda(I-B)^{-1}\kappa \\ V(\mathbf{y}^*) = \Lambda(I-B)^{-1}\Psi(I-B)^{-1}\Lambda' + \Theta \end{cases} \tag{4}$$

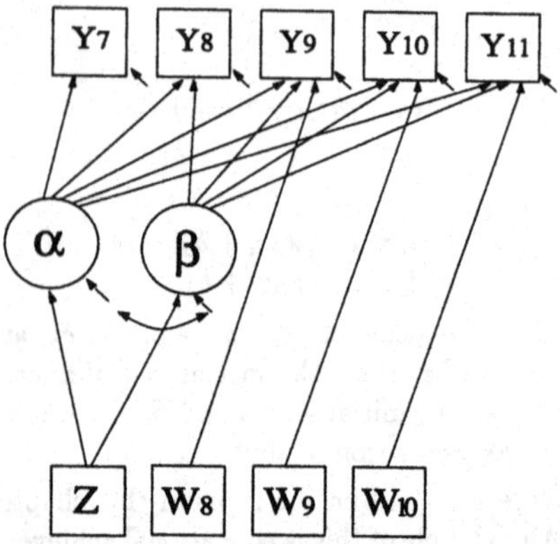

Figure 1: Latent Variable Growth Model Formulation

see, for example, Muthén (1989).

Translating Figure 1 into this general model, we have $\mathbf{y}^{*\prime} = (y_7, \cdots, y_{11}, z,\ w_8,\ w_9,\ w_{10})$, $\eta' = (\mathbf{y}^{*\prime}, \alpha, \beta)$,

$$\nu = 0,\ \Lambda = [\mathrm{I}\ 0],\ \ \epsilon = 0,\ \ B = \begin{bmatrix} B_{11} & 1 & \mathrm{x} \\ B_{21} & 0 & 0 \end{bmatrix} \Psi = \begin{bmatrix} \Psi_{y^*} & 0 \\ 0 & \Psi_{\alpha,\beta} \end{bmatrix}$$

and $\kappa' = (0, \cdots, 0, E(z), E(w_8), E(w_9), E(w_{10}), E(\alpha), E(\beta))$.

The growth model imposes a structure on the mean vector and covariance matrix. It is clear that the Figure 1 model can be easily generalized to applications with multiple indicators of latent variable constructs instead of single outcome measurements y at each time point. The covariates may also be latent variables with multiple indicators. Muthén (1983) showed that binary outcome measures y can also be handled when the $y*$ variables are measured dichotomously. In the continuous variable case, maximum-likelihood estimation is the usual estimator in SEM. Estimates may also be obtained for the individual growth curves by estimating the individual values of the intercept and slope factors α and β. This relates to Empirical Bayes estimation in the conventional growth literature (see, e.g. Bock, 1989).

4 A three-level hierarchical model

We will now add the complication of cluster sampling to the growth modeling. Here, data are obtained on individuals observed within groups. Such hierarchical data are naturally obtained in educational studies where students are sampled within randomly sampled schools. It is well-known that ignoring the hierarchical nature of the data and applying techniques developed for simple random samples distorts the standard errors of estimates and the chi-square test of model fit (see, e.g., Muthén & Satorra, 1993). Standard errors are deflated and chi-square values are inflated. This is particularly pronounced when the intraclass correlations are large and the number of individuals within each group is large. The following latent variable methodology draws on Muthén (1994a).

Conventional random-coefficient modeling for such three-level data is described e.g. in Goldstein (1987) and Bryk and Raudenbush (1992). For

Individual	:	$i = 1, 2, ..., n$
Time	:	$t = 1, 2, ..., T$
Group (School)	:	$g = 1, 2, ..., G$

consider the growth model, again expressed for the special case of $x_{it} = x_t$,

$$y_{itg} = \alpha_{ig} + x_t \beta_{ig} + \zeta_{itg} \tag{5}$$

where for simplicity there are no covariates

$$\begin{cases} \alpha_{ig} = \alpha_g + \delta_{\alpha ig} \\ \beta_{ig} = \beta_g + \delta_{\beta ig} \end{cases} \tag{6}$$

$$\begin{cases} \alpha_g = \alpha + \delta_{\alpha g} \\ \beta_g = \beta + \delta_{\beta g} \end{cases} \tag{7}$$

5 A two-level formulation of multilevel growth

In the case of growth modeling using a simple random sample of individuals, it was possible to translate the growth model from a two-level model to a one-level model by considering a $T \times 1$ vector of outcome variable for each individual. Analogously, we may reduce the three-level model of the previous section to two levels as follows.

$$\mathbf{y}_{ig} = \begin{pmatrix} y_{i1g} \\ \vdots \\ y_{iTg} \end{pmatrix} = [1\ \mathbf{x}] \begin{pmatrix} \alpha_{ig} \\ \beta_{ig} \end{pmatrix} + \zeta_{ig} \tag{8}$$

which may be expressed in five terms

$$\mathbf{y}_{ig} = [1\ \mathbf{x}] \begin{pmatrix} \alpha \\ \beta \end{pmatrix} + [1\ \mathbf{x}] \begin{pmatrix} \delta_{\alpha g} \\ \delta_{\beta g} \end{pmatrix} + \zeta_g^* + [1\ \mathbf{x}] \begin{pmatrix} \delta_{\alpha ig} \\ \delta_{\beta ig} \end{pmatrix} + \zeta_{ig}^* \tag{9}$$

The first term represents the mean as a function of the mean of the initial status and the mean of the growth rate. The second and third terms correspond to between-group (school) variation. The fourth and fifth terms correspond to within-group variation.

6 Latent variable modeling of two-level growth data

Assume $g = 1, 2, ..., G$ independently observed groups with $i = 1, 2, ..., N_g$ individual observations within group g. Let z and y represent group- and individual-level variables, respectively. Arrange the data vector for which independent observations are obtained as

$$\mathbf{d}_g' = (\mathbf{z}_g', \mathbf{y}_{g1}', \mathbf{y}_{g2}', ..., \mathbf{y}_{gN_g}') \tag{10}$$

where we note that the length of $\mathbf{d}_g$ varies across groups. The mean vector and covariance matrix are,

$$\mu_{\mathbf{d}_g}' = [\mu_{\mathbf{z}}', \mathbf{1}_{N_g}' \otimes \mu_{\mathbf{y}}'] \tag{11}$$

$$\Sigma_{\mathbf{d}_g} = \begin{pmatrix} \Sigma_{\mathbf{zz}} & \text{symmetric} \\ \mathbf{1}_{N_g} \otimes \Sigma_{\mathbf{yz}} & \mathbf{I}_{N_g} \otimes \Sigma_w + \mathbf{1}_{N_g}\mathbf{1}_{N_g}' \otimes \Sigma_B \end{pmatrix} \tag{12}$$

Assuming multivariate normality of $\mathbf{d}_g$, the ML estimator minimizes the function

$$F = \sum_{g=1}^{G} \{\log |\Sigma_{\mathbf{d}_g}| + (\mathbf{d}_g - \mu_{\mathbf{d}_g})' \Sigma_{\mathbf{d}_g}^{-1} (\mathbf{d}_g - \mu_{\mathbf{d}_g})\} \tag{13}$$

which may be simplified as (cf. McDonald & Goldstein, 1989; Muthén, 1989, 1990)

$$\begin{aligned} F = & \textstyle\sum_d^D G_d\{\ln |\Sigma_{B_d}| + tr[\Sigma_{B_d}^{-1}(S_{B_d} + N_d(\bar{v}_d - \mu)(\bar{v}_d - \mu)')]\} \\ & +(N - G)\{\ln |\Sigma_W| + tr[\Sigma_W^{-1} S_{PW}]\} \end{aligned} \tag{14}$$

where

$$\Sigma_{B_d} = \begin{pmatrix} N_d\Sigma_{zz} & \text{symmetric} \\ N_d\Sigma_{yz} & \Sigma_W + N_d\Sigma_B \end{pmatrix}$$

$$S_{B_d} = N_d G_d^{-1} \sum_{k=1}^{G_d} \begin{pmatrix} z_{dk} - \bar{z}_d \\ \bar{y}_{dk} - \bar{y}_d \end{pmatrix} [(z_{dk} - \bar{z}_d)'(\bar{y}_{dk} - \bar{y}_d)']$$

$$\bar{v}_d - \mu = \begin{pmatrix} \bar{z}_d - \mu_z \\ \bar{y}_d - \mu_y \end{pmatrix}$$

$$S_{PW} = (N - G)^{-1} \sum_{g=1}^{G} \sum_{i=1}^{N_g} (y_{gi} - \bar{y}_g)(y_{gi} - \bar{y}_g)'$$

Here, D denotes the number of groups of a distinct size, d is an index denoting a distinct group size category with group size N_d, G_d denotes the number of groups of that size, S_{B_d} denotes a between-group sample covariance matrix, and S_{PW} is the usual pooled-within sample covariance matrix.

Muthén (1989,1990) pointed out that the minimization of the ML fitting function defined by equation 14 can be carried out by conventional structural equation modeling software, apart from a slight modification due to the possibility of singular sample covariance matrices for groups with small G_d values. A multiple-group analysis is carried out for $D+1$ groups, the first D groups having sample size G_d and the last group having sample size $N-G$. Equality constraints are imposed across the groups for the elements of the parameter arrays μ, Σ_{zz}, Σ_{yz}, Σ_B, and Σ_W. To obtain the correct chi-square test of model fit, a separate H_1 analysis needs to be done (see Muthén, 1990 for details).

Muthén (1989, 1990) also suggested an ad hoc estimator which considered only two groups,

$$\begin{aligned} F' = \; & G\{\ln|\Sigma_{B_c}| + tr[\Sigma_{B_c}^{-1}(S_B + c(\bar{v}-\mu)(\bar{v}-\mu)')]\} \\ & +(N-G)\{\ln|\Sigma_W| + tr[\Sigma_W^{-1} S_{PW}]\} \end{aligned} \tag{15}$$

where the definition of the terms simplifies relative to equation 14 due to ignoring the variation in group size, dropping the d subcript, and using $D=1$, $G_d=G$, and $N_d=c$, where c is the average group size (see Muthén, 1990 for details). When data are balanced, i.e. the group size is constant for all groups, this gives the ML estimator. Experience with the ad hoc estimator for unbalanced data indicates that the estimates, and also the standard errors and chi-square test of model fit, are quite close to those obtained by the true ML estimator. This experience, however, is limited to models that do not have a mean structure and is therefore not directly applicable to growth models.

In line with Muthén (1989, 1990), Figure 2 shows a path diagram which is useful in implementing the estimation using F or F'. The figure corresponds to the case of no covariates given in equations 5 - 7 and 9. It shows how the covariance structure

$$\Sigma_W + N_d \Sigma_B \tag{16}$$

can be represented by latent variables, introducing a latent between-level variable for each outcome variable y. These latent between-level variables may also be related to observed between-level variables z_g. The between-level α and β factors correspond to the δ_{α_g} and δ_{β_g} residuals of equation 7. The within-level α and β factors correspond to the $\delta_{\alpha_{ig}}$ and $\delta_{\beta_{ig}}$ residuals of equation 6. From equation 9 it is clear that the the influence from these two factors is the same on the between side as it is on the within side. In Figure 2, the Σ_B structure is identical to the Σ_W structure. A strength of the latent variable approach is that this equality assumption can easily be relaxed. For example, it may not be necessary include between-group variation in the growth rate.

Specific to the growth model is the mean structure imposed on μ in equation 14, where μ represents the means of group- and individual-level variables. In the specific growth model shown in Figure 2, the mean structure arises from the five observed variable means being expressed as functions of the means of the α and β factors, here applied on the between side, see equation 9. Equation 14 indicates that the means need to be included on the between side of Figure 2 given that the mean term of F is scaled by

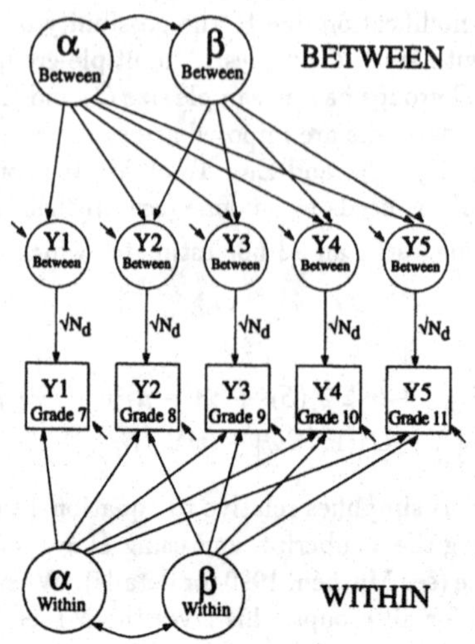

Figure 2: Two-level Latent Variable Growth Model Formulation

N_d, while the means on the within side are fixed at zero. This implies that dummy zero means are entered for the within group. The degrees of freedom for the chi-square test of model fit obtained in conventional software then needs to be reduced by the number of y variables.

Further details and references on latent variable modeling with two-level data are given in Muthén (1994b), also giving suggestions for analysis strategies. Software is available from the author for calculating the necessary sample statistics, including intraclass correlations.

7 Analysis of longitudinal achievement data

Mathematics achievement data from four time points will be used to illustrate the methodology described above. Data are from grades 7-10 of The Longitudinal Study of American Youth (LSAY) and were collected every Fall starting in 1987. The data we will analyze consists of a total sample of 2,488 students in 50 schools. The average school size is therefore 49. The intraclass correlations for math achievement for the four grades are estimated as 0.19, 0.16, 0.16, and 0.14.

The ML estimator of equation 14 will be reported. As a comparison, the ad hoc estimator of equation 15 will also be given for the final model. As a further comparison, the conventional ML estimator assuming a simple random sample will also be reported.

As an initial model, the linear growth model using $x_t = 0, 1, 2, 3$ was used, resulting in a chi-square value of 18.69 with 5 df ($p < 0.01$). It is interesting to note that conventional modeling gives a higher chi-square value with fewer degrees of freedom, 41.85 with 2 df. This illustrates the inflation of chi-square values when ignoring the clustering so that proper models might be inadvertently rejected.

To investigate if the misfit is due to the assumption of equal covariance structure on the between and within sides, the Σ_B structure was relaxed and an unrestricted Σ_B matrix was used (the latent between-level variables are then freely correlated). This improved the fit, but the improvement was not significant on the 1% level (the chi-square value was 9.17 with 2 df and $p = 0.01$; the chi-square difference test has a chi-square of 9.52 with 3 df, $p > 0.02$).

It was decided to retain the assumption of equal covariance structure on the between and within sides and instead relax the linearity assumption, $x_t = 0, 1, 2, 3$ by letting the last two values be free to be estimated. This non-linear growth model gave a strongly significant improvement over the initial model (the chi-square value was 6.27 with 3 df and p=0.10; the chi-square difference value was 12.43 with 2 df and $p < 0.001$). The estimates from this model are given in Table 1 below.

Table 1 gives estimates from three procedures: using the incorrect, conventional ML estimator, using the correct multilevel ML estimator, and using the ad hoc multilevel estimator. Growth scores refers to the values of x_t.

The multilevel ML approach shows that the estimated growth score for grade 9 is 2.505 which is larger than the linear growth value of 2.0. This means that growth is accelerated during grade eight when many new topics are introduced. The growth rate mean is positive as expected. The variation in both the initial status and the growth rate are significantly different from zero on the within (student) level, but that is not the case for the growth rate variation on the between (school) level. This indicates that schools vary more with respect to the intake of students than how the students develop in achievement over time. Variation in intake may depend on socio-economic neighborhood factors. The estimates show that about 18 % of the total variation in initial status is due to across-school variation. This is in line with the observed intraclass correlations. The within-level correlation between initial status and growth rate is 0.43.

Comparing the conventional ML approach with the multilevel ML approach, it is seen that the parameter estimates of the growth score and the growth rate mean are close. As expected, however, the standard errors for these estimates are too low for the conventional approach. The parameter estimates of initial status and growth rate variances and covariance are quite different. The conventional analysis estimates the total variance, adding between and within variation. The correlation between initial status and growth rate is 0.33, which is lower than the multilevel value of 0.43 for the within level.

Table 1: Estimates for three approaches to multilevel growth modeling

	Conventional ML Analysis		ML Multilevel Analysis		Ad Hoc Multilevel Analysis	
	Estimate	**SE**	**Estimate**	**SE**	**Estimate**	**SE**
Initial Status Mean	51.311	0.073	51.125	0.120	51.343	0.129
Growth scores						
Grade 8	1.00*	-	1.00*	-	1.00*	-
Grade 9	2.499	0.095	2.508	0.184	2.541	0.202
Grade10	3.636	0.153	3.635	0.290	3.691	0.318
Growth rate mean	2.443	0.112	2.433	0.229	2.400	0.244
Variance of						
Initial Status	87.032	4.795	66.202	4.277	66.248	4.241
Growth Rate	1.678	0.601	1.519	0.553	1.465	0.536
Covariance of						
Initial Status, Growth Rate	3.928	1.406	4.352	1.357	4.338	1.337
Residual variance of y						
Grade 7	12.614	4.127	15.118	3.858	15.272	3.827
Grade 8	15.964	2.210	18.964	2.085	19.099	2.099
Grade 9	25.956	2.844	22.460	2.839	22.474	2.886
Grade 10	43.231	5.393	38.983	5.531	38.833	5.683
Between Variance of						
Initial status			14.444	3.348	15.406	3.560
Growth Rate			0.094	0.068	0.112	0.077
Between Covariance of						
Initial status, Growth Rate			0.839	0.339	0.909	0.369
Between Residual variance of y						
Grade 7			1.133	0.439	1.342	0.501
Grade 8			0.718	0.278	0.850	0.317
Grade 9			0.220	0.232	0.331	0.271
Grade 10			0.563	0.443	0.766	0.513

*Fixed

Comparing the ad hoc multilevel approach with the ML multilevel approach, it is seen that the parameter estimates are on the whole reasonably close. For example, the ratio of between to total variance in initial status is 19 % compared to 18 % and the within-level correlation between initial status and growth rate is 0.44 compared to 0.43. The standard errors also seem close enough to serve as a rough approximations. The approximation to the chi-square value is 6.51 compared to 6.27 for the ML multilevel approach.

8 Conclusions

This chapter has shown that for an important special case it is possible to use the framework of latent variable modeling to carry out quite general growth modeling of three-level data. Maximum-likelihood estimation under normality assumptions can be carried out with existing structural equation modeling software. A simpler, ad hoc, estimator appears to work well and may be useful at least for initial model exploration. This provides a useful tool for educational research with longitudinal data on students observed within schools and other data with similar structures.

References

Bock, R. D. (1989). Multilevel analysis of educational data. San Diego, CA: Academic Press.

Bryk, A. S. and Raudenbush, S. W. (1992). Hierarchical linear models: Applications and data analysis methods. Newbury Park, CA: Sage Publications.

Goldstein, H. I. (1987). Multilevel models in educational and social research. London: Oxford University Press.

Jöreskog, K. G. and Sörbom, D. (1977). Statistical models and methods for analysis of longitudinal data. In D. J. Aigner and A. S. Goldberger (eds.), Latent variables in socio-economic models (pp. 285-325. Amsterdam: North-Holland

Laird, N.M., and Ware, J.,H. (1982). Random-effects models for longitudinal data. *Biometrics*, 38, 963-974.

McArdle, J.J. and Epstein, D. (1987). Latent growth curves within developmental structural equation models. *Child Development*, 58, 110-133.

McDonald and Goldstein (1989). Balanced versus unbalanced designs for linear structural relations in two-level data. *British Journal of Mathematical and Statistical Psychology*, 42, 215-232.

Meredith, W. and Tisak, J. (1984). "Tuckerizing" curves. Paper presented at the Psychometric Society annual meetings, Santa Barbara, CA.

Meredith, W. and Tisak, J. (1990). Latent curve analysis. *Psychometrika*, 55, 107-122.

Muthén, B. (1983). Latent variable structural equation modeling with categorical data. *Journal of Econometrics*, 22, 43-65.

Muthén, B. (1989). Latent variable modeling in heterogeneous populations. *Psychometrika*, 54, 557-585.

Muthén, B. (1990). Mean and covariance structure analysis of hierarchical data. Paper presented at the Psychometric Society meeting in Princeton, New Jersey. UCLA Statistics Series #62, August 1990.

Muthén, B. (1991). Analysis of longitudinal data using latent variable models with varying parameters. In L. Collins and J. Horn (eds.) *Best Methods for the Analysis of Change. Recent Advances, Unanswered Questions, Future Directions* (pp. 1-17). Washington D.C.: American Psychological Association.

Muthén, B. (1993). Latent variable modeling of growth with missing data and multilevel data. In C.R. Rao and C. M. Cuadras (eds.) *Multivariate Analysis: Future Directions 2* (pp. 199-210). Amsterdam: North-Holland.

Muthén, B. (1994a). Latent variable modeling of longitudinal and multilevel data. Invited paper for the annual meeting of the American Sociological Association, Section on Methodology, Showcase Session, Los Angeles, August 1994.

Muthén, B. (1994b). Multilevel covariance structure analysis. In J. J Hox and I. G. Kreft (eds.), Multilevel analysis methods, (pp. 376-391). Thousand Oakes, CA: Sage Publications.

Muthén, B. and Satorra, A. (1995). Complex sample data in structural equation modeling. Forthcoming in *Sociological Methodology* 1995.

Wheaton, B., Muthén, B., Alwin, D., and Summers. G. (1977). Assessing reliability and stability in panel models. In D.R. Heise (ed.) *Sociological Methodology* 1977 (pp. 84-136). San Francisco: Jossey-Bass.

Willett, J.B. and Sayer, A. G (1993). Using covariance structure analysis to detect correlates and predictors of individual change over time. Harvard University. *Psychological Bulletin*, 116(2), 363-381.

High-dimensional Full-information Item Factor Analysis

R. Darrell Bock Steven Schilling
University of Chicago

Item factor analysis is unexcelled as a technique of exploration and discovery in the study of behavior. From binary-scored responses to a multiple-item test or scale, it determines the dimensionality of individual variation among the respondents and reveals attributes of the items defining each dimension. In practical test construction and scoring, it is the best guide to item selection and revision, as well as an essential preliminary step in justifying unidimensional IRT item analysis. Because hundreds of items may be analyzed jointly, the detail and generality that may be achieved exceeds that of any other procedure for exploring relationships among responses.

Principal factor analysis of phi coefficients or of estimated tetrachoric correlation coefficients has never been very satisfactory for these purposes. The former introduces spurious "difficulty" factors, and the latter almost always results in nonpositive-definite correlation matrices that are, strictly speaking, unsuitable for factor analysis. The limited-information generalized least-squares item factor analysis method of Christoffersson (1975) and Muthén (1978) alleviates the latter problem, but the number of elements in the weight matrix of that method (of order n^4 in the number of variables) limits the number of items that can be analyzed.

The full-information method of Bock and Aitkin (1981) is more favorable in this respect. The computational demands are linear in the number of items, and the EM solution is easily extended to the several-hundred item range. As implemented in Bock, Gibbons, and Muraki (1988), however, this solution is limited computationally to no more than five factors. The barrier is the

Gauss-Hermite quadrature used to compute the expected values that appear in the maximum marginal likelihood equations for the factor loadings. Because the number of points in the factor space at which these expectations must be computed increases exponentially with the number of dimensions, the method is limited to perhaps five or six factors, even with fast computers.

The purpose of the present paper is to demonstrate that by the use of quadratures adapted to the location and dispersion of the posterior distributions in the factor space, given the corresponding answer pattern, evaluations at a properly selected fraction of points can result in quite satisfactory estimates of the factor loadings and marginal probabilities with very little increase in computing time as the number of factors increases. This approach puts high-dimensional item factor analysis within practical reach.

1 A Multiple-factor Item Response Model

Bock and Aitkin (1981) adapted Thurstone's (1947) multiple-factor model for test scores to item responses as follows:

For an n-item instrument, let $u = [u_j]$, $j = 1, 2, \ldots, n$ be a pattern of item scores, $u_j = 1$ if correct, and 0 otherwise. Assume a normal ogive response function, such that $P(u_j = 1|\theta, \zeta_j) = \Phi_j(\theta)$ and $P(u_{ij} = 0|\theta, \zeta_j) = 1 - \Phi_j(\theta)$, where $\Phi_j(\theta)$ is a standard normal distribution function, the argument of which is a function of an m-vector factor score, θ, and an $(m+1)$-vector parameter, ζ_j, of item j $(m < n)$. That is, $\Phi_j(\theta) = \Phi[z_j(\theta)]$, and

$$z_j(\theta) = (\gamma_j - \sum_k^m \alpha_{jk}\theta_k)/\sigma_j \,, \tag{1}$$

with

$$\sigma_j^2 = 1 - \sum_k^m \alpha_{jk}^2 \,, \tag{2}$$

and the elements of ζ_j are $\gamma_j, \alpha_{j2}, \alpha_{j3} \ldots \alpha_{jm}$, where γ_j is the threshold and the α_{jk} are the factor loadings of item j.

2 Maximum Marginal Likelihood Estimation of the Item Parameters

Suppose we have item response patterns from a random sample of N persons drawn from a population in which θ is distributed with density $g(\theta)$. Then the marginal probability of observing a pattern u_ℓ is, say,

$$\bar{P}_\ell = \int_{R(m)} L_\ell(\theta) g(\theta) d\theta \,,$$

where $L_\ell(\theta)$ is the conditional probability of pattern ℓ, given θ. We assume the item responses are independent, conditional on θ, so that

$$L_\ell(\theta) = \prod_j^n [P_j(\theta)]^{u_{\ell j}} [1 - P_j(\theta)]^{1-u_{\ell j}} \,. \tag{3}$$

For binary-scored items, the $\bar{P}_\ell$ are orthant probabilities of an assumed proper n-variate distribution; they sum to 1 over the full set of 2^n orthants. In the context of Bayes estimation of the factor score, θ, given a response vector u_ℓ, $L_\ell(\theta)$ is a likelihood and $g(\theta)$ an ordinate of the prior distribution.

If r_ℓ persons are observed to respond in pattern ℓ, the marginal likelihood of $\zeta = [\zeta_1, \zeta_2, \ldots \zeta_m]'$ is

$$L_n(\zeta) = \frac{N!}{r_1! r_2! \ldots r_s!} \bar{P}_1^{r_1} \bar{P}_2^{r_2} \ldots \bar{P}_s^{r_s} \,, \tag{4}$$

where s is the number of distinct patterns occurring in the sample, and $N = \Sigma_\ell^s r_\ell$.

For purposes of MML estimation, we reparameterize in ζ_j by setting

$$-\gamma_j/\sigma_j = c_j \alpha_{jk}/\sigma_j = a_{jk} \,, \tag{5}$$

for $\sigma_j > 0$. Bock, Gibbons, and Muraki (1988) refer to c_j as an item "intercept" and to the a_{jk} as item "slopes." So-called "Heywood" cases in which $\sigma_j = 0$ must be avoided either by omitting the offending variables or, preferably, imposing the constraint $\sum_k^m \alpha_{jk}^2 < 1$ on the estimator. For the latter, Bock, Gibbons, and Muraki (1988) employ a stochastic constraint based on the beta distribution.

Because of the rotational indeterminacy of the multiple-factor model, any rank m basis of the matrix of factor loadings may be chosen arbitrarily. We determine a basis by performing a MINRES principal factor solution on the tetrachoric correlation matrix (conditioned to be positive-definite) and transform the loadings so that $\alpha_{jk} = 0$, $k > j$. The transformation for this purpose is computed by modified Gram-Schmidt or Householder (1964) triangular orthogonalization of the m leading rows of the principal factor loadings, say A_{11}. The result is the so-called QL decomposition

$$A_{11} = LX \ , \ |A_{11}| \neq 0 \, ,$$

where L is lower triangular and X is orthogonal. Then

$$AX' = \begin{bmatrix} A_{11} \\ A_{21} \end{bmatrix} X' = \begin{bmatrix} L \\ A_{21}X' \end{bmatrix} ,$$

as required. We then transform the arbitrary solution to a preferred basis on either the principal-factor or Kaiser's varimax criterion.

Without loss of generality, we assume $g(\theta)$ to be multivariate normal with vector mean 0 and covariance matrix I in the population of respondents. These conventions fix the locations of the item thresholds and the scales of the factor loadings.

Subject to these constraints, the likelihood equations for the intercept and slope parameters of item j are (see Bock, 1989):

$$\sum_{\ell}^{s} r_\ell \int_{R(m)} \left(\frac{u_{\ell j} - P_j(\theta)}{P_j(\theta)[1 - P_j(\theta)]} \right) \frac{\partial P_j(\theta)}{\partial \zeta_j} \frac{L_\ell(\theta) g(\theta)}{\bar{P}_\ell} d\theta = 0 \, . \tag{6}$$

We observe that (5) is a weighted sum of posterior expectations of a nonlinear function of θ, given the response patterns, u_ℓ. For an EM solution of these equations, Bock and Aitkin (1981), and Bock, Gibbons, and Muraki (1988), sum under the integral sign and write (5) in the form of the likelihood equations for probit analysis with $m + 1$ independent variables:

$$\int_{R(m)} \left(\frac{\bar{r}_j - NP_j(\theta)}{P_j(\theta)[1 - P_j(\theta)]} \right) \frac{\partial P_j(\theta)}{\partial \zeta_j} d\theta = 0 \tag{7}$$

where

$$\bar{r}_j = \sum_{\ell}^{s} r_\ell u_{\ell_j} \frac{L_\ell(\theta) g(\theta)}{\bar{P}_\ell} \tag{8}$$

and

$$\bar{N} = \sum_{\ell}^{s} r_{\ell} \frac{L_{\ell}(\theta)g(\theta)}{\bar{P}_{\ell}} \tag{9}$$

The E-step of the EM algorithm then consists of item-by-item evaluations of (7) and (8), given a provisional value of ζ_j, and the M-step consists of a solution of (6) by m-fold gaussian quadrature and Fisher scoring.

For factor loadings in the typical range of 0.4 to 0.7, and number of items less than 20, the integrand in (7) is relatively smooth over the range of θ. The integral can then be evaluated with good accuracy by m-fold Gauss-Hermite quadrature with a small number of points per dimension. Bock, Gibbons, and Muraki (1988) report good recovery of item parameters from five-factor simulated data when the number of points per dimension is as small as 3. But when the number of items is large or some of the factor loadings near 1, the posterior distributions may become so concentrated that the integrand of (7) becomes "lumpy" and therefore difficult to evaluate by quadrature with a small number of points. Schilling (1993) demonstrated this problem by comparing the results of item factor analyses by 3-point quadratures in 5 dimensions with Gibbs sampling analysis of the same data.

The problem is even more severe when evaluating the marginal probabilities, $\bar{P}_{\ell}$. Their evaluation does not benefit from the sum over the answer patterns as does estimation of the item parameters. Good estimates of the $\bar{P}_{\ell}$ are critical, however, because they appear in the likelihood ratio criterion for terminating the stepwise addition of factors (see Example 2).

An alternative approach which avoids this difficulty is to make use of equation (5) and integrate adaptively before summing. That is, choose points that are optimal for quadrature of each posterior distribution rather than for the population distribution. Accurate quadrature of each term in the sum will be possible with relatively few points, as will the evaluation of the corresponding marginal probabilities. The computations will be somewhat heavier because the response functions, $P_j(\theta)$, must be evaluated at arbitrary points rather than at the fixed points of the Bock-Aitkin solution, and the M-step must be carried out with ungrouped data. These computational disadvantages are more than compensated, however, by the reduction of the number of quadrature points required in high dimensions when the quadrature points are chosen optimally for each posterior distribution.

In this paper, we examine the properties of solutions by adaptive quadrature that employ small numbers of favorably placed fixed points. Up to five dimensions, we use quadratures with three points per dimension. For five dimensions and above, we use certain fractions of these points as described below.

3 Adapting the Quadratures to the Posterior Distributions

In IRT, it is well-known that normal ogive and logistic response functions of tests with relatively few items yield posterior distributions of θ that are approximately normal, and the approximation improves rapidly as the number of items increases (see Bock & Mislevy, 1982). Applying this result to the multiple-factor case, we base the adaptation on estimated first and second-order moments of the posterior. Suppose the m-vector posterior mean given pattern is μ_ℓ and the corresponding covariance matrix is Σ_ℓ, $|\Sigma_\ell| > 0$.

To center a set of points, $X_h, h = 1, 2, \ldots, q^m$, on the posterior, we transform to

$$X^*_{\ell h} = T_\ell X_h + \mu_\ell \,, \tag{10}$$

where T_ℓ is the (lower triangular) Cholesky decomposition (see Bock, 1985, p. 85) of Σ_ℓ (that is, $\Sigma_\ell = T_\ell T'_\ell$).

To evaluate moments of the posterior and corresponding marginal probabilities of the response patterns by quadrature, we employ the multivariate version of the method introduced by Naylor and Smith (1982). The procedure is as follows. Suppose we have provisional estimates of the conditional mean and covariance matrix, say μ_ℓ and Σ_ℓ, of θ, given the pattern u_ℓ. For the evaluation of, say.

$$\begin{aligned} \bar{P}_\ell &= \int_{R(m)} f_\ell(\theta) d(\theta) \\ &= \int_{R(m)} L_\ell(\theta) g(\theta) d(\theta) \,, \end{aligned}$$

we observe that if the posterior density,

$$\frac{L_\ell(\theta)g(\theta)}{\tilde{P}_\ell(\theta)} \cong \mathcal{N}(\mu_\ell, \Sigma_\ell) ,$$

is approximately normal, we might expect the ratio $f_l(\theta)/\mathcal{N}(\mu_\ell, \Sigma_\ell)$ to be well-approximated by a low-degree polynomial. In that case,

$$\tilde{P}_\ell = \int_{R(m)} \left(\frac{f_\ell(\theta)}{\mathcal{N}(\mu_\ell, \Sigma_\ell)}\right) \mathcal{N}(\mu_\ell, \Sigma_\ell) d\theta$$

could be evaluated exactly with relatively few points per dimension by m-fold Gauss-Hermite quadrature, with the Hermite polynomials defined for a normal error function rather than the gaussian error function, $e^{-\theta^2}$. The quadrature formula is

$$\tilde{P}_\ell = (2\pi)^{m/2} |\Sigma_\ell|^{\frac{1}{2}} \sum_{h_m}^{q} \cdots \sum_{h_2}^{q} \sum_{h_1}^{q} L_\ell(X^*_{\ell h}) e^{\frac{1}{2} X'_{\ell h} X_{\ell h}} \prod_{j}^{m} A(X_{\ell h_j}) \, . \tag{11}$$

where $X^*_{\ell h}$ is a quadrature point in the transformed space and $A(X_{\ell h_k})$ is the corresponding weight. This formula has q^m terms. Moment functions, such as the terms in the likelihood equations (5), or μ_ℓ and Σ_ℓ, can be evaluated in a similar manner.

In the solution of the likelihood equations, we approximate the provisional μ_ℓ and Σ_ℓ by the (Newton-Gauss) posterior mode and inverse information matrix. These are obtained by a Fisher scoring solution of the MAP equation,

$$\sum_{j}^{n} \frac{u_{\ell j} - P_j(\tilde{\theta})}{P_j(\tilde{\theta})[1 - P_j(\tilde{\theta})]} \cdot \frac{\partial P_j(\tilde{\theta})}{\partial \theta} - \Sigma_\ell^{-1} \mu_\ell = 0 \tag{12}$$

using the information matrix

$$I(\tilde{\theta}_\ell) = \sum_{j}^{n} w_j a'_j a_j + \Sigma_\ell^{-1} , \tag{13}$$

where $\tilde{\theta}_\ell$ is the solution of (11) and $w_j = [\partial P_j(\tilde{\theta}_\ell)/\partial\theta]^2 / P_j(\tilde{\theta}_\ell)[1 - P_j(\tilde{\theta}_\ell)]$, for $a_j = [a_{j1}, a_{j2}, \ldots a_{jm}]$. Starting values for $\tilde{\theta}$ for the scoring iterations are the provisional values from the previous EM cycle. In the first EM cycle, $u_\ell = 0$ and $\sum_\ell = I$.

The final EM cycle, $\tilde{\theta}_\ell$ provides the estimated factor scores associated with pattern ℓ, and the square roots of the diagonal elements of the corresponding posterior covariance matrix serve as the standard errors of the respective scores.

The test of significance of an added factor is based on the difference of the likelihood ratio chi-square,

$$\chi^2_{LR} = -2 \sum_{\ell}^{s} r_\ell \log \frac{r_\ell}{\bar{P}_\ell}, \tag{14}$$

between an m and $m+1$ factor model. The degrees of freedom are $n - m$. If the sample size is sufficiently large that $s = 2^n$, or nearly so, (13) provides a test of the factor model against the general multinomial alternative.

4 Fractional Quadrature

The method of adaptive provides accurate evaluation of the likelihood equations and the marginal probabilities when the number of items is large, but it does not solve the problem of the number of quadrature points increasing exponentially with the number of factors. A similar difficulty occurs with the number of treatment combinations in factorial experimentation, however, and its solution by the introduction of fractional factorial designs has potential for full-information factor analysis. If the outcome of a factorial experiment is subject only to first- and second-order effects, a fractional design that leaves main effects and two-factor interactions unconfounded is capable of estimating them. To the extent that the posterior is determined by first- and second-order moments, an equivalent choice of points for adaptive quadrature should satisfactorily evaluate the marginal quantities in the likelihood equations and ratio.

An attractive set of designs for this purpose, based on three points per dimension, is available in Connor and Zelen (1956). Depending on the choice of defining contrasts, many such designs are possible, but their choices are perhaps as good as any. Five is the minimum number of dimensions for which a 1/3 replicate design (81 points) with unconfounded two-factor interactions is possible. In $d = 6$ or more dimensions, similar $1/3^{7-d}$ fractions can be constructed, always with 243 points. For MML estimation in item factor analysis, these designs effectively remove the exponential barrier to high-dimension numerical integration.

Points for these designs are given by Connor and Zelen (1956), but for computing purposes we prefer to generate them as needed from their defining contrasts.[1] The block confounding in these designs is not required in the present application.

5 Example 1

As a test of the procedure based on adaptive fractional quadrature, we simulated responses of 1000 subjects to a five-factor test with 32 items. The signs of the generating factor loadings were those of the first five columns of a 2^5 Hadamard matrix. The absolute values of the elements in columns 1 through 5 were 0.5, 0.37, 0.34, 0.31, 0.27, respectively. The item thresholds were all set at zero, and the factor scores were drawn from an $N(0, I)$ distribution.

To assess the accuracy of recovering the generating factor loadings, we performed full-information factor analyses of the simulated data by the following four methods:

1. Schilling's (1993) Monte Carlo EM method using an 81-point Gibbs sampler.
2. Bock-Aitkin (1981) nonadaptive Gauss-Hermite quadrature with 3 points per dimension (243 points in total).
3. Adaptive Gauss-Hermite quadrature with 3 points per dimension (243).
4. Fractional Gauss-Hermite quadrature with 3 points per dimension (one-third fraction; 81 points).

Table 1 shows the root mean square errors in recovering the loadings of each of the factors. The nonadaptive method (2) is appreciably less accurate than the others for all factors. Full Gauss-Hermite adaptive quadrature (3) has about the same accuracy as the Gibbs sampling method (1), but it requires three times as many points. Finally, the fractional Gauss-Hermite adaptive quadrature (4) has about the same accuracy as methods 1 and 3.

[1] We are indebted to Nickolay Trendafilov for supplying the computer routine for this purpose.

TABLE 1

Root Mean Square Errors of Recovering Generating Factor Loadings: Five-factor simulated data, 32-items, $N = 1{,}000$

Method	Factor 1	2	3	4	5
1. 81-point Monte Carlo EM	.0441	.0839	.1153	.0698	.0616
2. Nonadaptive Gauss-Hermite	.1373	.1400	.1117	.1346	.1061
3. Adaptive Gauss-Hermite	.0459	.0794	.1152	.0739	.0605
4. Fractional adaptive G-H	.0463	.1248	.1140	.0710	.0727

To compare the relative performance of fractional and full Gauss-Hermite quadrature in estimating the marginal probabilities of the patterns, we evaluated these probabilities by both methods for all patterns in the sample. Examining the distribution of the ratios of the former to the latter, we found the mean to be 1.007587 and the standard deviation to be 0.005726. The largest and smallest ratios were 1.0231 and 0.9932, respectively.

6 Example 2

To test MML estimation beyond the range beyond which full Gauss-Hermite quadrature is possible, we simulated a test of 128 items for which the factor loadings were the first eight columns of a Hadamard matrix of order 2^7. The magnitude of loadings for the successive factors were 0.5, 0.37, 0.34, 0.31, 0.27, 0.23, 0.19, and 0.15. Thus, the uniqueness for each variable was 0.2170. All thresholds were set a 0.

Ten thousand values of θ were drawn randomly from the eight variate unit normal distribution and the item responses simulated. Item factor analyses were carried out by Gibbs sampler EM with 243 randomly drawn points and with fractional Gauss-Hermite quadrature using a 1/27 replicate of the 3^8 factorial design. The root mean square errors in recovering the generating

TABLE 2
Root Mean Square Errors of Recovering Generating Factor Loadings: Eight-factor simulated data, 128-items, $N = 10{,}000$

Method	Factor 1 / 5	2 / 6	3 / 7	4 / 8
1. Monte Carlo EM	.0037	.0034	.0034	.0018
	.0016	.0015	.0015	.0011
2. Fractional adaptive G-H	.0069	.0053	.0028	.0015
	.0017	.0013	.0009	.0009

factor loadings are shown in Table 2.

The Monte Carlo results for the two largest principle factors are somewhat more accurate that those of the fractional adaptive Gauss-Hermite quadrature, but they are generally less accurate for the remaining factors. A possible reason of this ordering of the results is that the random points are more concentrated in the dimensions with greater variance, whereas the fractional design positions the quadrature points uniformly in all dimensions. These results seem favorable to exploratory analysis, where identifying minor dimensions is of primary interest.

The Gauss-Hermite quadrature also has the advantage of providing the estimates of the response pattern probabilities required in computing the likelihood ratio test of the significance of association accounted for by an added factor. For the eight-factor data, the change of negative log likelihood between the seventh and eighth factor was:

$$\begin{array}{r} 34745.46 \\ -31689.33 \\ \hline 3056.33 \end{array}$$

Twice this difference is distributed in large samples as a chi-square on degrees of freedom equal to the number of free parameters added to the model, 120 in this instance. With a sample size of 10,000, the analysis easily detected the eighth factor.

7 Discussion

Although numerous other marginalization methods have been proposed for Bayes and marginal maximum likelihood estimation, quadrature formulas have the advantage of being robust, nonstochastic, and better characterized mathematically. Up to now their use has been confined to low dimensional problems where the computational burden is manageable. For the large class of prolems where posterior (or predictive) distributions can be specified by first and second moments, however, the method of adaptive fractional quadrature described here overcomes the barrier of high dimensionality without appreciable loss of numerical accuracy. It has potential application to all nonlinear mixed and random effects models, of which item factor analysis is an important example.

REFERENCES

Bock, R. D. (1985 reprint). *Multivariate statistical methods in behavioral research.* Chicago: Scientific Software International.

Bock, R. D. (1989). Measurement of human variation: A two-stage model. In R. D. Bock (Ed.), *Multilevel analysis of educational data* (pp. 319–342). New York: Academic Press.

Bock, R. D. & Aitkin, M. (1981). Marginal maximum likelihood estimation of item parameters: Application of an EM algorithm. *Psychometrika,* **46**, 443–445.

Bock, R. D., Gibbons, R. D., & Muraki, E. (1988). Full information item factor analysis. *Applied Psychological Measurement,* **12**, (**3**), 261–280.

Bock, R. D., & Mislevy, R. J. (1982). Adaptive EAP estimation of ability in a microcomputer environment. *Applied Psychological Measurement,* **6**, 431–444.

Christoffersson, A. (1975). Factor analysis of dichotomized variables. *Psychometrika,* **40**, 5–32.

Connor, W. S. & Zelen, M. (1956). *Fractional factorial experimental designs for factors at three levels.* Washington, DC: U. S. Bureau of Standards.

Householder, A. S. (1964). *The theory of matrices in numerical analysis.* New York: Blaisdell.

Muthén, B. (1978). Contributions to factor analysis of dichotomized variables. *Psychometrika,* **43**, 551–560.

Naylor, J. C., & Smith, A. F. M. (1982). Applications of a method for the efficient computation of posterior distributions. *Applied Statistics,* **31**, 214–225.

Schilling, S. G. (1993). Advances in full-information factor analysis using the Gibbs sampler. University of Chicago, Department of Psychology. Unpublished dissertation.

Thurstone, L. L. (1947). *Multiple factor analysis.* Chicago: University of Chicago Press.

Dynamic Factor Models for the Analysis of Ordered Categorical Panel Data

Gerhard Arminger
Bergische Universität - GH Wuppertal

Abstract

Panel studies are conducted in the social and behavioral sciences to observe and to model structural relationships between variables in a population as well as the dynamic development of these variables over time. Examples of panel studies are the Panel Study of Income Dynamics (PSID), the German Socio-Economic Panel (GSOEP), the European Innovation Panel for the Manufacturing Industry and the Longitudinal Study of American Youth (LSAY). Data are typically collected using questionnaires yielding observed variables that may be continuous or limited dependent (censored, dichotomous, ordered or unordered categorical).

The models of Jöreskog and Molenaar are extended to deal with the specific problems of autoregressive models for panel data with ordered categorical outcomes. First, an autoregressive model is formulated for latent variables. Second, a dynamic factor model is given for measurements of the latent variables in the first step. Third, threshold models are used to take the ordered categorical data into account. Fourth, the special identification problems of such models such as equality restrictions on thresholds and factor loadings for ordered categorical variables are discussed. Fifth, an estimation procedure based on Muthén's estimation of conditional polychoric correlation coefficients is given. Finally, the model and the estimation method are illustrated using data on math scores and attitudes toward math from the LSAY panel.

*This paper has been prepared for the conference on causality and latent variables, March 18 - 19, 1994, University of California at Los Angeles. The author is grateful to the Graduate School of Education, UCLA, for the opportunity to be a visiting professor during the winter quarter 1994. Helpful comments and suggestions have been provided by Maia Berkane, Bengt Muthén and an unknown referee. The author is especially indebted to Ginger Nelson Goff who has prepared the data set analyzed in the paper.
The author's address is Gerhard Arminger, Bergische Universitt Wuppertal, Department of Economics (FB 6), D - 42097 Wuppertal, Germany.

1 Introduction

In the social and behaviorial sciences it is now common to collect data from a sample of n independent elements such as individuals, households or companies over a number of T times points, denoted as longitudinal or panel or repeated observation studies. The sample is often thought to be representative for a population, the number of time points is usually fairly small, ranging from $T = 2$ in intervention and $T = 20$ in prospective studies. The aim of collecting such data rather than cross-sectional data is to describe not only structural relationships between variables in the population but also to describe the development of variables over time. Additionally, one hopes that the observation and modelling of temporal development on one hand and the fact that at least in linear models unobserved individual heterogeneity can be eliminated by taking differences on the other hand yield more valid conclusion about possible causes than the analysis of cross-sectional data (a thorough discussion of the relationship between statistical models and causality is found in Sobel (1995). Examples for large scale panel studies are household panels such as the American Panel Study of Income Dynamics (PSID, Hill 1992), the German Socio-Economic Panel (GSOEP, Wagner, Schupp and Rendtel, 1991), enterprise panels such as the Community Innnovation Survey of the European Community (CIS, Eurostat 1994), the German Business Test Panel of the Munich IFO Institute (cf. Arminger and Ronning 1991) and panels in educational research such as the Longitudinal Study of American Youth (LSAY, Miller, Hoffer, Suchner, Brown and Nelson 1992). The variables collected (for instance, employment status, income, turnover, number of employees, test scores, attitudes toward school, aspirations) are usually a mixture of (quasi)continuous and limited dependent variables that may be censored, dichotomous, ordered or unordered categorical.

Standard statistical models often used for the analysis of panel data are found in the monographs of Chamberlain (1984), and Hsiao (1986) and the surveys of Hsiao (1995) and Hamerle and Ronning (1995). Most of the papers on analyzing panel data focus on the analysis of one dependent variable which may be directly observed over time. Already such analysis may prove theoretically and numerically difficult if the dependent variable is dichotomous (cf. Heckman 1981a,b, Chamberlain 1984) or unordered categorical (Fahrmeir and Tutz 1994, Keane 1994). In this paper, the focus is on descriptive models for multivariate outcomes in panel data such as item batteries used in the social sciences or errors in variable models in econometrics when more than one indicator is available. Such models have been proposed for continuous variables by Jöreskog for panel data (1978) and by Molenaar (1985) and Molenaar, De Gooijer and Schmitz (1992) for multivariate time series data in the context of LISREL model (Jöreskog and Sörbom 1993). These models are extended to deal with dichotomous, ordered categorical, censored metric and classified metric data based on the assumption of underlying multivariate normality as in Muthén (1984) and for panel models in Arminger (1994).

2 Model Specification

In this section, a special dynamic model is formulated which extends Molenaar's (1985) model to panel data.

2.1 An autoregressive model in the underlying variable

An autoregressive model for an underlying unobserved variable $\eta_{it} \sim p \times 1$ is given where i indexes the individual and t the time point:

$$\eta_{it} = \alpha^{(t)} + \sum_{j=1}^{r} B_j^{(t)} \eta_{i,t-j} + \Gamma^{(t)} x_{it} + \zeta_{it} \tag{1}$$

- $x_{it} \sim q \times 1$ is the vector of regressors which may include the former history of x.
- $\alpha^{(t)} \sim p \times 1$ is the vector of regression constants. In general, the regression constants may change over time and are therefore indexed by (t).
- $B_j^{(t)} \sim p \times p$ is the matrix of regression coefficients on the lagged values of $\eta_{i,t}$. Like the regression constants, $B_j^{(t)}$ may change over time. The index j denotes regression on $\eta_{i,t-j}$ with lag j. If j is set to 0, the model includes simultaneous relations between the components of $\eta_{i,t}$. The usefulness of such simultaneous relations may be in doubt and is therefore excluded.
- $\Gamma^{(t)} \sim p \times q$ is the matrix of regression coefficients on x_{it}. Again $\Gamma^{(t)}$ may change over time.
- $\zeta_{it} \sim p \times 1$ is the vector of disturbances with $E(\zeta_{it}) = 0$. $V(\zeta_{it}) = \Psi^{(t)}$, $E(\zeta_{it}\zeta_{jt}') = 0$ for $i \neq j$ $E(\zeta_{it}\zeta_{is}') = 0$ for $t \neq s$, and $E(\zeta_{it}\zeta_{js}') = 0$ for $i \neq j, t \neq s$. In addition, normality of $\zeta_{it} \sim \mathcal{N}(0, \Psi^{(t)})$ is assumed.

The model of (1) is a standard AR(r) model with exogenous variables x_{it} if $\alpha^{(t)}, B_j^{(t)}, \Gamma^{(t)}$ and $\Psi^{(t)}$ are not indexed by (t). If only time series data were available, the parameters in $\alpha^{(t)}, B_j^{(t)}, \Gamma^{(t)}$ and $\Psi^{(t)}$ could only be identified and estimated if they remain constant over time. However, in panel data, additional information is available because not only one individual but a whole sample is observed over time. In this model, homogeneity of coefficients over the sample at each time point is assumed. Therefore, consistent estimates of $\alpha^{(t)}, B_j^{(t)}, \Gamma^{(t)}$ and $\Psi^{(t)}$ may be obtained as n may go to infinity for each time point. Since $\alpha^{(t)}$ and $\Psi^{(t)}$ may vary over time, weak stationarity is not assumed. In intervention research, the possibility that the coefficients vary over time is particularly important, since an intervention may change the mean and the variance as well as regression relations for a dependent variable.

The model (1) assumes strict exogeneity of the error terms, therefore not allowing unobserved heterogeneity unless all coefficient matrices of $B_j^{(t)}$ are set to 0. This assumption may be relaxed to include unobserved heterogeneity across individuals in the form $\zeta_{it} = \nu_i + \delta_{it}$ or an autoregressive model for ζ_{it} if no state dependence is assumed (cf. Arminger 1994). State dependence exists if the latent variable η_{it} is connected to the observed variable y_{it} for instance by the threshold observation rule for dichotomous outcomes with $y_{it} = 1$ if $\eta_{it} > 0$ and $y_{it} = 0$ if $\eta_{it} \leq 0$ and the observed variable $y_{i,t-j}$ is included as a regressor in (1). (cf. Heckman 1981a,b).

2.2 A dynamic factor analytic measurement model

The unobserved variable η_{it} drives the multivariate time series $y_{it}^* \sim m \times 1$ in a dynamic factor model (Molenaar 1985):

$$y_{it}^* = \kappa^{(t)} + \sum_{l=0}^{s} \Lambda_l^{(t)} \eta_{i,t-l} + \epsilon_{it} \tag{2}$$

- $\kappa^{(t)} \sim m \times 1$ is the vector of regression constants of y_{it}^*. Again $\kappa^{(t)}$ may vary over time.
- $\Lambda_l^{(t)} \sim m \times p$ is the matrix of regression coefficients (or factor loadings) of y_{it}^* on $\eta_{i,t-l}$. The coefficients in $\Lambda_l^{(t)}$ may vary over time.
- $\epsilon_{it} \sim m \times 1$ is the vector of measurement errors with $E(\epsilon_{it}) = 0, E(\epsilon_{it}\epsilon_{it}') = \Theta^{(t,t)} =$ diag $\{\vartheta_{11}^{(t,t)}, \vartheta_{22}^{(t,t)} \ldots, \vartheta_{mm}^{(t,t)}\}, E(\epsilon_{it}\epsilon_{jt}') = 0$ for $i \neq j, E(\epsilon_{it}\epsilon_{is}') = \Theta^{(t,s)} =$ diag $\{\vartheta_{11}^{(t,s)}, \vartheta_{22}^{(t,s)} \ldots, \vartheta_{mm}^{(t,s)}\}$ and $E(\epsilon_{it}\epsilon_{js}') = 0$ for $i \neq j$ and $t \neq s$. The covariance matrices $\Theta^{(t,s)}$ may vary over time. The errors $\epsilon_i = \{\epsilon_{it}\}$ for all t follow a multivariate normal distribution.

The model in (2) is a simple extension of Molenaar's (1985) dynamic factor analysis model. Since Molenaar analyzes long individual time series which are assumed to be covariance stationary, $\kappa^{(t)}$ is set to 0 and $\Lambda_l^{(t)}$ and $\Theta^{(t,t)}$ do not depend on time. In addition $\Theta^{(t,s)}$ depends only on the lag $(t-s)$ and not on the specific time point t itself. If one uses panel data instead of time series data, the parameters $\kappa^{(t)}, \Lambda_l^{(t)}, \Theta^{(t,t)}$ and $\Theta^{(t,s)}$ may be identified and consistently estimated using the information from the sample of individuals. However, in most applications it will be desirable to restrict the parameters using substantive considerations such as equality of factor loadings over time. It should be noted that in the usual measurement model of y_{it}^* is given by:

$$y_{it}^* = \kappa^{(t)} + \Lambda_0^{(t)} \eta_{it} + \epsilon_{it} \tag{3}$$

This model corresponds to a classical measurement model for η_{it}. The dynamic factor model includes not only η_{it} but also the lagged variables $\eta_{i,t-l}$.$\Lambda_l^{(t)}$ can therefore be interpreted as a causal filter as in time series analysis.

2.3 The reduced form

Substituting model (1) into (2) yields a reduced form model for the vector $y_i^* = (y_{i1}^{*\prime}, y_{i2}^{*\prime}, \ldots, y_{iT}^{*\prime})'$ which is linear in $x_i \sim q^* \times 1 = (x_{i1}', x_{i2}', \ldots, x_{iT}')'$.

$$y_i^* = \gamma + \Pi x_i + \delta_i \tag{4}$$

The reduced form parameters $\gamma \sim mT \times 1$ and $\Pi \sim mT \times q^*$ are functions of $\alpha^{(t)}, B_j^{(t)}, \Gamma^{(t)}, \kappa^{(t)}$ and $\Lambda_l^{(t)}$. δ_i is a $mT \times 1$ random variable. Each component y_{ith}^*, that is the outcome of variable y_{ih}^* at time t for element i, may be written as

$$y_{ith}^* = \mu_{ith} + \delta_{ith} \tag{5}$$

where $\mu_{ith} = \gamma_{th} + \Pi_{th} x_i$. μ_{ith} is the fixed part of y_{ith}^* and δ_{ith} is a random variable consisting of linear combinations of ζ_{is} and ϵ_{is}, $s \leq t$. Therefore δ_{ith} is distributed as a normal random variable with $E(\delta_{ith}) = 0$, variance $V(\delta_{ith}) = \sigma_{hh}^{(t)}$ and covariances $Cov(\delta_{ith}, \delta_{isl}) = \sigma_{hl}^{(ts)}$. The covariance structure is a function of the parameters in (1) and (2). As an example, a simple first order autoregressive and dynamic factor-analytic model for panel data with three time points is considered. The autoregressive model in the latent variables η_{it} is given by:

$$\begin{aligned}
\eta_{i0} &= \zeta_{i0} \\
\eta_{i1} &= \alpha^{(1)} + B_1^{(1)} \eta_{i0} + \zeta_{i1} \\
\eta_{i2} &= \alpha^{(2)} + B_1^{(2)} \eta_{i1} + \zeta_{i2} \qquad (6) \\
\eta_{i3} &= \alpha^{(3)} + B_1^{(3)} \eta_{i2} + \zeta_{i2}
\end{aligned}$$

(7)

The dynamic factor-analytic measurement model is:

$$\begin{aligned}
y_{i1}^* &= \kappa^{(1)} + \Lambda_1^{(1)} \eta_{i0} + \Lambda_0^{(1)} \eta_{i1} + \epsilon_{i1} \\
y_{i2}^* &= \kappa^{(2)} + \Lambda_1^{(2)} \eta_{i1} + \Lambda_0^{(2)} \eta_{i2} + \epsilon_{i2} \qquad (8) \\
y_{i3}^* &= \kappa^{(3)} + \Lambda_1^{(3)} \eta_{i2} + \Lambda_0^{(3)} \eta_{i3} + \epsilon_{i3} \qquad (9)
\end{aligned}$$

(10)

The corresponding matrix formulation with variables $\eta_i = (\eta_{i0}', \eta_{i1}', \eta_{i2}', \eta_{i3}')'$, $x_i = (x_{i1}', x_{i2}', x_{i3}')'$ and $\zeta_i = (\zeta_{i0}', \zeta_{i1}', \zeta_{i2}', \zeta_{i3}')'$ is given by

$$\eta = \alpha + B\eta + \Gamma x + \zeta \tag{11}$$

with parameter vectors and matrices defined as

$$\alpha = \begin{pmatrix} 0 \\ \alpha^{(1)} \\ \alpha^{(2)} \\ \alpha^{(3)} \end{pmatrix}, \quad B = \begin{pmatrix} 0 & 0 & 0 & 0 \\ B_1^{(1)} & 0 & 0 & 0 \\ 0 & B_1^{(2)} & 0 & 0 \\ 0 & 0 & B_1^{(3)} & 0 \end{pmatrix} \tag{12}$$

$$\Gamma = \begin{pmatrix} 0 & 0 & 0 \\ \Gamma^{(1)} & 0 & 0 \\ 0 & \Gamma^{(2)} & 0 \\ 0 & 0 & \Gamma^{(3)} \end{pmatrix}, \quad \Psi = \begin{pmatrix} \Psi^{(0)} & 0 & 0 & 0 \\ 0 & \Psi^{(1)} & 0 & 0 \\ 0 & 0 & \Psi^{(2)} & 0 \\ 0 & 0 & 0 & \Psi^{(3)} \end{pmatrix} \tag{13}$$

The dynamic factor-analytic model for $y_i^* = (y_{i1}^{*\prime}, y_{i2}^{*\prime}, y_{i3}^{*\prime})'$ and $\epsilon_i = (\epsilon_{i1}', \epsilon_{i2}', \epsilon_{i3}')'$ may be written as

$$y_i^* = \kappa + \Lambda\eta_i + \epsilon_i \tag{14}$$

with parameter vectors and matrices

$$\kappa = \begin{pmatrix} \kappa^{(1)} \\ \kappa^{(2)} \\ \kappa^{(3)} \end{pmatrix}, \quad \Lambda = \begin{pmatrix} \Lambda_1^{(1)} & \Lambda_0^{(1)} & 0 & 0 \\ 0 & \Lambda_1^{(2)} & \Lambda_0^{(2)} & 0 \\ 0 & 0 & \Lambda_1^{(3)} & \Lambda_0^{(3)} \end{pmatrix}, \quad \Theta = \begin{pmatrix} \Theta^{(1,1)} & \Theta^{(1,2)} & \Theta^{(1,3)} \\ \Theta^{(2,1)} & \Theta^{(2,2)} & \Theta^{(2,3)} \\ \Theta^{(3,1)} & \Theta^{(3,2)} & \Theta^{(3,3)} \end{pmatrix} \tag{15}$$

The reduced form of the model is therefore given by:

$$y_i^* = \kappa + \Lambda(I-B)^{-1}\alpha + \Lambda(I-B)^{-1}\Gamma x_i + \Lambda(I-B)^{-1}\zeta_i + \epsilon_i \tag{16}$$

This formulation yields the conditional expected value

$$E(y_i^*|x_i) = \kappa + \Lambda(I-B)^{-1}\alpha + \Lambda(I-B)^{-1}\Gamma x_i \tag{17}$$

and the conditional covariance matrix

$$V(y_i^*|x_i) = \Lambda(I-B)^{-1}\Psi(I-B)^{-1\prime}\Lambda' + \Theta \tag{18}$$

The reduced form of the model formulation shows that — as in general mean and covariance structure models — restrictions are necessary to guarantee first and second order identifiability. The parameters in α and κ are not jointly identified. Therefore, either α or κ must be restriced. The covariance matrix Ψ^0 and the matrix $\Lambda_1^{(1)}$ of factor loadings cannot be identified because no indicator variables $y_{i,0}^*$ have been observed for the time point 0 before the first values are observed. Hence, assumptions about Ψ^0 and $\Lambda_1^{(1)}$ like $\Psi^0 = \Psi^1$ and $\Lambda_1^{(1)} = \Lambda_1^{(2)}$ must be made. These restrictions also solve the initial values problem because only latent variables appear as lagged variables in the model.

2.4 Threshold models for dependent variables

To include limited dependent variables, the following threshold models are used to connect the individual variable y^*_{ith} with the observed variable y_{ith} (cf. Muthén 1984, Ksters 1987). For convenience, the case index $i = 1, \ldots, n$ is omitted

- y_{th} is metric (identity relation):

$$y_{th} = y^*_{th} \tag{19}$$

- y_{th} is ordered categorical with unknown thresholds $\tau^{(t)}_{h,1} < \tau^{(t)}_{h,2} < \ldots < \tau^{(t)}_{h,K_h}$ and categories $y_{th} = 1, \ldots, K_h + 1$ (ordinal probit relation, McKelvey and Zavoina 1975):

$$y_{th} = l \Leftrightarrow y^*_{th} \in (\tau^{(t)}_{h,l-1}, \tau^{(t)}_{h,l}] \text{ with } \tau^{(t)}_{h,0} = -\infty \text{ and } (\tau^{(t)}_{h,K_h}, \tau^{(t)}_{h,K_h+1}] = (\tau^{(t)}_{h,K_h}, \infty] \tag{20}$$

  Additional identification restrictions are discussed in the next section. Since the thresholds $\tau^{(t)}_{h,l}$ define the categories of the observed variables, it may be necessary to restrict $\tau^{(t)}_{h,l}$ over time. Otherwise, the meaning of the categories of a Likert scale is assumed to change over time.

- Classified or grouped metric variables may be treated in analogy to the ordinal probit case with the difference that the boundaries for each class are now used as known thresholds (Stewart 1983). No identification restrictions are necessary.

- y_{th} is one-sided censored with a threshold $\tau^{(t)}_{h,1}$ known a priori (tobit relation, Tobin 1958).

$$y_{th} = \begin{cases} y^*_{th} & \text{if } y^*_{th} > \tau^{(t)}_{h,1} \\ \tau^{(t)}_{h,1} & \text{if } y^*_{th} \leq \tau^{(t)}_{h,1} \end{cases} \tag{21}$$

- y_{th} is double-sided censored with a priori known threshold values $\tau^{(t)}_{h,1} < \tau^{(t)}_{h,2}$ (two-limit probit relation, Rosett and Nelson 1975).

$$y_{th} = \begin{cases} \tau^{(t)}_{h,1} & \text{if } y^*_{th} \leq \tau^{(t)}_{h,1} \\ y^*_{th} & \text{if } \tau^{(t)}_{h,1} < y^*_{th} < \tau^{(t)}_{h,2} \\ \tau^{(t)}_{h,2} & \text{if } \tau^{(t)}_{h,2} \leq y^*_{th} \end{cases} \tag{22}$$

In applications, all of the threshold models may be used for each component y_{ith}, thereby allowing to deal simultaneously with continuous, censored, dichotomous and ordered categorical variables.

3 Identification restrictions for ordered categorical probit regression

For the moment a univariate ordered categorical variable y_{i1} with outcomes $k = 1, \ldots, K+1$ is considered which depends on x_{i1} through the linear model

$$y_{i1}^* = \gamma_1 + \Pi_{1 \cdot} x_{i1} + \delta_{i1}, \ \delta_{i1} \sim \mathcal{N}(0, \sigma_{11}) \tag{23}$$

and the observation rule $y_{i1} = k \Leftrightarrow \tau_{k-1}^{(1)} < y_{i1}^* \leq \tau_k^{(1)}, \ k = 1, \ldots, K+1$.
The probability that $y_{i1} = k$ is therefore given by

$$Pr(y_{i1} = k) = Pr(\tau_{k-1}^{(1)} < y_{i1}^* \leq \tau_k^{(1)}) = Pr(y_{i1}^* \leq \tau_k) - Pr(y_{i1}^* \leq \tau_{k-1}) \tag{24}$$

Considering only $Pr(y_{i1}^* \leq \tau_k)$ yields through the normality assumption:

$$Pr(y_{i1}^* \leq \tau_k^{(1)}) = Pr(\delta_{i1} \leq \tau_k^{(1)} - \gamma_1 - \Pi_{1 \cdot} x_{i1}) = \Phi\left(\frac{\tau_k^{(1)} - \gamma_1 - \Pi_{1 \cdot} x_{i1}}{\sqrt{\sigma_{11}}}\right) \tag{25}$$

This probability does not change if a scalar c is added to $\tau_k^{(1)}$ and γ_1 and if the numerator and the denominator are both multiplied by a scalar d. Therefore, only the difference $\tau_k^{(1)} - \gamma_1$ is identified and $\tau_k^{(1)}, \gamma_1$ and $\Pi_{1 \cdot}$ are identified only up to a scalar. For identification, usually $\tau_1^{(1)}$ is set to 0 and σ_{11} is set to 1. For dichotomous data, this identification rule yields the probit model.

If, in addition to y_{i1}, the same variable is observed a second time as y_{i2} and the linear model for y_{i2}^* is given by

$$y_{i2}^* = \gamma_2 + \Pi_{2 \cdot} x_{i2} + \delta_{i2}, \delta_{i2} \sim \mathcal{N}(0, \sigma_{22}) \tag{26}$$

then the probability $Pr(y_{i2}^* \leq \tau_k^{(2)})$ may be written as:

$$Pr(y_{i2}^* \leq \tau_k^{(2)}) = \Phi\left(\frac{\tau_k^{(2)} - \gamma_2 - \Pi_{2 \cdot} x_{i2}}{\sqrt{\sigma_{22}}}\right) \tag{27}$$

The same identification problems occur as for $\tau_k^{(1)}, \gamma_1$ and $\Pi_{1 \cdot}$. Even, if $\tau_1^{(2)}$ is set to 0 as before, equality constraints like $\tau_k^{(1)} = \tau_k^{(2)}, k > 2, \gamma_1 = \gamma_2$ and $\Pi_{1 \cdot} = \Pi_{2 \cdot}$ cannot be tested if σ_{22} is unknown. On the one hand, setting σ_{22} equal to $\sigma_{11} = 1$ may not be useful since the errors may show different variances over time. Then, only proportionality restrictions such as $\tau_k^{(2)} = \tau_k^{(1)}/\sqrt{\sigma_{22}}$, $\gamma_2 = \gamma_1/\sqrt{\sigma_{22}}$ and $\Pi_{2 \cdot} = \Pi_{1 \cdot}/\sqrt{\sigma_{22}}$ can be formulated and tested (cf. Sobel and Arminger 1992). On the other hand, the proportionality constant $\nu^{(2)} = 1/\sqrt{\sigma_{22}}$ can be estimated from the ordinal probit regression for y_{i2} if it is assumed that $\tau_k^{(2)} = \tau_k^{(1)}$ and/or $\gamma_2 = \gamma_1$ and/or $\Pi_{2\cdot} = \Pi_{1\cdot}$. In this case one finds:

$$Pr(y_{i2}^* \leq \tau_k^{(1)}) = \Phi\left(\frac{\tau_k^{(1)} - \gamma_1 - \Pi_{1\cdot}x_{i2}}{\sqrt{\sigma_{22}}}\right) = \Phi\left(\nu^{(2)}\tau_k^{(1)} - \nu^{(2)}\gamma_1 - \nu^{(2)}\Pi_{1\cdot}x_{i2}\right) \quad (28)$$

If more than two panel waves are observed, a proportionality coefficient $\nu^{(t)}$ for each wave $t > 1$ will be used. The assumption that σ_{11} and σ_{22} are equal and can therefore be set to 1 is meaningful if covariance stationarity is assumed. This assumption will therefore be important for time series of ordinal data. However, in panel data, the variances σ_{tt} may depend on the time point t because σ_{tt} here describes the conditional variance of the population at time t.

4 Sequential marginal ML estimation

The estimation of the parameters of the models discussed in section 2 is usually performed with full information or sequential marginal limited information maximum likelihood estimation methods. For general mean and covariance structures it is assumed that a $P \times 1$ vector y_i^* of latent dependent variables follows a multivariate normal distribution with conditional mean and covariance:

$$E(y_i^*|x_i) = \gamma(\vartheta) + \Pi(\vartheta)x_i, \quad V(y_i^*|x_i) = \Sigma(\vartheta) \quad (29)$$

In the analysis of panel data, P equals T if a univariate dependent variable is analyzed or P equals $m \cdot T$ if a multivariate dependent variable is analyzed. $\gamma(\vartheta)$ is a $P \times 1$ vector of regression constants and $\Pi(\vartheta)$ is a $P \times R$ matrix of reduced form regression coefficients. x_i is a $R \times 1$ vector of explanatory variables. $\Sigma(\vartheta)$ is the $P \times P$ covariance matrix of the errors of the reduced form. ϑ is the $\bar{q} \times 1$ vector of structural parameters to be estimated. The reduced form parameters $\gamma(\vartheta), \Pi(\vartheta)$ and $\Sigma(\vartheta)$ are continously differentiable functions of a common vector ϑ.

The estimation of the structural parameter vector and the threshold parameters collected in ϑ from the observed data vector y_i proceeds in three stages. (Muthén 1984, Küsters 1987). Computation of the estimates with the MECOSA program is described in Schepers and Arminger (1992).

1. In the first stage the threshold parameters τ, the reduced form coefficients γ and Π of the regression equation, and the reduced form error variance σ_l^2 of the lth equation are estimated using marginal maximum likelihood. Note that this first stage is the estimation of the mean structure without restrictions as in equation (29). The parameters to be estimated in the lth equation are the thresholds denoted by the vector

τ_l, the regression constant denoted by γ_l, the regression coefficients, i.e. the lth row of Π denoted by $\Pi_{l.}$ and the variance denoted by σ_l^2. The marginal estimation is performed using ordinary univariate regression, tobit and ordinal probit regression.

2. In the second stage the problem is to estimate the covariances of the error terms in the reduced form equations. Note that in this stage the covariances are estimated without parametric restrictions. Since the errors are assumed to be normally distributed and strongly consistent estimators of the reduced form coefficients have already been obtained in the first stage, the estimation problem reduces to maximizing the loglikelihood function

$$l_{lj}(\sigma_{lj}) = \sum_{i=1}^{n} \ln P(y_{il}, y_{ij}|x_i, \hat{\tau}_l, \hat{\gamma}_l, \hat{\Pi}_{l.}, \hat{\sigma}_l^2, \hat{\tau}_j, \hat{\gamma}_j, \hat{\Pi}_{j.}, \hat{\sigma}_j^2, \sigma_{lj}) \quad , \tag{30}$$

in which $P(y_{il}, y_{ij}|x_i, \hat{\tau}_l, \hat{\gamma}_l, \hat{\Pi}_{l.}, \hat{\sigma}_l^2, \hat{\tau}_j, \hat{\gamma}_j, \hat{\Pi}_{j.}, \hat{\sigma}_j^2, \sigma_{lj})$ is the bivariate probability of y_{il} and y_{ij} given x_i and the reduced form coefficients. A typical example of this bivariate probability is the case when y_l and y_j are both ordinal. Then the probability that $y_{il} = a$ and $y_{ij} = b$ is given by:

$$P(y_{il} = a, y_{ij} = b|x_i) = \int_{\hat{\tau}_{l,(a-1)}}^{\hat{\tau}_{l,(a)}} \int_{\hat{\tau}_{j,(b-1)}}^{\hat{\tau}_{j,(b)}} \varphi(y_l^*, y_j^*|\hat{\mu}_{il}, \hat{\sigma}_l^2, \hat{\mu}_{ij}, \hat{\sigma}_j^2, \sigma_{lj}) dy_j^* dy_l^* \tag{31}$$

in which $\hat{\mu}_{il} = \hat{\gamma}_l + \hat{\Pi}_{l.} x_i$, $\hat{\mu}_{ij} = \hat{\gamma}_j + \hat{\Pi}_{j.} x_i$ and $\varphi(y_l^*, y_j^*|\mu_l, \sigma_l^2, \mu_j, \sigma_j^2, \sigma_{lj})$ is the bivariate normal density function with covariance σ_{lj}. Note that in the ordinal case $\hat{\sigma}_l^2 = \hat{\sigma}_j^2 = 1$. Hence, σ_{lj} is a correlation coefficient which is called the polychoric correlation coefficient. The loglikelihood function $l_{lj}(\sigma_{lj})$ has to be modified accordingly if variables with other measurement levels are used. Note that the covariances σ_{lj} are the covariances of the error terms in the equations for y_l^*, $l = 1, \ldots, P$ conditional on x_i.

The estimated thresholds $\hat{\tau}_l$, the reduced form coefficients $\hat{\gamma}_l$ and $\hat{\Pi}_{l.}$, the variances $\hat{\sigma}_l^2$ and the covariances $\hat{\sigma}_{lj}$ from all equations are then collected in a vector $\hat{\omega}_n$ which depends on the sample size n. For the final estimation stage, a strongly consistent estimate of the asymptotic covariance matrix W of $\hat{\omega}_n$ is computed. This estimate is denoted by $\hat{W}_n$. The asymptotic covariance matrix W is difficult to derive since the estimates of $\hat{\sigma}_{lj}$ of the second stage depend on the estimated coefficients $\hat{\tau}_f, \hat{\gamma}_f, \hat{\Pi}_{f.}, \hat{\sigma}_f^2, f = l, j$ of the first stage. The various elements of the asymptotic covariance matrix W are given in Küsters (1987). The estimate $\hat{W}_n$ is computed in MECOSA by using analytical first order and numerical second order derivatives of the first and second stage loglikelihood function.

3. In the third stage the vector ω_n of thresholds, the reduced form regression coefficients and the reduced form covariance matrix is written as a function of the structural parameters of interest, collected in the parameter vector ϑ. The parameter vector ϑ is then estimated by minimizing the quadratic form

$$Q_n(\vartheta) = (\hat{\omega}_n - \omega(\vartheta))' \hat{W}_n^{-1} (\hat{\omega}_n - \omega(\vartheta)) \tag{32}$$

which is a minimum distance approach based on the asymptotic normality of the estimators of the reduced form coefficients. The vector $\hat{\omega}_n$ is asymptotically normal with expected value $\omega(\vartheta)$ and covariance matrix W. Since $\hat{W}_n$ is a strongly consistent estimate of W the quadratic form $Q_n(\vartheta)$ is centrally χ^2 distributed with $\tilde{p}-\tilde{q}$ degrees of freedom if the model is specified correctly and the sample size is sufficiently large. The number $\tilde{p}$ indicates the number of elements in $\hat{\omega}_n$ while $\tilde{q}$ is the number of elements in ϑ. The computation of $\hat{W}_n$ is quite cumbersome for models with many parameters. The function $Q_n(\vartheta)$ is minimized using the Davidon-Fletcher-Powell algorithm with numerical first derivatives.

The program MECOSA follows these three estimation stages. In the third stage, the facilities of GAUSS are fully exploited to be able to estimate parameters under arbitrary restrictions. The parameter vector $\omega(\vartheta)$ may be defined using the matrix language and the procedure facility of GAUSS. Consequently, arbitrary restrictions can be placed on $\tau(\vartheta), \gamma(\vartheta), \Pi(\vartheta)$ and $\Sigma(\vartheta)$ including the restrictions placed on $\Sigma(\vartheta)$ in the analysis of panel data.

The MECOSA program provides estimates of the reduced form parameters $\Pi(\vartheta)$ in the first stage by regressing all dependent variables in $y_i^* = (y_{i1}^*, \dots, y_{iT}^*)'$ on all regressors collected in x_i without the usual restriction that the dependence of $y_{i,t-j}$, $j \geq 1$ on x_{it} is zero in panel data. If these restrictions are necessary, one can trick the program into restricted estimation already in the first stage by running the first stage of MECOSA T times separately for each y_{it}^* which is then regressed on x_{it}. In the second stage, the standard batch input of MECOSA has to be corrected to deal with the T estimation results from the first stage. The polychoric covariances and/or correlation coefficients of the reduced form are then computed under the restrictions of the first stage. Algorithmic details are found in Arminger (1992).

5 Development of math scores and attitudes toward math in the Longitudinal Study of American Youth

To illustrate the models, the identification problems and the estimation method discussed in the section 2, 3, 4 and 5, data from the Longitudinal Study of American Youth (LSAY, citation) are used. The dependent variables of interests are math score (M) and attitude toward math (A). The math score is observed directly, the attitude is measured by three items, I1, I2 and I3. These items are five point Likert scales with I1 "how much do you like your math class", I2 "how clearly is your math class teacher explaining the material" and I3 "how difficult or easy is the math course for you" with the lower categories indicating likes and the higher categories indicating dislikes. The math scores were measured in fall 1987, fall 1988, fall 1989 and fall 1990 and are denoted by 1M, 3M, 5M and 7M. The attitude variables were collected in fall 1987, spring 1988, spring 1989, fall 1989, spring 1990 and fall 1990. These variables are denoted by 1I1, 1I2, 1I3, 2I1, 2I2, 2I3, 4I1, 4I2, 4I3, 5I1, 5I2, 5I3, 6I1, 6I2, 6I3 and 7I1, 7I2, 7I3. Note that the math scores were collected at time points 1, 3, 5 and 7 while the attitude variables were not collected at time points 1, 2, 4, 5, 6 and 7. At time point 3, no attitude variables were collected. The sample size of the data set for the analysis is 1519. As background variable, the parents socioeconomic status (SES) is used.

After some preliminary analysis, the following model has been used to describe the development of these variables over time. First the measurement model for the attitude variable $\eta_{i2}^{(t)}$ at time point 1 is discussed:

$$\begin{pmatrix} y_{i1}^{(1)*} \\ y_{i2}^{(1)*} \\ y_{i3}^{(1)*} \end{pmatrix} = \begin{pmatrix} \kappa_1^{(1)} \\ \kappa_2^{(1)} \\ \kappa_3^{(1)} \end{pmatrix} + \begin{pmatrix} 1 \\ \lambda_2 \\ \lambda_3 \end{pmatrix} \eta_{i2}^{(1)} + \begin{pmatrix} \epsilon_{i1}^{(1)} \\ \epsilon_{i2}^{(1)} \\ \epsilon_{i3}^{(1)} \end{pmatrix} \tag{33}$$

$y_{ij}^{(1)*}, j = 1,2,3$ denotes the underlying variables for the ordered categorical indicators 1I1, 1I2 and 1I3. $\eta_{i2}^{(1)}$ is the attitude variable at time 1, $\epsilon_{ij}^{(1)}, j = 1,2,3$ denotes the measurement errors for the three variables. The scale of $\eta_{i2}^{(1)}$ is set to the scale of $y_{i1}^{(1)*}$ by fixing the first factor loading to 1. λ_2 and λ_3 are the factor loadings for item 1I2 and 1I3. The parameters $\kappa_j^{(1)}, j = 1,2,3$ are the regression constants for the items. The variances of $\epsilon_{ij}^{(1)}, j = 1,2,3$ are not free, but fixed parameters determined by the fact that the reduced form for each equation is estimated under the identification restriction that the reduced form variance is equal to 1. The covariances between the measurement errors at the same time point are set to 0 according to the classical factor analytic model (Lawley and Maxwell 1971).

The measurement model for the time points 2, 4, 5, 6 and 7 is given by:

$$\begin{pmatrix} y_{i1}^{(t)*} \\ y_{i2}^{(t)*} \\ y_{i3}^{(t)*} \end{pmatrix} = \begin{pmatrix} \kappa_1^{(t)} \\ \kappa_2^{(t)} \\ \kappa_3^{(t)} \end{pmatrix} + \begin{pmatrix} \nu_1^{(t)} \\ \nu_2^{(t)} \cdot \lambda_2 \\ \nu_3^{(t)} \cdot \lambda_3 \end{pmatrix} \eta_{i2}^{(t)} + \begin{pmatrix} \epsilon_{i1}^{(t)} \\ \epsilon_{i2}^{(t)} \\ \epsilon_{i3}^{(t)} \end{pmatrix} \tag{34}$$

The random variables and parameters are constructed in the same way as in (33). Since the items are supposed to have the same connotation over time, it is asumed that the factor loadings are fixed over time. However, since the regression parameters are only identified up to scale in the reduced form, the assumption of equality has to be modified by multiplying the factor loadings with a proportionality constant $\nu_j^{(t)}$ for each j for the time points $t > 1$. The regression constants are allowed to vary over time. The error variables $\epsilon_{ij}^{(t)}$ are assumed to be correlated with $\epsilon_{ij}^{(s)}$ over all time points for $j = 1,2,3$. The assumption of equality is made not only for the factor loadings but also for the thresholds in the ordered categorical threshold model. If the thresholds for item $j, j = 1,2,3$ at time 1 are denoted by $\tau_2^{(j)}, \tau_3^{(j)}, \tau_4^{(j)}$ for the five point Likert scales with $\tau_0^{(j)} = -\infty, \tau_1^{(j)} = 0$ and $\tau_5^{(j)} = +\infty$, then the corresponding thresholds for time point $t > 1$ are given by multiplying the thresholds $\tau_k^{(j)}$, with the proportionality constant $\nu_j^{(t)}, t > 1$.

The factor loadings of $y_{ij}^{(t)*}$ on $\eta_{i2}^{(t-j)}, j > 1$ have been set to 0. Hence, a simple measurement model without autoregressive structure is assumed. The autoregression is captured in the structural equation model for the variable $\eta^{(i1}$ for math scores and the variable $\eta_{i2}^{(t)}$ for the attitude variable. The autoregressive structure is depicted graphically as follows:

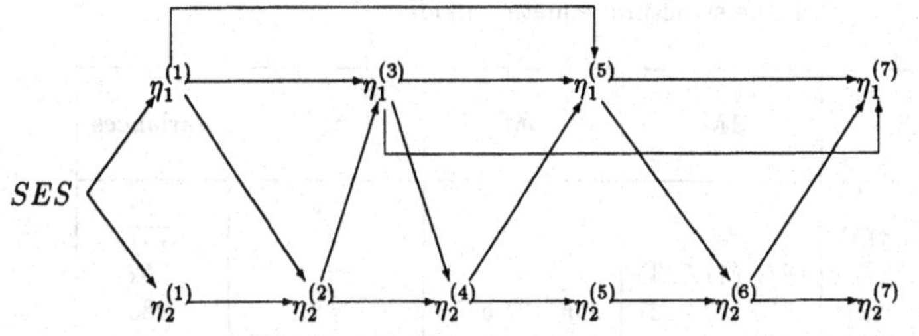

FIGURE 1. Autoregressive structure of math scores and attitude toward math.

It is assumed that the regression coefficients of $\eta_1^{(1)}$ and $\eta_2^{(t)}$ on SES are non-zero for all $t > 1$. In this autoregressive model, SES is a background variable for both math score and attitude at all time points. (These arrows are not indicated in figure 1.) The math score variable is assumed to follow a second order autoregressive process. The attitude variable is assumed to follow a first order autoregressive process. In addition, it is assumed that the score and the attitude variable are dependent on each other, however they are related through the lagged variables rather than by a contemporaneous relationship. The errors are assumed to be uncorrelated over both variables and the time points. Since no specific model for the initial values has been assumed except for dependence on SES, the regression coefficients have to be interpreted given the outcomes at time 1,2 and 3. Hence, no additional restrictions have been made on the parameters. The results for the structural equation model are given in Table 1.

Table 1 Part I: Regression coefficients, z-values and error variances in the structural equation model

	SES	1M	3M	5M	Variances
1M	.399 (13.315)				.757
3M	.003 (.132)	.871 (47.723)			.252
5M	.058 (2.982)	.375 (17.252)	.606 (27.635)		.286
7M	.085 (3.711)		.357 (22.985)	.634 (37.567)	.449
1A	-.042 (-1.303)				.705
2A	.031 (1.125)	-.073 (-3.322)			.321
4A	.044 (1.479)		-.013 (-0.616)		.389
5A	.054 (1.756)				.518
6A	-.011 (.444)			.002 (.126)	.293
7A	0.06 (2.189)				.514

NOTE: The dependent variables are in the row, the explanatory variables are in the column of the table. The variances are added as the last column. The z-values are given in parentheses. M stands for the math score, A for attitude. The time points are again denoted by 1, 2, 3, 4, 5, 6 and 7.

Table 1 Part II: Regression coefficients and z-values in the structural equation model

	1A	2A	4A	5A	6A
1M					
3M		-.013(-0.705)			
5M			-.004(-0.225)		
7M					-0.064(-3.153)
1A					
2A	.645(25.557)				
4A		.578(19.514)			
5A			.523(15.769)		
6A				.735(31.624)	
7A					.404(14.895)

The estimated regression coefficients reveal strong autocorrelation for each of the dependent variables. For the math score, both the first order and the second order autoregressive coefficients are highly significant (test level $\alpha = 0.05$) and positive. The regression coefficients are of equal size allowing to add further restrictions. For the attitude variable, the first order coefficients are high and of about equal magnitude. The parameter from 2A to 4A has to be interpreted differently since time point 3 in missing for η_2. The relationship between math score and attitude, as captured in the lagged regression coefficients of 1M to 2A, 3M to 4A, 5M to 6A, 2A to 3M, and so forth is practically negligeable. Therefore, no dependence of math score and attitude on each other over time is shown by the model. The regression coefficients for SES are strong only at time 1 for the math score. Given 1M, all other regression coefficients for SES are quite small. The highest error variances are found for the initial values 1M and 1A as is to be expected.

The threshold parameters for the ordered categorical variables are given in Table 2.

Table 2: Threshold parameters for the ordered categorical items

	τ_2	τ_3	τ_4
1I1	.677	1.171	1.418
1I2	.618	1.051	1.337
1I3	.722	1.319	1.666

The differences in the threshold parameters indicate that the distance between threshold $\tau_1 = 0$ and τ_2 are about twice as large as the distance between τ_2 and τ_3 and τ_3 and τ_4 implying that the use of the coding 1 to 5 for the ordered categorical items might lead to model misspecification.

The proportionality constants are given in Table 3.

Table 3: Proportionality constants $\nu_j^{(t)}, j = 1, 2, 3$. z-values for $H_0: \nu_j^{(t)} = 1$ are given in parentheses

Times	I1	I2	I3
2	1.016(0.475)	.964(-0.946)	.939(-2.055)
4	1.039(1.026)	1.020(0.483)	.957(-1.410)
5	.991(-0.248)	.956(-1.144)	.912(-2.832)
6	1.071(1.842)	1.042(1.060)	.944(-1.766)
7	1.121(3.045)	1.040(0.990)	1.009(0.280)

Of the 15 z-values given in this table, only 3 are different from 1 on a $\alpha = 0.05$ test level. Hence, it may be surmised that the proportionality constants could be set to 1 indicating that the variance of the measurement errors for the attitude items does not vary over time, given that thresholds and factor loadings are assumed not to change over time. The factor loadings are given in Table 4.

Table 4: Factor loadings. z-values in parentheses

Item	Loading
1I1	1
1I2	0.818 (26.538)
1I3	0.776 (28.697)

The values for the regression constants are not of interest and are therefore not given. The error variances of the measurement model vary between 0.213 and 0.678 indicating a low overall measurement error for the items given that t he reduced form variances are set to 1. The serial covariances between the measurement errors for each items are very low (0.028 to .293).

The number of parameters in the reduced form for 23 variables is 335, the number of estimated parameters in the model is 131 yielding 204 degrees of freedom. The model was estimated using MECOSA (Schepers and Arminger 1992). The chi-square minimum distance statistic is 725.872 indicating lack of fit. However, the root mean square error of approximation (RMSEA) (Browne and Cudeck 1993) is equal to 0.041. Using the rule of thumb that RMSEA should be less than 0.05 the model provides a sufficently good approximation to the true vector of reduced form coefficients.

References

Arminger, G., and Ronning, G. (1991), Ein Strukturmodell für Preis-, Produktions- und Lagerhaltungsentscheidungen von Firmen, *IFO-STUDIEN, Zeitschrift für empirische Wirtschaftsforschung*, 37, 229-254.

Arminger, G. (1992), Analyzing Panel Data with Non-Metric Dependent Variables: Probit Models, Generalized Estimating Equations, Missing Data and Absorbing States, *Discussion Paper No. 59*, Deutsches Institut für Wirtschaftsforschung, Berlin.

Arminger, G. (1994), Probit-Models for the Analysis of Non-Metric Panel Data, *Allgemeines Statistisches Archiv*, Vol. 78, No. 1, 125 - 140.

Browne, M.E., and Cudeck, R., (1993), Alternative Ways of Assessing Model Fit, in Bollen, K.A., and Long, J.S. (eds.), *Testing Structural Equation Models*, Newbury Park, Sage, 136-162.

Chamberlain, G., (1984), Panel Data, in Z. Griliches and M.D. Intriligator (eds.), *Handbook of Econometrics*, Vol. II, North-Holland, Amsterdam, 1246 - 1318.

Eurostat (1994), *REDIS News, The Newsletter of Research and Development and Innovation Statistics, Vol. 2, No. 2*, Luxemburg.

Fahrmeir, L., and Tutz, G., (1994), *Multivariate Statistical Modelling Based on Generalized Linear Models* forthcoming in Springer-Verlag, Heidelberg.

Hamerle, A., and Ronning, G. (1995), Panel Analysis for Qualitative Variables, forthcoming in G. Arminger, C. C. Clogg and M. E. Sobel (eds.), *Handbook of Statistical Modeling for the Behavioral Sciences*, New York: Plenum.

Heckman, J. J. (1981a), Statistical Models for Discrete Panel Data, in C. F. Manski and D. McFadden (eds), *Structural Analysis of Discrete Data with Econometric Applications*, 114-178.

Heckman, J. J. (1981b), The Incidental Parameters Problem and the Problem of Initial Conditions in Estimating a Discrete Time-Discrete Stochastic Process, in C. F. Manski and D. McFadden (eds), *Structural Analysis of Discrete Data with Econometric Applications*, 179-195.

Hill, M. (1992), *The Panel Study of Income Dynamics – A User's Guide*, Newbury Park, Sage.

Hsiao, C. (1986), *Analysis of Panel Data*,Cambridge, Mass.: Cambridge University Press.

Hsiao, C. (1995), Panel Analysis for Metric Data, forthcoming in G. Arminger, C.C. Clogg and M.E. Sobel (eds.), *Handbook of Statistical Modeling in the Social and Behavioral Sciences*, New York, Plenum.

Jöreskog, K. G. (1978), An Econometric Model for Multivariate Panel Data, *Annales de l'INSEE*, 30/31, 355 - 366.

Jöreskog, K. G., and Sörbom, D. (1993), *LISREL 8: Structural Equation Modeling with the SIMPLIS Command Language*, Hillsdale, Lawrence Erlbaum.

Keane, M.P., (1994), A Computationally Practical Simulation Estimator for Panel Data, *Econometrica*, Vol. 62, No. 1, 95 - 116.

Küusters, U. (1987), *Hierarchische Mittelwert- und Kovarianzstrukturmodelle mit nichtmetrischen endogenen Variablen.* Heidelberg: Physica Verlag.

McKelvey, R. D., and Zavoina, W. (1975), A Statistical Model for the Analysis of Ordinal Level Dependent Variables. *Journal of Mathematical Sociology 4*, 103–120.

Miller, J.D., Hoffer, T., Suchner, R.W., Brown, K.G., and Nelson, C., (1992), *LSAY Codebook: Student, Parent, and Teacher Data for Cohort Two for Longitudinal Years One through Four (1987 - 1991)*, Vol. 2, Northern Illinois University, De Kalb, 60115 -

2854.

Molenaar, P. C. M., (1985), A Dynamic Factor Model for the Analysis of Multivariate Time Series, *Psychometrika*, Vol. 50, No. 2, 181 - 202.

Molenaar, P. C. M., De Gooijer, J. G., and Schmitz, B. (1992), Dynamic Factor Analysis of Nonstationary Multivariate Time Series, *Psychometrika*, Vol. 57, No. 3, 333 - 349.

Muthén, B. (1984), A General Structural Equation Model with Dichotomous, Ordered Categorical, and Continuous Latent Variable Indicators. *Psychometrika 49*, 115–132.

Rosett, R.N. and Nelson, F.D. (1975). Estimation of the two-limit probit regression model. *Econometrica 43*, 141–146.

Schepers, A. and Arminger, G. (1992), MECOSA: *A Program for the Analysis of General Mean- and Covariance Structures with Non-Metric Variables, User Guide*, SLI-AG, Züricher Str. 300, CH-8500 Frauenfeld, Switzerland.

Sobel, M. E. (1995), Causal Inference in the Social and Behavioral Sciences, forthcoming in G. Arminger, C.C. Clogg and M.E. Sobel (eds.), *Handbook of Statistical Modeling in the Social and Behavioral Sciences*, New York, Plenum.

Sobel, M. and Arminger, G. (1992), Modeling Household Fertility Decisions: A Nonlinear Simultaneous Probit Model, *Journal of the American Statistical Association*, 87, 38-47.

Stewart, M. B. (1983), On Least Squares Estimation when the Dependent Variable is Grouped. *Review of Economic Studies L*, 737–753.

Tobin, J. (1958), Estimation of Relationships for Limited Dependent Variables. *Econometrica 26*, 24–36.

Wagner, G., Schupp, J., and Rendtel, U. (1991). The Socio-Economic Panel (SOEP) for Germany - *Methods of Production and Management of Longitudinal Data*, Discussion Paper No. 31a, DIW Berlin.

Model fitting procedures for nonlinear factor analysis using the errors-in-variables parameterization

Yasuo Amemiya and Ilker Yalcin
Iowa State University

Abstract

Traditional factor analysis and structural equation modeling use models which are linear in latent variables. Here, a general parametric nonlinear factor analysis model is introduced. The identification problem for the model is discussed, and the errors-in-variables parametrization is proposed as a solution. Two general procedures for fitting the model are described. Tests for the goodness of fit of the model are also proposed. The usefulness and comparison of the model fitting procedures are studied based on a simulation.

1 Introduction

Factor analysis and more general structural equation modeling have been popular statistical tools in social and behavioral sciences. Traditional models in these types of analyses are linear in underlying factors or latent variables. However, practical need for more general nonlinear models have been pointed out, especially in behavioral sciences. Capabilities to fit and assess models nonlinear in unobservable factors greatly enrich the available tools for exploratory multivariate analysis. When the subject-matter theory suggests nonlinear relationships, fitting such nonlinear models is much more desirable than the use of ad-hoc linear models providing only unsatisfactory approximate explanation. In addition, consideration of nonlinear models may solve the often-cited problem in behavioral and social sciences that a proposed linear model often fail to fit the data well or fit the data only with more factors than expected.

There has been a limited literature on nonlinear factor analysis. The development of nonlinear factor analysis was started by McDonald (1962,1967a,1967b). The work was followed up by Etezadi-Amoli and McDonald (1983) and Moijaart and Bentler (1986). This series of work was concerned with polynomial or power relationships, and its model fitting procedures were based on sample moments. Statistical properties of these estimation methods have not been discussed in the literature. Amemiya (1993a,b) introduced a more general nonlinear model and discussed its identification problem. He also developed an instrumental variable technique for fitting a single equation at a time in the system of the general nonlinear factor analysis equations. He used the so-called small$-\sigma$ asymptotics to develop the theory supporting the method and to motivate an estimator which corrects for the bias due to the nonlinearity. See also Yalcin and Amemiya (1993) and Yalcin (1995).

For individuals $t = 1, 2, ..., n$, let $\mathbf{Z}_t$ be a $p \times 1$ vector of observations. A general parametric nonlinear factor analysis model for $\mathbf{Z}_t$ can be written as

$$\mathbf{Z}_t = \mathbf{g}^*(\mathbf{f}_t^*; \beta^*) + \epsilon_t, \qquad t = 1, 2, ..., n, \tag{1}$$

where $\mathbf{f}_t^*$ is a $k \times 1$ underlying unobservable random factor vector, $\mathbf{g}^*$ is a p–variate function of $\mathbf{f}_t^*$ indexed by an unknown parameter vector β^*, and ϵ_t is a $p \times 1$ unobservable error vector with mean zero. As in linear factor analysis we will assume that $\mathbf{f}_t^*$'s and ϵ_t's are independent, and that the p–components of ϵ_t are independent. Thus, $\Psi = \mathrm{Var}\{\epsilon_t\}$ is diagonal.

Some examples of a typical component, say the ith component, of such a function $\mathbf{g}^*$ are

$$\begin{aligned} g_i^*(\mathbf{f}_t^*, \beta^*) &= \beta_{0i}^* + \frac{\beta_{1i}^*}{\beta_{2i}^* + e^{\left(\beta_{3i}^* - \beta_{4i}^* f_{1t}^* - \beta_{5i}^* f_{2t}^*\right)}}, \\ g_i^*(\mathbf{f}_t^*, \beta^*) &= \beta_{0i}^* + \beta_{1i}^* e^{\beta_{2i}^* f_t^*}, \\ g_i^*(\mathbf{f}_t^*, \beta^*) &= \beta_{0i}^* + \beta_{1i}^* f_{1t}^* + \beta_{2i}^* f_{1t}^{*2} + \beta_{3i}^* f_{2t}^* + \beta_{4i}^* f_{2t}^{*2} + \beta_{5i}^* f_{1t}^* f_{2t}^*. \end{aligned}$$

An identification problem exists for model (1) and is discussed in Section 2. For an identifiable version of model (1), two model fitting procedures are introduced in Section 3 and 4. A simulation study is presented in Section 5.

2 Errors-in-variables parametrization

As in the linear model, the nonlinear factor analysis model (1) has an identification problem. The $k \times 1$ factor vector $\mathbf{f}_t^*$ can be transformed by a one-to-one linear transformation $\mathbf{f}_t^{**} = \mathbf{h}(\mathbf{f}_t^*; \beta^*)$, possibly depending on β^*. Then, (1) can be written in an equivalent form

$$\mathbf{Z}_t = \mathbf{g}^*\left(\mathbf{h}^{-1}(\mathbf{f}_t^*; \beta^*)\right) + \epsilon_t = \mathbf{g}^{**}\left(\mathbf{f}_t^{**}; \beta^*\right) + \epsilon_t, \tag{2}$$

with $\mathbf{f}_t^{**}$ as the factor vector. Models (1) and (2) are equivalent nonlinear factor models with two different factor vectors. For model fitting and estimation, this indeterminacy has to be removed. Because of this nonlinear indeterminacy of $\mathbf{f}_t^*$, imposing a particular distributional form for $\mathbf{f}_t^*$ makes little sense. Also, the nonlinear model is generally considered for data with complex appearance and features. Thus, we proceed without a particular distributional assumption on $\mathbf{f}_t^*$ and ϵ_t. This approach follows the asymptotic robustness study for the linear model as given in, e.g., Anderson and Amemiya (1988), Browne and Shapiro (1988), and Amemiya and Anderson (1990).

As discussed in Amemiya (1993a), an identified model without the indeterminacy in (2) is the one written in the errors-in-variables parameterization. The indeterminacy can be removed if there exists a one-to-one nonlinear functional relationship between $\mathbf{f}_t^*$ and some k components of $\mathbf{g}^*$ in the model (1). In such a case, the factor vector $\mathbf{f}_t$ is identified as the "true values" measured by the k corresponding components of $\mathbf{Z}_t$. No generality is lost by placing such k components as the last k components, and so we write $\mathbf{Z}_t = (\mathbf{Y}_t', \mathbf{X}_t')'$ for $(p-k) \times 1$ $\mathbf{Y}_t$ and $k \times 1$ $\mathbf{X}_t$, and write (1) in the form

$$\begin{aligned} \mathbf{Y}_t &= \mathbf{g}(\mathbf{f}_t; \beta) + \mathbf{e}_t, \\ & \qquad\qquad\qquad\qquad t = 1, 2, ..., n, \\ \mathbf{X}_t &= \quad \mathbf{f}_t \quad + \mathbf{u}_t, \end{aligned} \tag{3}$$

where $\mathbf{f}_t$ is the $k \times 1$ factor identified in this way, $\epsilon_t = (\mathbf{e}_t', \mathbf{u}_t')'$, $\mathbf{g}$ is a $(p-k)$-valued nonlinear function of $\mathbf{f}_t$ resulting from this definition of $\mathbf{f}_t$, and $\boldsymbol{\beta}$ is the parameter for such a function $\mathbf{g}$. The original $\boldsymbol{\beta}^*$ may be unidentifiable, but the parameter $\boldsymbol{\beta}$ is assumed to be written with no redundancy so that it is identified. For future reference, we also write $\boldsymbol{\Psi} = \mathrm{Var}\{\epsilon_t\} = \textit{block diag}\ (\boldsymbol{\Psi}_{ee}, \boldsymbol{\Psi}_{uu})$, where $\boldsymbol{\Psi}_{ee}$ and $\boldsymbol{\Psi}_{uu}$ are diagonal, and let $\boldsymbol{\psi}$ be the $p \times 1$ vector consisting of the diagonal elements of $\boldsymbol{\Psi}_{ee}$ and $\boldsymbol{\Psi}_{uu}$. For the general model (1), we only consider the models that can be transformed to those given in (3). This restricts the form of the nonlinear models that can be fitted and estimated. However, the errors-in-variables parameterization allows a straightforward interpretation of the nonlinear model in addition to providing a clearly identified model. Also, for the exploratory and model building purposes, various plotting techniques such as a scatter plot matrix of observed variables can suggest models in such a parametrization. Throughout our development, we also assume that $(p-k)(p-k+1)/2 \geq p$. This condition assumes that the dimension k of the factor vector can not be very large in comparison to the number of observed variables p. This is identical to the requirement for the identification of the linear unrestricted (exploratory) factor analysis. We assume this condition, because the general nonlinear factor model (1) includes such a linear model as a special case, and because it plays a key role in our model fitting and checking procedures. There is no restriction on the dimension of the parameter vector $\boldsymbol{\beta}$ in $\mathbf{g}$. Thus, for example, even with a small k, a polynomial model of any order can be fitted and estimated with parameterization (3).

3 Extended linear maximum likelihood method

An approach to developing model fitting is to use the idea behind a well established and accepted procedure for the linear case and to extend it to the nonlinear model. For the linear model, the most widely used method is the maximum likelihood estimation under the normality assumption. For the linear model such a method produces estimators and test statistics with good properties even for nonnormal data. See, e.g., Anderson and Amemiya (1988), Browne and Shapiro (1988), and Amemiya and Anderson (1990). Hence, we present a fitting procedure for the nonlinear model (3) which reduces to an algorithm for the maximum normal likelihood if the model happens to be linear in factors. The nonlinear model is considered usually for nonnormal data, and a specific distributional assumption makes little sense. Also, even under some assumed distribution for $\mathbf{f}_t$ and ϵ_t, the distribution of observation $\mathbf{Y}_t$ is generally complicated and usually lacks an explicit form of the likelihood. Thus, instead of simply applying the maximum likelihood method, we extend the essential idea underlying a maximum likelihood algorithm for the linear model by taking advantage of the specific structure of the factor analysis model in general. The algorithm mentioned here for the linear model is that suggested in Jöreskog (1967, 1969, 1977) and Pantula and Fuller (1986). The algorithm finds the estimator of the relationship parameter (factor loadings) for a given value of the error variance estimate, obtains the error variance estimate given the relationship parameter by the generalized least squares applied to a transformed moment matrix, and iterates these two steps. We apply this idea to the nonlinear model with a careful development of an appropriate extension and with the use of an approximation to the nonlinear function. Of course, the linear approximation is inappropriate, and a certain form of quadratic approximation is utilized. For this approximation, and due to the nature of the derivatives of a nonlinear function, estimation of factor scores is needed in the iterative process. The estimation of the factor covariance matrix is of no particular interest for the nonlinear model, and is not a part of the procedure. But, an estimator of the factor covariance matrix can be obtained easily once the other intrinsic parameters have been estimated.

To start the procedure iterated over estimates of $\mathbf{f}_t$, $\boldsymbol{\beta}$, and $\boldsymbol{\Psi}$, we need initial values. An initial estimate $\boldsymbol{\beta}^{(0)}$ of $\boldsymbol{\beta}$ can be obtained by combining Amemiya's (1993a, b) single-equation instrumental variable estimates or by minimizing

$$\sum_{t=1}^{n}[\mathbf{Y}_t - \mathbf{g}(\mathbf{X}_t;\boldsymbol{\beta})]'[\mathbf{Y}_t - \mathbf{g}(\mathbf{X}_t;\boldsymbol{\beta})]. \tag{4}$$

An initial estimate $\boldsymbol{\Psi}^{(0)}$ can be obtained by applying the ordinary least squares to

$$\mathbf{m}^{(0)} = \frac{1}{n}\sum_{t=1}^{n}\mathbf{r}_t^{(0)}\mathbf{r}_t^{(0)\prime},$$

where

$$\mathbf{r}_t^{(0)} = \mathbf{Y}_t - \mathbf{g}(\mathbf{X}_t;\boldsymbol{\beta}^{(0)}).$$

That is, for the $p \times 1$ vector $\boldsymbol{\psi}^{(0)}$ satisfying vec $\boldsymbol{\Psi}^{(0)} = \mathbf{L}_p\boldsymbol{\psi}^{(0)}$ with the $p^2 \times p$ known transformation matrix $\mathbf{L}_p$ with zero and one elements,

$$\boldsymbol{\psi}^{(0)} = \left(\mathbf{C}^{(0)\prime}\mathbf{C}^{(0)}\right)^{-1}\mathbf{C}^{(0)\prime}\text{vech}\,\mathbf{m}^{(0)}, \tag{5}$$

where

$$\begin{aligned}
\mathbf{C}^{(0)} &= \mathbf{K}_{p-k}^{+}\{\mathbf{L}_{p-k}, \frac{1}{n}\sum_{t=1}^{n}\left(\mathbf{G}(\mathbf{X}_t;\boldsymbol{\beta}^{(0)})\otimes\mathbf{G}(\mathbf{X}_t;\boldsymbol{\beta}^{(0)})\right)\mathbf{L}_k\},\\
\mathbf{G}(\mathbf{f};\boldsymbol{\beta}) &= \frac{\partial}{\partial\mathbf{f}'}\mathbf{g}(\mathbf{f};\boldsymbol{\beta}),\\
\mathbf{K}_{p-k}^{+} &= \left(\mathbf{K}_{p-k}'\mathbf{K}_{p-k}\right)^{-1}\mathbf{K}_{p-k}',
\end{aligned} \tag{6}$$

and $\mathbf{K}_{p-k}$ is the transformation matrix such that vec$\mathbf{A} = \mathbf{K}_{p-k}$vech$\mathbf{A}$ for a $(p-k)\times(p-k)$ symmetric $\mathbf{A}$. Any negative element of $\boldsymbol{\psi}^{(0)}$ is replaced by zero, and the remaining elements are re-estimated by the reduced regression. An additional modification is performed to assure the upper bound for the elements of $\boldsymbol{\psi}^{(0)}$. We will also use $\mathbf{f}_t^{(0)} = \mathbf{X}_t$ as the initial value for the factor scores.

The i−th step of the iterative procedure consists of the following:

(E1) The $\mathbf{f}_t^{(i)}$ is computed as

$$\mathbf{f}_t^{(i)} = \mathbf{X}_t + \boldsymbol{\Psi}_{uu}^{(i-1)}\hat{\mathbf{G}}_t^{(i-1)\prime}\left(\hat{\boldsymbol{\Sigma}}_{tt}^{(i-1)}\right)^{-1}\mathbf{v}_t^{(i-1)},$$

where

$$\hat{\mathbf{G}}_t^{(i-1)} = \mathbf{G}(\mathbf{f}_t^{(i-1)};\boldsymbol{\beta}^{(i-1)}),$$

$$\hat{\boldsymbol{\Sigma}}_{tt}^{(i-1)} = \frac{n-1}{n}\{\boldsymbol{\Psi}_{ee}{}^{(i-1)} + \hat{\mathbf{G}}_t^{(i-1)}\boldsymbol{\Psi}_{uu}{}^{(i-1)}\hat{\mathbf{G}}_t^{(i-1)\prime}\} + \frac{1}{n}\left[\mathbf{I}_{p-k}, -\hat{\mathbf{G}}_t^{(i-1)}\right]\mathbf{m}_{ZZ}\left[\mathbf{I}_{p-k}, -\hat{\mathbf{G}}_t^{(i-1)}\right]',$$

$$\mathbf{m}_{ZZ} = \frac{1}{n-1}\sum_{t=1}^{n}\left(\mathbf{Z}_t - \bar{\mathbf{Z}}\right)\left(\mathbf{Z}_t - \bar{\mathbf{Z}}\right)', \tag{7}$$

$$\bar{\mathbf{Z}} = \frac{1}{n}\sum_{t=1}^{n}\mathbf{Z}_t,$$

$$\mathbf{v}_t^{(i-1)} = \mathbf{Y}_t - \mathbf{g}(\mathbf{X}_t;\boldsymbol{\beta}^{(i-1)}) + \frac{1}{2}\mathbf{H}\left(\mathbf{f}_t^{(i-1)};\boldsymbol{\beta}^{(i-1)}\right)\mathrm{vec}\boldsymbol{\Psi}_{uu}{}^{(i-1)},$$

$$\mathbf{H}(\mathbf{f};\boldsymbol{\beta}) = [\mathbf{h}_1(\mathbf{f};\boldsymbol{\beta}), \mathbf{h}_2(\mathbf{f};\boldsymbol{\beta}), \ldots, \mathbf{h}_{p-k}(\mathbf{f};\boldsymbol{\beta})]', \tag{8}$$

$$\mathbf{h}_i(\mathbf{f};\boldsymbol{\beta}) = \mathrm{vec}\left[\frac{\partial^2 g_i(\mathbf{f};\boldsymbol{\beta})}{\partial\mathbf{f}\partial\mathbf{f}'}\right],$$

and g_i is the i–th component of $\mathbf{g}$.

(E2) The $\boldsymbol{\beta}^{(i)}$ is the value of $\boldsymbol{\beta}$ that minimizes

$$\sum_{t=1}^{n}\hat{\mathbf{v}}_t'(\boldsymbol{\beta})\hat{\boldsymbol{\Sigma}}_{tt}^{-1}(\boldsymbol{\beta})\hat{\mathbf{v}}_t(\boldsymbol{\beta}),$$

where

$$\hat{\mathbf{v}}_t(\boldsymbol{\beta}) = \mathbf{Y}_t - \mathbf{g}(\mathbf{X}_t;\boldsymbol{\beta}) + \frac{1}{2}\mathbf{H}\left(\mathbf{f}_t^{(i)};\boldsymbol{\beta}\right)\mathrm{vec}\boldsymbol{\Psi}_{uu}{}^{(i-1)},$$

$$\hat{\boldsymbol{\Sigma}}_{tt}(\boldsymbol{\beta}) = \left[\mathbf{I}_{p-k}, -\mathbf{G}(\mathbf{f}_t{}^{(i)};\boldsymbol{\beta})\right]\left(\boldsymbol{\Psi}^{(i-1)} + \frac{1}{n}\mathbf{m}_{ZZ}\right)\left[\mathbf{I}_{p-k}, -\mathbf{G}(\mathbf{f}_t^{(i)};\boldsymbol{\beta})\right]'.$$

(E3) The $\boldsymbol{\Psi}^{(i)}$ is obtained by the generalized least squares applied to

$$\boldsymbol{\gamma}^{(i)} = \mathrm{vech}\,\frac{1}{n}\sum_{t=1}^{n}\hat{\mathbf{v}}_t(\boldsymbol{\beta}^{(i)})\hat{\mathbf{v}}_t'(\boldsymbol{\beta}^{(i)}).$$

For the $p\times 1$ $\boldsymbol{\psi}^{(i)}$ satisfying $\mathrm{vec}\boldsymbol{\Psi}^{(i)} = \mathbf{L}_p\boldsymbol{\psi}^{(i)}$,

$$\boldsymbol{\psi}^{(i)} = \left(\hat{\mathbf{C}}^{(i)\prime}\hat{\mathbf{V}}^{-1}\hat{\mathbf{C}}^{(i)}\right)^{-1}\hat{\mathbf{C}}^{(i)\prime}\hat{\mathbf{V}}^{-1}\boldsymbol{\gamma}^{(i)},$$

where

$$\hat{\mathbf{C}}^{(i)} = \mathbf{K}_q^{+}\{\mathbf{L}_q, \frac{1}{n}\sum_{t=1}^{n}\left(\hat{\mathbf{G}}_t^{(i)}\otimes\hat{\mathbf{G}}_t^{(i)}\right)\mathbf{L}_k\},$$

$$\hat{\mathbf{G}}_t^{(i)} = \mathbf{G}\left(\mathbf{f}_t{}^{(i)};\boldsymbol{\beta}^{(i)}\right),$$

$$\hat{\mathbf{V}} = \frac{2}{n^2}\mathbf{K}_{p-k}^{+}\sum_{t=1}^{n}(\hat{\mathbf{Q}}_{tt}\otimes\hat{\mathbf{Q}}_{tt})\mathbf{K}_{p-k}^{+\prime},$$

$$\hat{\mathbf{Q}}_{tt} = \left[\mathbf{I}_{p-k}, -\hat{\mathbf{G}}_t^{(i)}\right]\left(\boldsymbol{\Psi}^{(i-1)} + \frac{1}{n}\mathbf{m}_{ZZ}\right)\left[\mathbf{I}_{p-k}, -\hat{\mathbf{G}}_t^{(i)}\right]'.$$

(E4) Update $i = i + 1$.

This procedure should be iterated for a few iterations. See, e.g., Carroll and Ruppert (1982). Observe that the minimization in (E2) itself is usually carried out using some iterative procedure. The final step produces the extended linear maximum likelihood (ELM) estimates $\hat{\beta}$, $\hat{\Psi}$, and $\hat{\mathbf{f}}_t$. If desired, the factor covariance matrix $\Phi = \mathrm{Var}\{\mathbf{f}_t\}$ can be estimated by a simple unweighted estimator

$$\hat{\Phi} = \frac{1}{n}\sum_{t=1}^{n}\left[\left(\hat{\mathbf{f}}_t - \bar{\mathbf{f}}\right)\left(\hat{\mathbf{f}}_t - \bar{\mathbf{f}}\right)' + \hat{\Psi}_{uu}\hat{\mathbf{G}}_t'\hat{\Sigma}_{tt}^{-1}\hat{\mathbf{G}}_t\hat{\Psi}_{uu}\right] - \hat{\Psi}_{uu},$$

where

$$\begin{aligned}
\bar{\mathbf{f}} &= \frac{1}{n}\sum_{t=1}^{n}\hat{\mathbf{f}}_t, \\
\hat{\mathbf{G}}_t &= \mathbf{G}\left(\hat{\mathbf{f}}_t; \hat{\beta}\right), \\
\hat{\Sigma}_{tt} &= [\mathbf{I}_{p-k}, -\mathbf{G}_t]\left(\frac{n-1}{n}\hat{\Psi} + \frac{1}{n}\mathbf{m}_{ZZ}\right)[\mathbf{I}_{p-k}, -\mathbf{G}_t]',
\end{aligned}$$

and the modification of Amemiya (1985) should be incorporated to assume the nonnegative definiteness of $\hat{\Phi}$.

In step (E3), any negative element of $\psi^{(i)}$ is replaced by zero, and the remaining elements are re-estimated by the reduced generalized least squares. Also, an upper bound modification is incorporated. This is based on the observation that for the general nonlinear model with no linear relationship the sample covariance matrix $\mathbf{m}_{ZZ}$ estimates the sum of Ψ and a positive definite matrix. First the largest root $\hat{\lambda}$ of $|\Psi^{(i)} - \lambda\mathbf{m}_{ZZ}| = 0$ is obtained. If $\hat{\lambda} < 1 + \frac{1}{n}$, then $\Psi^{(i)}$ is unchanged. If $\hat{\lambda} \geq 1 + \frac{1}{n}$, then we replace $\Psi^{(i)}$ with

$$\frac{1}{\hat{\lambda} - n^{-1}}\Psi^{(i)}. \tag{9}$$

With this modification, $\Psi^{(i)}$ is not "too large" in the sense that the variability due to the factors and nonlinear relationships remains in $\mathbf{m}_{ZZ}$. The same upper bound modification is used for the initial estimate $\Psi^{(0)}$ in (5).

The $n^{-1}\mathbf{m}_{ZZ}$ modification in (E1), (E2), and (E3) are for numerical stability, relevant particularly in small samples. For the linear unrestricted factor model, $\mathbf{f}_t^{(i)}$ is not needed in (E2) and (E3), and so the iteration is only over β and Ψ. For such a linear case, except for the $n^{-1}\mathbf{m}_{ZZ}$ modification and the $\Psi^{(i)}$ upper bound modification, $\hat{\beta}$, $\hat{\Psi}$, and $\hat{\Phi}$ iterated to convergence are in fact the maximum likelihood estimators. Note that the step (E2) for a given value of ψ gives the maximum likelihood estimator of β in the linear models for either fixed or normally distributed factors. The expression for $\hat{\mathbf{v}}_t(\beta)$ in (E2) includes the second derivative bias adjustment term which vanishes for the linear case.

The step (E3) is a generalized least squares applied to a pseudo moment matrix. It is possible to apply the generalized least squares to vech $\hat{\mathbf{v}}_t(\beta^{(i)})\hat{\mathbf{v}}_t'(\beta^{(i)})$ with an individual specific weight. We recommend the pseudo moment approach because of better numerical stability, of much simpler computation, and of the availability of a corresponding goodness of fit test statistic. The generalized least squares residual sum of squares in (E3) at the final iteration serves as a test statistic for the fit of the model, and can be compared to the upper percentage points of the chi-square distribution with $d = \frac{1}{2}(p-k)(p-k+1)-p$ degrees of freedom. This test statistic reduces to the likelihood ratio test statistic for the model fit in the normal linear model case. The originally assumed condition that $d \geq 0$ was used in (E3).

4 Approximate conditional likelihood method

Although the fitting method developed in the previous section makes intuitive sense, the computation is extensive due to the estimation of the factor scores for all individuals at each iteration step. Such individual factor score computation and the evaluation of various quantities at the estimated factor scores may potentially introduce unstability of the algorithm as well as unwelcome variability in small samples. As an alternative, a somewhat simpler fitting method using an approximate conditional likelihood is considered in this section. The method also uses an expansion of the nonlinear function around the observed $\mathbf{X}_t$ instead of $\mathbf{f}_t$ to avoid the computation of factor score estimates. Although we intend to develop a method useful for a large class of distributions, we first assume hypothetically in model (3) that $\mathbf{f}_t \sim N(\boldsymbol{\mu}, \boldsymbol{\Phi})$ and $\boldsymbol{\epsilon}_t \sim N(\mathbf{0}, \boldsymbol{\Psi})$. Since $\mathbf{Y}_t$ is nonnormal for any $\mathbf{g}$ nonlinear in $\mathbf{f}_t$, the likelihood based on $\mathbf{Z}_t$ cannot be written down explicitly. To obtain a workable form of the likelihood, we use a quadratic approximation to the function $\mathbf{g}$ around the observable $\mathbf{X}_t$ as given by

$$\mathbf{g}(\mathbf{f}_t;\boldsymbol{\beta}) \doteq \mathbf{g}(\mathbf{X}_t;\boldsymbol{\beta}) - \mathbf{G}(\mathbf{X}_t;\boldsymbol{\beta})\mathbf{u}_t + \frac{1}{2}\mathbf{H}(\mathbf{X}_t;\boldsymbol{\beta})\text{vec } \boldsymbol{\Psi}_{uu}, \tag{10}$$

where $\mathbf{G}(\mathbf{f};\boldsymbol{\beta})$ and $\mathbf{H}(\mathbf{f};\boldsymbol{\beta})$ are defined in (6) and (8), respectively. The approximation in (10) is based on ignoring the terms of order higher than two and on replacing $\mathbf{u}_t\mathbf{u}_t'$ with its expectation $\boldsymbol{\Psi}_{uu}$. With this approximation, the equation for $\mathbf{Y}_t$ in (3) becomes

$$\mathbf{Y}_t = \mathbf{q}(\mathbf{X}_t;\boldsymbol{\beta}) + \boldsymbol{\eta}_t, \tag{11}$$

where

$$\begin{aligned} \mathbf{q}(\mathbf{X}_t;\boldsymbol{\beta}) &= \mathbf{g}(\mathbf{X}_t;\boldsymbol{\beta}) + \frac{1}{2}\mathbf{H}(\mathbf{X}_t;\boldsymbol{\beta})\text{vec } \boldsymbol{\Psi}_{uu}, \\ \boldsymbol{\eta}_t &= \mathbf{e}_t - \mathbf{G}(\mathbf{X}_t;\boldsymbol{\beta})\mathbf{u}_t. \end{aligned}$$

Using (11) and the conditional distribution of $\boldsymbol{\eta}_t$ given $\mathbf{X}_t$, the approximate conditional distribution of $\mathbf{Y}_t$ given $\mathbf{X}_t$ is

$$N(\boldsymbol{\nu}_t, \boldsymbol{\Gamma}_{tt}),$$

where

$$\begin{aligned} \boldsymbol{\nu}_t &= \mathbf{q}(\mathbf{X}_t;\boldsymbol{\beta}) + \boldsymbol{\Sigma}_{uvt}(\boldsymbol{\Phi} + \boldsymbol{\Psi}_{uu})^{-1}\left(\mathbf{X}_t - \bar{\mathbf{X}}\right), \\ \boldsymbol{\Gamma}_{tt} &= \boldsymbol{\Sigma}_{vvt} - \boldsymbol{\Sigma}_{vut}(\boldsymbol{\Phi} + \boldsymbol{\Psi}_{uu})^{-1}\boldsymbol{\Sigma}_{vut}', \\ \boldsymbol{\Sigma}_{vut} &= -\mathbf{G}(\mathbf{X}_t;\boldsymbol{\beta})\boldsymbol{\Psi}_{uu}, \\ \boldsymbol{\Sigma}_{vvt} &= \boldsymbol{\Psi}_{ee} + \mathbf{G}(\mathbf{X}_t;\boldsymbol{\beta})\boldsymbol{\Psi}_{uu}\mathbf{G}(\mathbf{X}_t;\boldsymbol{\beta})'. \end{aligned}$$

The approximate likelihood function is a product of the densities of $\mathbf{Z}_t$'s, and each density is the product of the marginal density of $\mathbf{X}_t$ and the conditional density of $\mathbf{Y}_t$ given $\mathbf{X}_t$. Thus, the approximate log likelihood is, except for a multiplier $(-1/2)$ and an additive constant,

$$l = l_1 + l_2,$$

where

$$\begin{aligned} l_1 &= \sum_{t=1}^{n} \left[log|\mathbf{\Gamma}_{tt}| + (\mathbf{Y}_t - \boldsymbol{\nu}_t)' \mathbf{\Gamma}_{tt}^{-1} (\mathbf{Y}_t - \boldsymbol{\nu}_t) \right], \\ l_2 &= \sum_{t=1}^{n} \left[log|\mathbf{\Phi}^*| + (\mathbf{X}_t - \boldsymbol{\mu})' \mathbf{\Phi}^{*-1} (\mathbf{X}_t - \boldsymbol{\mu}) \right], \\ \mathbf{\Phi}^* &= \mathbf{\Phi} + \mathbf{\Psi}_{uu}. \end{aligned}$$

We observe that the normal marginal likelihood l_2 is valid only for normal $\mathbf{X}_t = \mathbf{f}_t + \mathbf{u}_t$. But, simple unbiased estimators

$$\begin{aligned} \hat{\boldsymbol{\mu}} &= \bar{\mathbf{X}} = \frac{1}{n} \sum_{t=1}^{n} \mathbf{X}_t, \\ \hat{\mathbf{\Phi}}^* &= \mathbf{m}_{\mathbf{XX}} = \frac{1}{n-1} \sum_{t=1}^{n} (\mathbf{X}_t - \bar{\mathbf{X}})(\mathbf{X}_t - \bar{\mathbf{X}})', \end{aligned}$$

are reasonable estimators of $\boldsymbol{\mu}$ and $\mathbf{\Phi}^*$ regardless of the distributional forms of random $\mathbf{f}_t$ and $\mathbf{u}_t$. Thus, to estimate $\boldsymbol{\beta}$ and $\mathbf{\Psi}$, parameters of our primary interest, we consider l_1 with $\hat{\boldsymbol{\mu}}$ and $\hat{\mathbf{\Phi}}^*$ substituted. This is the so-called pseudo likelihood approach. The l_1 which is also a normal likelihood does not directly depend on the normality of $\mathbf{f}_t$, but can be justified if the distribution of $\mathbf{e}_t$ and the conditional distribution of $\mathbf{u}_t$ given $\mathbf{X}_t$ are approximately normal. Note that l_1 is only an approximate likelihood and that, in particular $\mathbf{\Gamma}_{tt}$ is not the exact conditional covariance matrix. The use of such an approximate expression for a covariance matrix in a role other than a (possibly incorrect) weight for a quadratic form (to be minimized) is known to produce estimators with potentially poor properties. See, e.g., van Houwelingen (1988). Hence, instead of minimizing l_1 with respect to $\boldsymbol{\beta}$ and $\mathbf{\Psi}$, we separate the two parameters, and iterate between the two.

To start the procedure, $\boldsymbol{\beta}^{(0)}$ and $\mathbf{\Psi}^{(0)}$ defined in (4) and (5) can be used as initial estimates for $\boldsymbol{\beta}$ and $\mathbf{\Psi}$, respectively. The i–th step estimates of $\boldsymbol{\beta}$ and $\mathbf{\Psi}$ are obtained as follows:

(A1) The $\boldsymbol{\beta}^{(i)}$ is obtained by minimizing

$$\sum_{t=1}^{n} \left[\mathbf{Y}_t - \boldsymbol{\nu}_t^{(i-1)}(\boldsymbol{\beta}) \right]' \mathbf{\Gamma}_{tt}^{(i-1)-1} \left[\mathbf{Y}_t - \boldsymbol{\nu}_t^{(i-1)}(\boldsymbol{\beta}) \right],$$

where

$$\begin{aligned} \boldsymbol{\nu}_t^{(i-1)} &= \mathbf{g}(\mathbf{X}_t; \boldsymbol{\beta}) + \frac{1}{2} \mathbf{H}(\mathbf{X}_t; \boldsymbol{\beta}) \mathrm{vec} \mathbf{\Psi}_{\mathrm{uu}}^{(i-1)} - \mathbf{G}(\mathbf{X}_t; \boldsymbol{\beta}) \mathbf{\Psi}_{\mathrm{uu}}^{(i-1)} \mathbf{m}_{\mathbf{XX}}^{-1} (\mathbf{X}_t - \bar{\mathbf{X}}), \\ \mathbf{\Gamma}_{tt}^{(i-1)} &= \mathbf{\Psi}_{\mathrm{ee}}^{(i-1)} + \tilde{\mathbf{G}}_t^{(i-1)} \left[\mathbf{\Psi}_{\mathrm{uu}}^{(i-1)} - \mathbf{\Psi}_{\mathrm{uu}}^{(i-1)} \mathbf{m}_{\mathbf{XX}}^{-1} \mathbf{\Psi}_{\mathrm{uu}}^{(i-1)} \right] \tilde{\mathbf{G}}_t^{(i-1)'} \\ &\quad + \frac{1}{n} \left[\mathbf{I}_{p-k}, -\tilde{\mathbf{G}}_t^{(i-1)} \right] \mathbf{m}_{\mathbf{ZZ}} \left[\mathbf{I}_{p-k}, -\tilde{\mathbf{G}}_t^{(i-1)} \right]', \\ \tilde{\mathbf{G}}_t^{(i-1)} &= \mathbf{G}(\mathbf{X}_t; \boldsymbol{\beta}^{(i-1)}), \end{aligned}$$

and $\mathbf{m}_{ZZ}$ is defined in (7).

(A2) The $\mathbf{\Psi}^{(i)}$ is obtained by the generalized least squares applied to

$$\mathbf{m}^{(i)} = \frac{1}{n}\sum_{t=1}^{n}\left[\mathbf{r}_t^{(i)}\mathbf{r}_t^{(i)\prime} + \mathbf{A}_t^{(i)}\right],$$

where

$$\begin{aligned}
\mathbf{r}_t^{(i)} &= \mathbf{Y}_t - \boldsymbol{\nu}_t^{(i-1)}(\boldsymbol{\beta}^{(i)}),\\
\mathbf{A}_t^{(i)} &= \tilde{\mathbf{G}}_t^{(i)}\mathbf{\Psi}_{\mathrm{uu}}^{(i-1)}\mathbf{m}_{XX}^{-1}\mathbf{\Psi}_{\mathrm{uu}}^{(i-1)}\tilde{\mathbf{G}}_t^{(i)\prime}\\
&\quad + \mathbf{D}_t^{(i)}\left[\sum_{s=1}^{n}\mathbf{D}_s^{(i)\prime}\mathbf{\Gamma}_{ss}^{(i-1)-1}\mathbf{D}_s^{(i)}\right]^{-1}\mathbf{D}_t^{(i)\prime},\\
\mathbf{D}_t^{(i)} &= \frac{\partial \mathbf{g}\left(\mathbf{X}_t;\boldsymbol{\beta}^{(i)}\right)}{\partial \boldsymbol{\beta}'}.
\end{aligned}$$

For the $p \times 1$ $\boldsymbol{\psi}^{(i)}$ satisfying $\mathrm{vec}\mathbf{\Psi}^{(i)} = \mathbf{L}_p\boldsymbol{\psi}^{(i)}$,

$$\boldsymbol{\psi}^{(i)} = \left(\tilde{\mathbf{C}}^{(i)\prime}\tilde{\mathbf{V}}^{-1}\tilde{\mathbf{C}}^{(i)}\right)^{-1}\tilde{\mathbf{V}}^{-1}\tilde{\mathbf{C}}^{(i)\prime}\mathrm{vech}\,\mathbf{m}^{(i)},$$

where

$$\begin{aligned}
\tilde{\mathbf{C}}^{(i)} &= \mathbf{K}_q^{+}\{\mathbf{L}_q, \frac{1}{n}\sum_{t=1}^{n}\left(\tilde{\mathbf{G}}_t^{(i)}\otimes\tilde{\mathbf{G}}_t^{(i)}\right)\mathbf{L}_k\},\\
\tilde{\mathbf{V}} &= \frac{2}{n^2}\mathbf{K}_{p-k}^{+}\sum_{t=1}^{n}\left(\tilde{\mathbf{Q}}_{tt}\otimes\tilde{\mathbf{Q}}_{tt}\right)\mathbf{K}_{p-k}^{+\prime},\\
\tilde{\mathbf{Q}}_{tt} &= \mathbf{\Psi}_{\mathrm{ee}}^{(i-1)} + \tilde{\mathbf{G}}_t^{(i)}\left[\mathbf{\Psi}_{\mathrm{uu}}^{(i-1)} - \mathbf{\Psi}_{\mathrm{uu}}^{(i-1)}\mathbf{m}_{XX}^{-1}\mathbf{\Psi}_{\mathrm{uu}}^{(i-1)}\right]\tilde{\mathbf{G}}_t^{(i)\prime}\\
&\quad + \frac{1}{n}\left[\mathbf{I}_{p-k}, -\tilde{\mathbf{G}}_t^{(i-1)}\right]\mathbf{m}_{ZZ}\left[\mathbf{I}_{p-k}, -\tilde{\mathbf{G}}_t^{(i-1)}\right]'.
\end{aligned}$$

(A3) Update $i = i + 1$.

This iterative procedure should be continued for a few iterations only, instead of searching for convergence. But, in our experience, the estimates in some cases converge after a reasonable number of iterations. Note that the step (A1) itself generally utilizes an iterative algorithm. The final step produces the approximate conditional likelihood (ACL) estimates $\tilde{\boldsymbol{\beta}}$ and $\tilde{\mathbf{\Psi}}$. In step (A2) above, any negative element of $\boldsymbol{\psi}^{(i)}$ is replaced by zero, and the remaining elements are re-estimated by the reduced generalized least squares. The additional upper bound modification as given in (9) is also incorporated in computation of $\boldsymbol{\psi}^{(i)}$. The terms involving $n^{-1}\mathbf{m}_{ZZ}$ in (A1) and (A2) are included as small order adjustment to guarantee numerical stability of the inverse matrices. An estimator of the factor covariance matrix $\mathbf{\Phi}$ can be obtained by $\mathbf{m}_{XX} - \tilde{\mathbf{\Psi}}_{\mathrm{uu}}$ with modification suggested in Amemiya (1985). The minimization in (A1) is similar to that in (E2) for the extended linear maximum likelihood procedure. But, (A1) is based on the approximate conditional distribution and uses the expansion around $\mathbf{X}_t$ instead of around $\mathbf{f}_t$ in (E2). Thus, e.g., the direction of the second derivative bias adjustment based on the similar idea looks opposite in the two approaches. The generalized least squares in (A2) is applied to a moment matrix as in (E3), but is based on the conditional distribution. The generalized least squares residual sum of squares in (A2) at the final iteration can be compared to the chi-square distribution with degrees of freedom $d = \frac{1}{2}(p-k)(p-k+1) - p$ to test the fit of the model. This test is based on the conditional distribution and differs from that in the previous section. For a special case of the linear model with normally distributed $\mathbf{f}_t$ and ϵ_t, the procedure based on (A1)–(A3) iterated to convergence does not reduce to the maximum likelihood estimation. This is because the information on $\mathbf{\Psi}_{ee}$ is ignored in step (A1) and is not used in obtaining $\boldsymbol{\beta}^{(i)}$, and because in turn some information about $\mathbf{\Phi}$ contained in l_1 is ignored. However, for such a normal linear model, the resulting estimator is consistent and efficient, although not fully efficient. The general difficulty dealing with the likelihood based on a nonlinear function of a random factor and the relatively simple form of the approximate conditional likelihood make this approach appealing for the general nonlinear model.

5 A simulation study

To study finite-sample properties of the estimators, a Monte-Carlo simulation was conducted. Model (3) with $p = 4$ observed variables was considered, where for $t = 1, 2, ..., n$,

$$\begin{aligned} Y_{1t} &= \beta_1 + \frac{\beta_2}{1 + e^{\beta_3 - \beta_4 f_t}} + e_{1t}, \\ Y_{2t} &= \beta_5 + \beta_6 f_t + \beta_7 f_t^2 + e_{2t}, \\ Y_{3t} &= \beta_8 + \beta_9 f_t + \beta_{10} f_t^2 + e_{3t}, \\ X_t &= f_t + u_t. \end{aligned}$$

The true parameter values for the coefficients are

$$\boldsymbol{\beta}' = (\beta_1, \beta_2, ..., \beta_{10}) = (0\ , 7\ , 10\ , 0.5\ , 50\ , -5\ , 0.1\ , 65\ , -7\ , 0.2)\,.$$

We generated random samples assuming

$$\begin{aligned} f_t &\sim N(20, 36)\,, \\ \begin{pmatrix} \mathbf{e}_t \\ u_t \end{pmatrix} &\sim N(\mathbf{0}\ , \ diag(\psi_{ee11}, \psi_{ee22}, \psi_{ee33}, \psi_{uu})). \end{aligned}$$

To study the effect of the error variance size relative to the total variance, we considered two sets of the error variances. We set

$$\begin{aligned} \psi_{eeii} &= \delta\, Var\{Y_{it}\},\ i = 1, 2, 3, \\ \psi_{uu} &= \delta\, Var\{X_t\}, \end{aligned}$$

and considered two cases $\delta = 0.1$ and $\delta = 0.2$ corresponding to the cases with error variances approximately 10% and 20% of the total variances. (The numerical approximation was used for $\mathrm{Var}\{Y_{1t}\}$). For the sample size n, two choices $n = 300$ and 500 were considered. For each of the four combinations of δ and n, 1000 Monte Carlo samples were generated. From each sample, initial estimates for $\boldsymbol{\beta}$ and $\boldsymbol{\psi}$ were obtained by using (4) and (5), respectively. Starting from these initial estimates, a step consisting of (E1)–(E4) in Section 3 was iterated twice to obtain the extended linear maximum likelihood (ELM) estimates of $\boldsymbol{\beta}$ and $\boldsymbol{\psi}$. Also, using the same initial values, a step consisting of (A1)–(A3) in Section 4 was iterated twice to obtain the approximate conditional likelihood (ACL) estimates of $\boldsymbol{\beta}$ and $\boldsymbol{\psi}$.

The results for some selected parameters are summarized in Table 1. In the table, RMSE and RB stand for the square root of the mean squared error and the relative bias (bias / true value), respectively. Although the estimators of $\boldsymbol{\beta}$ may not possess moments, these tables provide a useful summary. In addition, some boxplots are given in Figure 1 and Figure 2 representing general patterns of empirical distributions of the estimates. In the boxplots, whiskers are drawn to the nearest value not exceeding $1.5 \times$ (*interquartile range*) from the quartiles. There were one or two outliers outside the range of the boxplots, but none of them were extreme outliers.

For estimation of $\boldsymbol{\beta}$, the initial estimator is the ordinary nonlinear least squares estimator defined in (4) which ignores the error in $\mathbf{X}_t$. As can be seen in the table and figures, the initial estimator of $\boldsymbol{\beta}$ has a large bias which increases with δ and does not decrease with n. In fact, the initial estimator is so biased that in some cases all 1000 samples give the values on one side (either larger or smaller) of the true value. Thus, the ordinary nonlinear least squares estimator of $\boldsymbol{\beta}$ is unpractical unless the error variances are very small. On the other hand, the ELM and ACL estimators of $\boldsymbol{\beta}$ have small biases, and the ACL is nearly median unbiased. This fact and their nearly symmetric empirical distributions around the true values are two reasonable features of the ELM and ACL estimators for practical use. The ELM estimator of $\boldsymbol{\beta}$ takes some outlying values when the relative size δ of the error variance is large or the sample size n is small. This also affects the RMSE and RB, and may possibly indicate the nonexistence of moments in some cases. If we exclude few outlying values, the ELM estimator of $\boldsymbol{\beta}$ generally tends to have slightly smaller variability than the ACL, but the difference is small.

For estimation of $\boldsymbol{\psi}$, the initial estimator (5) is an unweighted estimator, while the ELM and ACL estimators are weighted estimators (with bias adjustment). As a result, the ELM and ACL estimators improve largely over the initial estimator in terms of variability. The performances of the ELM and ACL estimators are similar to each other, except that the ELM tends to have a smaller bias but a slightly larger number of outlying values. As in estimation of $\boldsymbol{\beta}$, the outlying values are not extreme, and the number of samples containing them is very small. The estimation of $\boldsymbol{\psi}$ becomes increasingly more difficult when the error variances become large relative to the total variances.

Overall, the computationally simpler ACL fitting procedure is generally more stable than the ELM, and seems to produce estimators with at least similar efficiency. Since the normality of the factor and error vectors used in this simulation corresponds to the ideal case for the ACL, and since the ELM estimator is more efficient in general for the linear models, the relative efficiency of the ACL against ELM may decrease for other distributional situations. But, for the studied situation, the factor score estimation and the repeated evaluation of functions at the score estimates in the ELM procedure seem to result in the finite sample instability producing more outliers than the ACL. The ELM

Table 1: Root mean squared error and relative bias for three estimators

$n = 300,\ \delta = 0.1$						
	Initial		ELM		ACL	
Parameter	RMSE	RB	RMSE	RB	RMSE	RB
β_4	0.1132	-0.2093	0.0701	-0.0848	0.0073	0.0331
β_5	8.7747	-0.1667	4.1434	-0.0176	2.7802	-0.0006
β_{10}	0.0352	-0.1683	0.0155	-0.0140	0.0110	0.0036
ψ_{ee11}	0.2792	-0.0853	0.1474	-0.0873	0.1534	-0.0555
ψ_{ee22}	2.0480	-0.0900	1.3914	-0.0212	1.2358	0.0054
ψ_{uu}	0.9302	0.0475	0.5045	-0.0196	0.4576	-0.0449

$n = 500,\ \delta = 0.1$						
	Initial		ELM		ACL	
Parameter	RMSE	RB	RMSE	RB	RMSE	RB
β_4	0.1115	-0.2120	0.0546	-0.0789	0.0563	0.0251
β_5	8.5938	-0.1666	2.4003	-0.0193	2.3242	-0.0012
β_{10}	0.0348	-0.1693	0.0087	-0.0158	0.0088	0.0033
ψ_{ee11}	0.2182	-0.0799	0.1114	-0.0747	0.1165	-0.0416
ψ_{ee22}	1.6591	-0.0903	0.9675	-0.0092	0.9386	0.0073
ψ_{uu}	0.7485	0.0781	0.3742	-0.0185	0.3724	-0.0437

$n = 300,\ \delta = 0.2$						
	Initial		ELM		ACL	
Parameter	RMSE	RB	RMSE	RB	RMSE	RB
β_4	0.2015	-0.3868	0.0546	-0.2424	0.0563	-0.0826
β_5	18.0742	-0.3544	9.9081	-0.0993	6.0836	0.0110
β_{10}	0.0730	-0.3590	0.0343	-0.0407	0.0269	0.0306
ψ_{ee11}	0.7565	-0.2245	0.3564	-0.0611	0.3487	0.0645
ψ_{ee22}	6.8998	-0.2275	3.8092	-0.0165	3.1516	0.0400
ψ_{uu}	2.5832	0.1325	1.3115	-0.0257	1.4179	-0.1119

$n = 500,\ \delta = 0.2$						
	Initial		ELM		ACL	
Parameter	RMSE	RB	RMSE	RB	RMSE	RB
β_4	0.2018	-0.3941	0.1461	-0.2303	0.1001	-0.0908
β_5	18.0186	-0.3562	7.7465	-0.0878	4.7030	0.0029
β_{10}	0.0725	-0.3593	0.0246	-0.0683	0.0208	0.0227
ψ_{ee11}	0.6835	-0.2342	0.2608	-0.0452	0.2836	0.0711
ψ_{ee22}	6.3766	-0.2361	2.6675	0.0013	2.4453	0.0397
ψ_{uu}	2.3628	0.1784	1.0265	-0.0256	1.2965	-0.1128

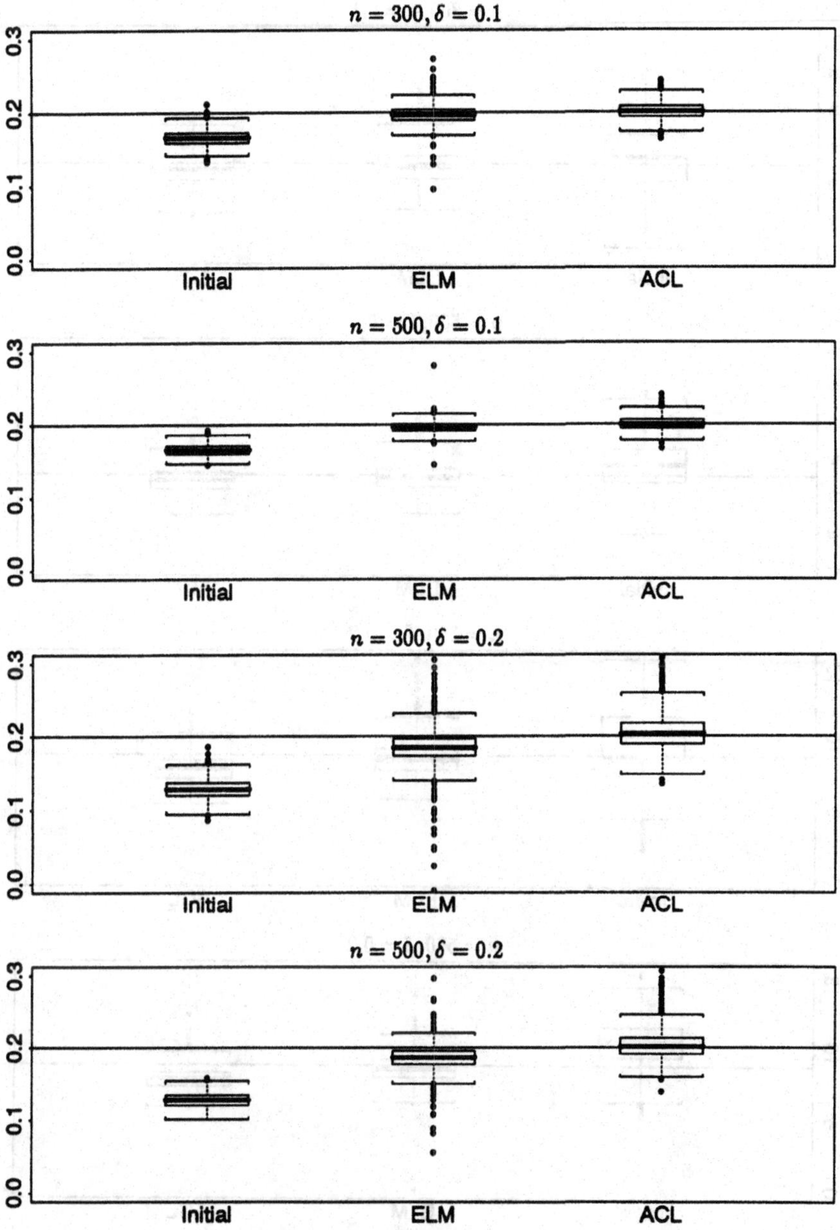

Figure 1: Boxplots for estimates of β_{10}

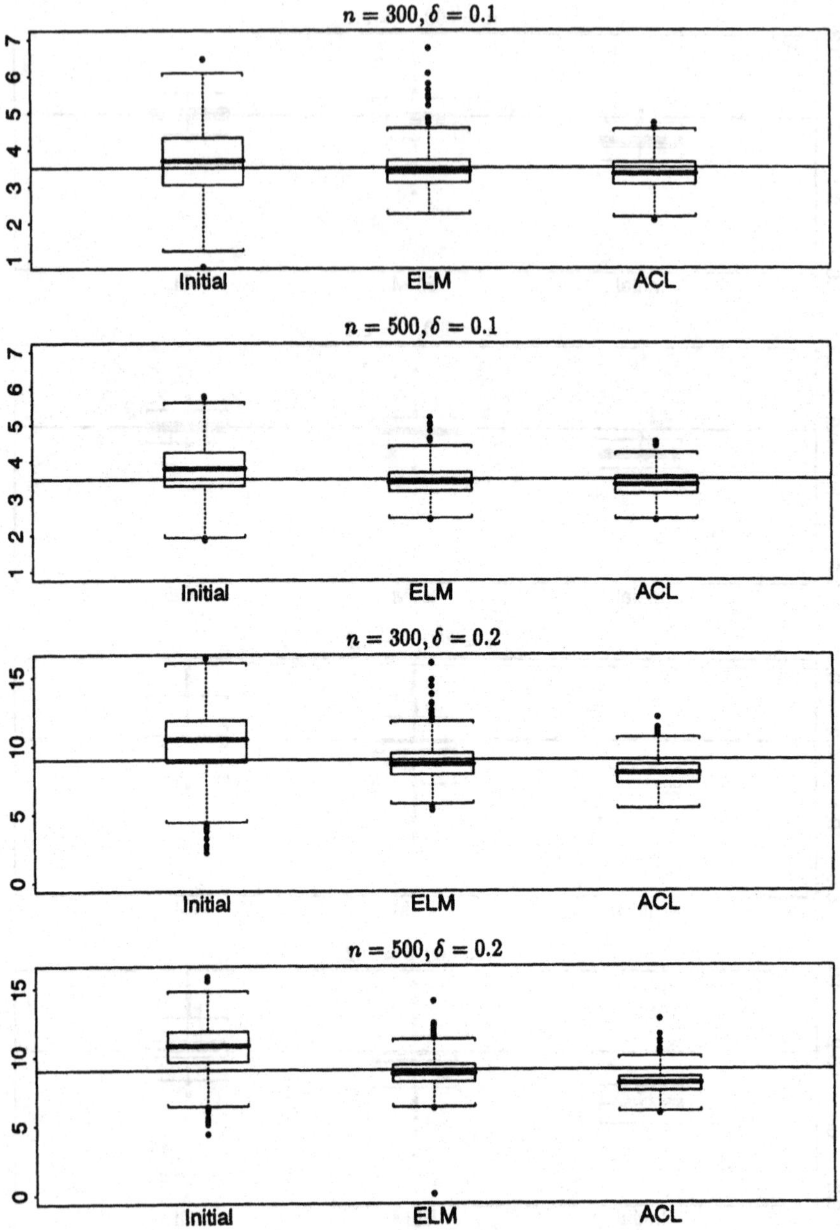

Figure 2: Boxplots for estimates of ψ_{uu}

procedure could be further modified to reduce the occurrence of outlying estimates and to improve the finite sample properties. These and a number of other methodological and theoretical questions regarding nonlinear factor analysis are being investigated. However, the results of this study are very promising for developing more refined practical model fitting procedures based on the ELM and ACL approaches.

6 References

Amemiya, Y. (1985). What should be done when an estimated between-group covariance matrix is not nonnegative definite? *American Statisti.* **39** 112-117.

Amemiya, Y. (1993a). Instrumental variable estimation for nonlinear factor analysis. In Multivariate Analysis: Future Directions 2. (C.M. Cuadras, C.R. Rao, Ed.), pp. 113-129, North-Holland Series in Statistics and Probability, Amsterdam.

Amemiya, Y. (1993b). On nonlinear factor analysis. *Proceedings of ASA, Social Statistics Section* 290-294.

Amemiya, Y. and Anderson, T. W. (1990). Asymptotic chi-square tests for a large class of factor analysis models. *Ann. Statist.* **18** 1453-1463.

Anderson, T. W. and Amemiya, Y. (1988). The asymptotic normal distribution of estimators in factor analysis under general conditions. *Ann. Statist.* **16** 759-771.

Browne, M. W. and Shapiro, A. (1988). Robustness of normal theory methods in the analysis of linear latent variate models. *British J. Math. Statist. Psych.* 41 193-208.

Carroll, R. J. and Ruppert, D. (1982). A comparison between maximum likelihood and generalized least squares in a heteroscedastic linear model. *Jour. of Amer. Stat. Assoc.* **77** 878-882.

Etezadi-Amoli, J. and McDonald, R. P. (1983). A second generation nonlinear factor analysis. *Psychometrika* **48** 315-342.

Jöreskog, K. G. (1967). Some contributions to maximum likelihood factor analysis. *Psychometrika* **32** 443-482.

Jöreskog, K. G. (1969). A general approach to confirmatory maximum likelihood factor analysis. *Psychometrika* **34** 183-202.

Jöreskog, K. G. (1977). Factor analysis by least squares and maximum likelihood methods. In *Statistical Methods for Digital Computers,* Vol. 13, K.Enslein, R. Ralston, and S.W. Wilf (eds.). Wiley, New York.

McDonald, R. P. (1962). A general approach to nonlinear factor analysis. *Psychometrika* **27** 397-415.

McDonald, R. P. (1967a). Numerical methods for polynomial models in nonlinear factor analysis. *Psychometrika* **32** 77-112.

McDonald, R. P. (1967b). Factor interaction in nonlinear factor analysis. *British J. Math. Statist. Psyc.* **20** 205-215.

Mooijaart, A. and Bentler, P. (1986). Random polynomial factor analysis. In *Data Analysis and Informatics, IV* (E. Diday et al., Ed.), pp. 241-250. Elsevier Science Publishers B.V., New York.

Pantula, S. G. and Fuller, W. A. (1986). Computational algorithms for the factor model. *Commun. Statist. Part B* **15**, 227-259.

van Houwelingen, J. C. (1988). Use and abuse of variance models in regression. *Biometrics* **44** 1073-1081.

Yalcin, I. (1995). Nonlinear factor analysis. Ph.D. dissertation. Iowa State University.

Yalcin, I. and Amemiya, Y. (1993). Fitting of a general nonlinear factor analysis model. *Proceedings of ASA, Statistical Computing Section* 92-96.

Multivariate Regression with Errors in Variables: Issues on Asymptotic Robustness

Albert Satorra
Departament d'Economia
Universitat Pompeu Fabra, Barcelona

Abstract

Estimation and testing of functional or structural multivariate regression with errors in variables, with possibly unbalanced design for replicates, and not necessarily normal data, is developed using only the sample cross-product moments of the data. We give conditions under which normal theory standard errors and an asymptotic chi-square goodness-of-fit test statistic retain their validity despite non-normality of constituents of the model. Assymptotic optimality for a subvector of parameter estimates is also investigated. The results developed apply to methods that are widely available in standard software for structural equation models, such as LISREL or EQS.

1 Introduction

Consider the following multivariate linear regression model with errors in variables, with independent replicates and multiple samples,

$$\begin{cases} Y_{ig} &= \alpha_g + B_g x_{ig} + v_{ig} \\ X_{ig} &= F_g x_{ig} + u_{ig} \\ W_{ig} &= H_g x_{ig} + w_{ig} \end{cases} \tag{1}$$

for $i = 1, \ldots, n_g$; $g = 1, \ldots, G$. Here G is the number of samples or groups; $Y_{ig}\,(p_g \times 1)$, $X_{ig}\,(q_g \times 1)$ and $W_{ig}\,(r_g \times 1)$ are vectors of observable variables; $F_g\,(q_g \times q)$ and $H_g\,(r_g \times q)$ are selection matrices (identity matrices with some of the rows suppressed); $x_{ig}\,(q \times 1)$ is a vector of unobservable regressors; $v_{ig}\,(p_g \times 1)$ is a disturbance term independent of the x_{ig}; $u_{ig}\,(q_g \times 1)$ and $w_{ig}\,(r_g \times 1)$ are vectors of measurement errors associated with the variables X_{ig} and W_{ig} ; and $\alpha_g\,(p_g \times 1)$ and $B_g(p_g \times q)$ are the intercept vector and slope matrix for group g. In order to have an identified model, we assume that (at least) for one value of g, say $g^\star$, $F_{g^\star} = H_{g^\star} = I_{q\times q}$. We define the augmented vector of observable

*A preliminary version of this paper was presented at the conference "Causality and Latent Variables", University of California, Los Angeles, March 19-20th, 1994. Computer assistance of Jan Graffelman, of University Pompeu Fabra, is acknowledged. Work supported by the Spanish DGICYT grants PB91-0814 and PB93-0403

variables $z_{ig} \equiv (Y'_{ig}, X'_{ig}, W'_{ig}, 1)'$, with the last component being constant to one. Note that the number of components c_g $(= p_g + q_g + r_g + 1)$ of z_{ig} is allowed to vary with g. For further use, let $c^*_g = c_g(c_g + 1)/2$, and $c^* = \sum_{g=1}^{G} c^*_g$. Some of the values of p_g, q_g or r_g may equal zero, in which case the corresponding equation vanishes.

We further assume that v_{ig}, u_{ig} and w_{ig} are mutually independent centered vector variables, with variance matrices $\mathcal{E}v_{ig}v'_{ig} = \Xi_g$, $\mathcal{E}u_{ig}u'_{ig} = \Psi_g$ and $\mathcal{E}w_{ig}w'_{ig} = H_g\Psi_v H'_g$, where Ξ_g $(p_g \times p_g)$ and Ψ_g $(q \times q)$ are non-negative semidefinite matrices of unknown coefficients. Moreover, the terms u_{ig} and w_{ig} are assumed to be *normally distributed.*

Throughout the paper, we distinguish two types of models. The functional model (FM), where for $g = 1, \ldots, G$, $\{x_{ig}; i = 1, \ldots, n_g\}$ is viewed as a set of fixed (across hypothetical sample replications) unknown vectors, and the structural model (SM) where $\{x_{ig}; i = 1, \ldots, n_g\}$ are regarded as iid random vectors with common mean vector μ_g and covariance matrix Φ_g. In the FM, we let

$$\lim_{n_g \to \infty} {n_g}^{-1} \sum_{i=1}^{n_g} x_{gi} = \mu_g, \quad \lim_{n_g \to \infty} {n_g}^{-1} \sum_{i=1}^{n_g} x_{gi}x'_{gi} = \Phi_g, \tag{2}$$

where μ_g and Φ_g are assumed to be finite.

As an example, consider the case where $G = 1$ and $p_1 = q_1 = 1$. In this case, model (1) is a simple regression model with errors in variables, with two replicates for the true unknown regressor vector x_{ig}. This model is usually estimated under the assumption that $\{z_{ig}; i = 1, \ldots, n_g\}$ is iid normally distributed (Fuller, 1987). When $G = 2$, model (1) encompasses the case for example where in the first sample ($g = 1$) we only observe Y and X, while in the second sample (g=2) we observe only the variables X and W; i.e., the case of $p_1 = q_1 = 1$, $r_1 = 0$ and $p_2 = 0$, $q_2 = r_2 = 1$. The variables X_{ig} and W_{ig} can be viewed as two different "replicates" of a common vector of underlying variables x_{ig}.

In the present paper we are concerned with correct inferences for model (1) under minimal assumptions with regard to x_{gi} and v_{gi}, and when the estimation method used involve only the sample cross-product moments of the data. Maximum likelihood under the assumption that $\{z_{ig}; i = 1, \ldots, n_g\}$ is iid normally distributed is one of such estimation methods. Standard errors and chi-square goodness of fit test can be developed using or not this normality assumption on the $\{z_{ig}$s. The major import of the present paper will be to show that some of the normal theory inferences retain their validity under the weak assumption of just finiteness of the second-order moments of x_{gi} and v_{gi}. The theoretical results of the paper apply to normal theory methods that are widely implemented in commercial computer software such as LISREL (Jöreskog & Sörbom, 1989), EQS (Bentler, 1989), LISCOMP (Muthén, 1989), among others.

The approach we develop extends easily to model (1) expanded with additional equations for replicates, i.e. in the case of $W^{(m)}_{ig} = H^{(m)}_g x_{ig} + w^{(m)}_{ig}; m = 1, \ldots, M_g$, with $M_g \geq 2$, instead of only the equation for W_{ig}. [1]

In applications, the latent predictor x_{ig} could refer, for example, to a continuous average over times of measurements (e.g., pollution concentrations, blood pressures), then observations at a finite numer of randomly selected times can be regarded as independent replications of measurements of x_{ig}. Our model allows for unbalanced designs on the

[1] For the implementation of the model using standard software, we require however that the number of replicates M_g is not too large.

number of replicates (i.e. different number of replicates for different groups of cases). Missing values could also generate such unbalanced designs. See for example Gleser (1992) and Hasabelnaby, Ware and Fuller (1989) for examples of the use of replicates to assess reliability in regression models with errors in variables. A major advantage of the approach we advocate, is that an efficient analysis is guaranteed without requiring the development of customized software.

The validity of normal theory inferences when the data are non-normal has been called asymptotic robustness (Anderson, 1987). Asymptotic robustness in the case of single-sample analysis has been investigated by Amemiya (1985), Anderson and Amemiya (1988), Amemiya and Anderson (1990), Browne (1987, 1990), Browne and Shapiro (1987), Anderson (1987, 1989), Kano (1993), Satorra and Bentler (1990, 1991), Mooijaart and Bentler (1991), Browne (1990), Satorra (1992) and Satorra and Neudecker (1994). Asymptotic robustness in single-sample analysis of regression with errors in variables was first investigated by Dahm and Fuller (1986).

Asymptotic robustness in the case of multi-sample analysis has been investigated by Satorra (1993a,1993b), and more recently by Satorra (1995), where a general approach that covers functional and structural latent variables is encompassed. In the present paper, the results of Satorra (1995) will be specialized to the case of model (1).

In multi-sample analysis of structural equation models, independent samples from several populations (groups) are analyzed simultaneously under a common model, with parameter restrictions across populations; interest focuses in studying the between-population variation of model parameters (Jöreskog, 1971). It has been shown that the presence of missing data in structural equation models can also be treated using a multi-sample analysis approach (Allison, 1987; Arminger and Sobel, 1989; Muthén, Kaplan and Hollis, 1987).

The structure of the paper is as follows. In Section 2 we deduce the multivariate linear relation and moment structure associated with model (1). In Section 3 we review general minimun distance estimation and inferences for moment structures. In Section 4 we apply the results on asymptotic robustness of linear latent variable models, as given in Satorra (1995), to the model set-up of (1). In Section 5, a small Monte Carlo study is conducted to illustrates the performance in finite samples of the asymptotic robustness results explained in Section 4.

The following notation will be used. Given a set of vectors $\{a_i; i = 1, \ldots, I\}$, $\text{vec}\{a_i; i = 1, \ldots, I\}$ denotes the vector formed by vertical stacking of the a_i. When the index variation is clear from the context, $\{a_i; i = 1, \ldots, I\}$ is simplified to $\{a_i\}$. Given a set of square matrices $\{A_i; i = 1, \ldots, I\}$, $\text{diag}\{A_i; i = 1, \ldots, I\}$ denotes the block-diagonal matrix with the A_i as diagonal blocks. For any matrix A, $\text{vec}\,(A)$ denotes the column vector formed by vertical stacking of the columns of A, one below the other. When A is a symmetric matrix, $\text{v}\,(A)$ is the vector obtained from $\text{vec}\,(A)$ by eliminating the duplicated elements associated with the symmetry of A. We use the "duplication" and "elimination" matrices D and D^+ (Magnus and Neudecker, 1993), such that $\text{vec}\,(A) = D\text{v}\,(A)$ and $\text{v}\,(A) = D^+\text{vec}\,(A)$, for a symmetric matrix A. The matrices D and D^+ are of varying order determined by the context.

2 Multivariate Linear Relations and Moment Structures

In this section we will first express model (1) as a multivariate linear relation, and then we deduce the moment structure.

We write model (1) as

$$\begin{pmatrix} Y_{ig} \\ X_{ig} \\ W_{ig} \\ 1 \end{pmatrix} = \begin{pmatrix} I_{p_g} & B_g & \alpha_g & 0 & 0 \\ 0 & I_q & 0 & 1 & 0 \\ 0 & H_g & 0 & 0 & 1 \\ 0 & 0 & 1 & 0 & 0 \end{pmatrix} \begin{pmatrix} v_{ig} \\ x_{ig} \\ 1 \\ u_{ig} \\ w_{ig} \end{pmatrix}, \tag{3}$$

in the case of the functional model (FM), and as

$$\begin{pmatrix} Y_{ig} \\ X_{ig} \\ W_{ig} \\ 1 \end{pmatrix} = \begin{pmatrix} I_{p_g} & B_g & (\alpha_g + B_g\mu_g) & 0 & 0 \\ 0 & I_q & \mu_g & 1 & 0 \\ 0 & H_g & H_g\mu_g & 0 & 1 \\ 0 & 0 & 1 & 0 & 0 \end{pmatrix} \begin{pmatrix} v_{ig} \\ x_{ig} - \mu_g \\ 1 \\ u_{ig} \\ w_{ig} \end{pmatrix}, \tag{4}$$

in the case of the structural model (SM). . In both cases, we obtain what is called a *multivariate linear relation* (Anderson, 1987)

$$z_{ig} = \Pi_g \xi_{ig}, \tag{5}$$

where

$$z_{ig} \equiv \begin{pmatrix} Y_{ig} \\ X_{ig} \\ W_{ig} \\ 1 \end{pmatrix}$$

is the augmented vector of observable variables and where in the FM:

$$\Pi_g \equiv \begin{pmatrix} I_{p_g} & B_g & \alpha_g & 0 & 0 \\ 0 & I_q & 0 & 1 & 0 \\ 0 & H_g & 0 & 0 & 1 \\ 0 & 0 & 1 & 0 & 0 \end{pmatrix} \qquad \xi_{ig} \equiv \begin{pmatrix} v_{ig} \\ x_{ig} \\ 1 \\ u_{ig} \\ w_{ig} \end{pmatrix} \tag{6}$$

and in the SM:

$$\Pi_g \equiv \begin{pmatrix} I_{p_g} & B_g & (\alpha_g + B_g\mu_g) & 0 & 0 \\ 0 & I_q & \mu_g & 1 & 0 \\ 0 & H_g & H_g\mu_g & 0 & 1 \\ 0 & 0 & 1 & 0 & 0 \end{pmatrix} \qquad \xi_{ig} \equiv \begin{pmatrix} v_{ig} \\ x_{ig} - \mu_g \\ 1 \\ u_{ig} \\ w_{ig} \end{pmatrix} \tag{7}$$

are, respectively, a parameter matrix and a vector of latent constituents of the model.

For the developments to be presented below, we write the above multivariate linear relation as

$$z_{ig} = \sum_{\ell=0}^{L} \Pi_{g,\ell}\xi_{ig,\ell}, \tag{8}$$

where $\ell = 0, \ldots, L$ indexes the subvector of ξ, and
in the FM:

$$L = 2,\ \xi_{ig,0} = (x'_{ig}, 1)', \quad \xi_{ig,1} = v_{ig}, \quad \xi_{ig,2} = (u'_{ig}, w'_{ig})';$$

in the SM:

$$L = 3,\ \xi_{ig,0} = 1, \quad \xi_{ig,1} = v_{ig}, \quad \xi_{ig,2} = x_{ig} - \mu_g, \quad \xi_{ig,3} = (u'_{ig}, w'_{ig})'.$$

Consider now the following matrix of cross-product moments

$$S_{\xi\xi,g} \equiv n_g^{-1} \sum_{i=1}^{n_g} \xi_{ig}\xi'_{ig},$$

where we assume $\lim_{n_g \to \infty} S_{\xi\xi,g} = \Sigma_{\xi\xi,g}$, a finite matrix. Clearly, under the above set-up,

$$\Sigma_{\xi\xi,g} = \begin{pmatrix} \Xi_g & 0 & 0 & 0 & 0 \\ 0 & \Phi_g & \phi_g & 0 & 0 \\ 0 & \phi'_g & \nu_g & 0 & 0 \\ 0 & 0 & 0 & \Psi_g & 0 \\ 0 & 0 & 0 & 0 & H_g\Psi_g H'_g \end{pmatrix}, \tag{9}$$

where for SM, $\phi_g = 0$. Note that we have $\phi_g = \lim_{n_g \to \infty} n^{-1} \sum_{i=1}^{n_g} x_{ig}$, and that ν_g will be equal to 1. The above matrix (9) is block-diagonal, where the diagonal blocks are
in the FM:

$$\Xi_g, \quad \begin{pmatrix} \Phi_g & \phi_g \\ \phi'_g & \nu_g \end{pmatrix}, \quad \begin{pmatrix} \Psi_g & 0 \\ 0 & H_g\Psi_g H'_g \end{pmatrix};$$

in the SM:

$$\Xi_g, \quad \Phi_g, \quad \nu_g, \quad \begin{pmatrix} \Psi_g & 0 \\ 0 & H_g\Psi_g H'_g \end{pmatrix}.$$

Let us introduce the following vectors ψ_g and τ, where
in the FM:

$$\psi_g = (\phi'_g, \mathrm{v}'(\Xi_g), \mathrm{v}'(\Phi_g), \nu_g)', \quad \tau = (\alpha', \mathrm{vec}'(B), \mathrm{v}'(\Psi))'; \tag{10}$$

in the SM:

$$\psi_g = (\mathrm{v}'(\Xi_g), \mathrm{v}'(\Phi_g), \nu'_g)', \quad \tau = (\alpha', \mathrm{vec}'(B), \mathrm{v}'(\Psi), \mu'_g)'. \tag{11}$$

For both, FM and SM, let

$$\psi = (\psi'_1, \ldots, \psi'_g, \ldots, \psi'_G)'$$

and

$$\vartheta = \begin{pmatrix} \tau \\ \psi \end{pmatrix}.$$

We denote by t the dimension of τ, and by t^* the dimension of the whole parameter vector ϑ.

Consider now the usual sample (uncentered) moment matrix

$$S_g = n_g^{-1} \sum_{i=1}^{n_g} z_{ig} z'_{ig},$$

and assume

$$\lim_{n\to\infty} S_g = \Sigma_g,$$

a finite matrix. Clearly, given the multivariate linear relation (5), we have

$$\Sigma_g = \Pi_g \Sigma_{\xi\xi,g} \Pi'_g, \tag{12}$$

where $\Pi_g = \Pi_g(\tau)$ and $\Sigma_{\xi\xi,g} = \Sigma_{\xi\xi,g}(\vartheta)$ are twice continuously differentiable matrix-valued functions of τ and ϑ respectively; thus,

$$\Sigma_g = \Sigma_g(\vartheta), \quad g = 1, \ldots, G, \tag{13}$$

where $\Sigma_g(.)$ is a twice continuously differentiable matrix-valued function of the parameter vector ϑ. The equation (13) is called a *moment structure*.

Assume now that the samples from the different groups are independent, so the moment matrices S_g are mutually independent across groups; denote by $n = \sum_{g=1}^{G} n_g$ the overall sample size.

Statistical inferences for ϑ will be undertaken by fitting simultaneously the sample moment matrices S_g's to the moment structures $\Sigma_g(\vartheta)$'s. To that aim, let $s \equiv \text{vec}\{s_g;\, g = 1, \ldots, G\}$, where $s_g \equiv \text{v}(S_g)$, and $\sigma \equiv \text{vec}\{\sigma_g;\, g = 1, \ldots, G\}$, where $\sigma_g \equiv \text{v}(\Sigma_g)$; thus, we can express (13) in the compact form of

$$\sigma = \sigma(\vartheta). \tag{14}$$

A minimum discrepancy (MD) function approach will be used to fit the moment structure $\sigma(\vartheta)$ to the vector of sample moments s.

2.1 Minimum discrepancy function analysis

Before describing the MD approach, let us first introduce the asymptotic variance matrix Γ of the (scaled) vector of sample moments $\sqrt{n}s$.

Since the S_g are mutually independent, the asymptotic variance matrix Γ of $\sqrt{n}s$ is block diagonal;

$$\Gamma = \text{diag}\left\{\frac{n}{n_g}\Gamma_g; g = 1, \ldots, G\right\}, \tag{15}$$

where Γ_g is the asymptotic variance matrix of $\sqrt{n_g}s_g$.

Clearly, when the Γ_g are finite, an (unbiased) consistent estimate of Γ is

$$\hat{\Gamma} \equiv \text{diag}\{\frac{n}{n_g}\hat{\Gamma}_g\}, \tag{16}$$

where

$$\hat{\Gamma}_g \equiv \frac{1}{n_g - 1} \sum_{i=1}^{n_g} (d_{gi} - s_g)(d_{gi} - s_g)' \tag{17}$$

and $d_{gi} \equiv \mathrm{v}(z_{gi} z_{gi}')$. The matrix (16) will be called the asymptotic robust (AR) estimator of Γ, to distinguish it from the estimator of Γ based on the NA assumption that will be introduced below.

A general method of estimation is weighted least-squares (WLS) (see, e.g., Browne, 1984, Chamberlain, 1982, Satorra, 1989), where the estimator of ϑ is defined as

$$\hat{\vartheta} \equiv \operatorname{argmin} [s - \sigma(\vartheta)]' \hat{V} [s - \sigma(\vartheta)], \tag{18}$$

with a matrix $\hat{V}$ that converges in probability to V, a positive definite $c^* \times c^*$ matrix. Define $\hat{\sigma} \equiv \sigma(\hat{\vartheta})$, the vector of fitted moments, and let $\hat{\tau}$ be the corresponding subvector of $\hat{\vartheta}$. We consider the Jacobian matrix $\Delta \equiv \partial\sigma(\theta)/\partial\sigma'$, and let $\hat{\Delta}$ be the matrix Δ evaluated at $\vartheta = \hat{\vartheta}$.

As has been shown elsewhere (e.g., Satorra, 1989), under standard regularity conditions, the estimator $\hat{\vartheta}$ is consistent and asymptotically normal. The asymptotic variance (avar) matrix $\hat{\vartheta}$ is given by

$$\operatorname{avar}(\hat{\vartheta} \mid V, \Gamma) = n^{-1} (\Delta' V \Delta)^{-1} \Delta' V \Gamma V \Delta (\Delta' V \Delta)^{-1}. \tag{19}$$

When $V\Gamma V = V$, or $\Gamma V \Gamma = \Gamma$ and Δ is in the column space of Γ (see Satorra and Neudecker, 1994), then clearly the variance matrix (19) simplifies to

$$\operatorname{avar}(\hat{\vartheta} \mid V) = n^{-1} (\Delta' V \Delta)^{-1}. \tag{20}$$

In this case, we say that the corresponding WLS estimator is asymptotically optimal. See Satorra and Neudecker (1994) for detail discussion on the asymptotic optimality of alternative WLS estimators in the context of linear latent variable models.

When in (19) we replace $\hat{\Gamma}$, $\hat{V}$ and $\hat{\Delta}$ by Γ, V and Δ respectively, we obtain the so-called asymptotic robust (AR) estimator of the variance matrix of $\hat{\vartheta}$. In practice, however, one wants to avoid the use of $\hat{\Gamma}$, since it involves computing the fourth-order sample moments of the data. Further, the use of $\hat{\Gamma}$ raises concern about robustness of the inferences to small samples.

The following normality assumption (NA) avoids totally the use of the fourth-order moment matrix $\hat{\Gamma}$, and it is often used to draw statistical inferences in structural equation modeling.

Assumption 1 (Normality assumption, NA) *For $g = 1, \ldots, G$, and $\ell = 0, \ldots, L$ equation (8) holds with $\{\xi_{ig,\ell}; i = 1, \ldots, n_g\}$ iid normally distributed.*

Remark Clearly, NA implies that $\{z_{ig}; i = 1, \ldots, n_g\}$ are iid normally distributed. Inferences derived under this assumption will be called NA inferences. In fact, the major importance of the present paper is to develop results showing that some of the NA inferences are valid even though NA does not hold. This is of practical relevance since under NA the computation of the sample fourth-order moments of the data can be avoided.

Now we review the formulae for asymptotic inferences under NA. The asymptotic variance matrix Γ_g of $\sqrt{n_g}s_g$ based on NA will be denoted as $\Gamma_g^\star$; and that of Γ as $\Gamma^\star$. We have

$$\Gamma^\star = \text{diag}\{\frac{n}{n_g}\Gamma_g^\star;\ g = 1,\ldots,G\}.$$

When NA holds the asymptotic variance matrix of $\hat{\vartheta}$ is given by (19) with Γ replaced by $\Gamma^\star$.

Define

$$\Omega \equiv \text{diag}\left\{\frac{n}{n_g}\Omega_g;\ g = 1,\ldots,G\right\}, \tag{21}$$

where

$$\Omega_g \equiv 2D^+(\Sigma_g \otimes \Sigma_g)D^{+\prime}, \tag{22}$$

and let $\hat{\Omega}$ denote the estimator of Ω obtained by substituting S_g for Σ_g in (22).

Under assumption NA, the matrix Γ can be expressed as (see, for example, Satorra 1992b)

$$\Gamma^\star = \Omega - \Upsilon, \tag{23}$$

where

$$\Upsilon \equiv \text{diag}\{\frac{n}{n_g}\Upsilon_g;\ g = 1,\ldots,G\}, \tag{24}$$

$$\Upsilon_g \equiv 2D^+(\mu_g{\mu_g}' \otimes \mu_g{\mu_g}')D^{+\prime},$$

and μ_g is the expectation of z_g; $\mu_g \equiv \mathcal{E}(z_g)$.

Define also

$$V^\star \equiv \text{diag}\left\{\frac{n_g}{n}V_g^\star;\ g = 1,\ldots,G\right\}$$

where

$$V_g^\star \equiv \frac{1}{2}D'({\Sigma_g}^{-1} \otimes {\Sigma_g}^{-1})D, \tag{25}$$

and denote by $\hat{V}^\star$ the estimator of $V^\star$ obtained by substituting S_g for Σ_g in (25). The use of $\hat{V} = \hat{V}^\star$ in (18) produces the so-called NA-WLS estimators, which have been shown to be asymptotically equivalent to the maximum likelihood estimators under NA (e.g., Satorra, 1992b). Note that $V^\star = \Omega^{-1}$ ($\hat{V}^\star = \hat{\Omega}^{-1}$).

When NA holds and Δ is in the column space of $\Gamma^\star$, the asymptotic variance matrix (19) of the NA-WLS estimator reduces to

$$\text{avar}(\hat{\vartheta} \mid V = V^\star) = (\Delta' V^\star \Delta)^{-1}, \tag{26}$$

since $\Gamma^\star V^\star \Gamma^\star = \Gamma^\star$ (see Lemma 2 of Satorra, 1992b), and hence the reduced expression (20) is attained. The standard errors extracted from (26) will be called the NA standard errors. The NA standard errors may not be correct when NA does not hold.

When the z_g are normally distributed, it can be shown (e.g., by direct application of results of Meredith and Tisak, 1990) that the log-likelihood function is an affine transformation of

$$F_{ML} \equiv \sum_{g=1}^{G} \frac{n_g}{n}[\log \mid \Sigma_g(\vartheta) \mid + \text{trace}\{S\Sigma_g(\vartheta)^{-1}\} - \log \mid S_g \mid - p_g]; \tag{27}$$

thus, the minimization of $F_{ML} = F_{ML}(\vartheta)$ yields the maximum likelihood (ML) estimator of ϑ. Since $\frac{\partial^2 F_{ML}}{\partial\sigma\partial\sigma'} = V^\star$ (see, e.g., Neudecker and Satorra, 1991), ML estimation is asymptotically equivalent to NT-WLS (Shapiro, 1985; Newey, 1988).

In addition to parameter estimation, we are interested in testing the goodness of fit of the model. Consider the following goodness-of-fit test statistic

$$T_V^\star = n(s-\hat{\sigma})'(\hat{P}\hat{V}^\star\hat{P}')^+(s-\hat{\sigma}), \tag{28}$$

where

$$\hat{P} \equiv I - \hat{\Delta}(\hat{\Delta}'\hat{V}\hat{\Delta})^{-1}\hat{\Delta}'V. \tag{29}$$

Let P denote the asymptotic limit of $\hat{P}$. It will be shown below that under MLR and MS, $T_V^\star$ is asymptotically chi-square distributed with

$$r = \text{rank}\{PV^\star P'\} \tag{30}$$

degrees of freedom (df). Clearly, when $V = V^\star$, an alternative expression of $T^{\star\star} \equiv T_{V^\star}^\star$ is

$$T^{\star\star} = n(s-\hat{\sigma})'(\hat{V}^\star - \hat{V}^\star\hat{\Delta}(\hat{\Delta}'\hat{V}^\star\hat{\Delta})^{-1}\hat{\Delta}'\hat{V}^\star)(s-\hat{\sigma}).$$

When Γ is finite, and the AR estimate $\hat{\Gamma}$ (see (16)) is available, an alternative goodness-of-fit test statistic is

$$T_V = n(s-\hat{\sigma})'(\hat{P}\hat{\Gamma}\hat{P}')^+(s-\hat{\sigma}). \tag{31}$$

Under a correct model, the statistic T_V can be shown to be asymptotically chi-square distributed with the degrees of freedom of (30) regardless of whether NA holds or not (e.g., Satorra, 1993b). Due to this, we call T_V the AR goodness-of-fit test statistic. Under NA single-group covariance structure analysis, T_V is Browne's (1984) residual-based chi-square goodness-of-fit statistic. The test statistic $T^{\star\star}$ is asymptotically equivalent to the usual goodness-of-fit test statistics developed under NA using the likelihood ratio principle (Satorra, 1992b, 1993b).

In the next section we show that some of the inferences derived under NA retain their validity even when the assumption NA is violated.

2.2 Asymptotic robustness of normal theory methods

In this section we review the results on asymptotic robustness as discussed in Satorra (1995), and we specialize them to the model context described above. We first present the assumptions MLR, MS and I of Satorra (1995). Given the equalities, (8), (9), (12) and (13), these assumptions are clearly verified within the model set-up described in Section 1.

Assumption 2 (Multivariate linear relation, MLR) *For $g = 1, \ldots, G$,*

$$z_{ig} = \sum_{\ell=0}^{L} \Pi_{g\ell}\xi_{ig,\ell}, \tag{32}$$

where the $\xi_{ig,\ell}$ are vectors of (observed or latent) variables and the $\Pi_{g\ell}$ are matrices of coefficients. Further,

1. *$\{\xi_{ig,0}; i = 1, \ldots, n_g\}$ are fixed vectors such that for each component u_i of $\xi_{ig,0}$, $\lim_{n_g \to \infty} \frac{1}{n_g} \sum_{i=1}^{n_g} u_i^2 = \phi_u$ is finite.*

2. *for $\ell = 1, \ldots, L$, the $\{\xi_{ig,\ell}; i = 1, \ldots, n_g\}$ are iid and mutually independent across $\ell = 1, \ldots, L_g$.*

3. *$\{\xi_{ig,L}; i = 1, \ldots, n_g\}$ are iid normally distributed.*

When $\xi_{ig,\ell}$ is stochastic, it has zero mean and finite variance matrix, say $\Phi_{g\ell}$.

ASSUMPTION 3 (MOMENT STRUCTURE, MS) *For $g = 1, \ldots, G$,*

$$\Sigma_g = \sum_{\ell=0}^{L} \Pi_{g\ell} \Phi_{g\ell} \Pi_{g\ell}{}', \tag{33}$$

where the matrices $\Pi_{g\ell}$ and $\Phi_{g\ell}$ are of conformable dimensions and $\Pi_{g\ell} = \Pi_{g\ell}(\tau)$ and $\Phi_{gL_g} = \Phi_{gL_g}(\tau)$ are twice continuously differentiable matrix-valued functions of a $t-$dimensional parameter vector τ.

Note that MS does not restrict the variance matrices $\Phi_{g\ell}$, except for Φ_{gL}. It should be noted that equality (33) is implied already by MLR. Assumption MS, however, structure the matrices $\Pi_{g\ell}$ and Φ_{gL} to be functions of τ. Clearly, MS can be fitted without the necessity of MLR to hold.

In the model set-up of Section 1, the following identification assumption is satisfied (recall that $F_{g\star} = H_{g\star} = I_{q\times q}$).

ASSUMPTION 4 (IDENTIFICATION, I) *When $s = \sigma$ and $\hat{V} = V$ the optimization problem of (18) has (locally) a unique solution ϑ that lies in the interior of Θ. Furthermore, the matrix $\Delta' V \Delta$ is nonsingular.*

The following theorem summarizes the major results on asymptotic robustness.

THEOREM 1 (cf. , Satorra, 1995) *Under MLR, MS and I,*

1. *$\hat{\vartheta}$ is a consistent estimator of ϑ.*

2. *$\sqrt{n}(s - \hat{\sigma})$ is asymptotically normally distributed, with zero mean and variance matrix determined only by τ, V and $\Gamma^\star$.*

3. *$\sqrt{n}(\hat{\tau} - \tau)$ is asymptotically normally distributed with $avar(\hat{\tau}) = avar(\hat{\tau} \mid V, \Gamma^\star)$ (i.e. the corresponding submatrix of (19) with $\Gamma = \Gamma^\star$).*

4. *when $V = V^\star$, then $avar(\hat{\tau})$ is the $t \times t$ leading principal submatrix of $\Delta' V^\star \Delta$, and $\hat{\tau}$ is an efficient estimator within the class of WLS estimators of (18).*

5. *the asymptotic distribution of the goodness-of-fit test statistic $T_V^\star$ of (28) is chi-square with degrees of freedom given by (30).*

Note that NA is not listed in the conditions of the theorem; i.e., MLR and MS are enough to guaranty that NA inferences concerning the residual vector $(s - \hat{\sigma})$, the subvector τ of ϑ, and the goodness-of-fit test statistic $T_V^\star$, remain asymptotically valid whether NA holds or not. Furthermore, under MLR and MS we attain valid inferences when Γ is replaced by Ω, an incorrect asymptotic variance matrix for $\sqrt{n}s$ even under NA. As it is illustrated in the next section, the validity of using Ω for statistical inferences allows the use of standard software of covariance structure analysis (such as LISREL of Jöreskog & Sörbom (1989), or EQS of Bentler (1989)) to analyze mean and covariance structures.

The theorem guarantees also that when MLR and MS hold, the NA standard errors of $\hat{\tau}$ coincide (asymptotically) with the AR standard errors based on the matrix $\hat{\Gamma}$ of fourth-order sample moments; the NA-WLS estimator $\hat{\tau}$ is asymptotically efficient; and T_V^* has the same asymptotic distribution as the AR test statistic T_V.

In the next section we present a Monte Carlo illustration showing the performance in finite samples of the asymptotic statments of Theorem 1.

3 Monte Carlo illustration

This section presents a small Monte Carlo study where we investigate the performance in finite samples of NA-WLS inferences in the context of a multi-sample regression with error in variables, and data deviating from the normality assumption.

We consider the following regression model with errors in variables:

$$\left\{ \begin{array}{lcl} Y & = & \alpha + \beta x + \zeta \\ X_1 & = & x + \epsilon_1 \\ X_2 & = & x + \epsilon_2 \end{array} \right. \tag{34}$$

Two-sample data is considered with the variable X_2 missing in the second subsample. This is a regression model in which Y is the dependent variable and X_1 and X_2 are two indicators of the "true" regressor x. We distinguish between the cases of x assumed "fixed" across replications (the functional model) or considered to be random (the structural model). Note that this model could serve in an applicaton of a regression with error in variables where two replicates X_1 and X_2 are available for a subsample of cases (group $g = 1$), while only one replicate is available in the rest of the sample (group $g = 2$).

We use the estimation method NA-WLS described in Section 3. In the illustration, we are concerned with correct asymptotic inferences when no distributional assumptions are imposed on the true regressor x, or on the disturbance regression term ζ. However, normality will be assumed for the distribution of the measurement error variables ϵ_1 and ϵ_2. The measurement error variances and the parameters α and β are set equal across groups. The second-order moments of the random constituents of the model are assumed to be finite. In the functional model case, we need to assume also that the uncentered second order moments of x converge as $n \to +\infty$ (Condition 1 in MLR).

3.1 The Functional Model

The equations describing the two-sample model are a measurement model for the first group

$$z_{i1} \equiv \begin{pmatrix} 1 \\ X_{i1,1} \\ X_{i1,2i} \\ Y_{i1} \end{pmatrix} = \begin{pmatrix} 1 & 0 & 0 \\ 0 & 1 & 0 \\ 0 & 1 & 0 \\ 0 & 0 & 1 \end{pmatrix} \begin{pmatrix} 1 \\ x_{i1} \\ Y_{i1} \end{pmatrix} + \begin{pmatrix} 0 \\ \epsilon_{i1,1} \\ \epsilon_{i1,2} \\ 0 \end{pmatrix}, \quad i = 1, \ldots, n_1, \tag{35}$$

a measurement model for the second group

$$z_{i2} \equiv \begin{pmatrix} 1 \\ X_{i2,1} \\ Y_{i2} \end{pmatrix} = \begin{pmatrix} 1 & 0 & 0 \\ 0 & 1 & 0 \\ 0 & 0 & 1 \end{pmatrix} \begin{pmatrix} 1 \\ x_{i2} \\ Y_{i2} \end{pmatrix} + \begin{pmatrix} 0 \\ \epsilon_{i2,1} \\ 0 \end{pmatrix}, \ i = 1, \ldots, n_2, \tag{36}$$

and a structural equation that is common to both groups

$$\begin{pmatrix} 1 \\ x_{ig} \\ Y_{ig} \end{pmatrix} = \begin{pmatrix} 0 & 0 & 0 \\ 0 & 0 & 0 \\ \alpha & \beta & 0 \end{pmatrix} \begin{pmatrix} 1 \\ x_{ig} \\ Y_{ig} \end{pmatrix} + \begin{pmatrix} 1 \\ x_{ig} \\ \zeta_{ig} \end{pmatrix}, \quad i = 1, \ldots, n_g, \quad g = 1, 2. \tag{37}$$

Clearly, assumptions MLR hold with

$$\Pi_1 = \left[\Lambda_1 (I - B)^{-1} \mid \begin{matrix} 0_{2\times 2} \\ I_{2\times 2} \end{matrix} \right], \tag{38}$$

$$\Pi_2 = \left[\Lambda_2 (I - B)^{-1} \mid \begin{matrix} 0_{2\times 2} \\ I_{1\times 2} \end{matrix} \right], \tag{39}$$

$$\Lambda_1 = \begin{pmatrix} 1 & 0 & 0 \\ 0 & 1 & 0 \\ 0 & 1 & 0 \\ 0 & 0 & 1 \end{pmatrix}, \Lambda_2 = \begin{pmatrix} 1 & 0 & 0 \\ 0 & 1 & 0 \\ 0 & 0 & 1 \end{pmatrix}, B = \begin{pmatrix} 0 & 0 & 0 \\ 0 & 0 & 0 \\ \alpha & \beta & 0 \end{pmatrix},$$

$$\xi_{i1} = (1, x_{i1}, \zeta_{i1}, \epsilon_{i1,1}, \epsilon_{i1,2})'$$

and

$$\xi_{i2} = (1, x_{i2}, \zeta_{i2}, \epsilon_{i2,1})'.$$

Thus we have the correspondence $\xi_{ig,0} = (1, x_{ig})$, $\xi_{ig,1} = \zeta_{ig}$,for $g = 1, 2$, and $\xi_{i1,2} = (\epsilon_{i1,1}, \epsilon_{i1,2})'$ and $\xi_{i2,2} = \epsilon_{i2,1}$.

Note also that assumption MS holds with the matrices Π_g of (5) and (39) being functions of β and α, and

$$\Phi_g = \text{diag}\,(\Phi_{g,00}, \Phi_{g,11}, \Phi_{g,22}),$$

where

$$\Phi_{g,00} = \begin{pmatrix} \nu_g & \phi_{g,01} \\ \phi_{g,01} & \phi_{g,00} \end{pmatrix}, \quad \Phi_{g,11} = \begin{pmatrix} \phi_{g,11} \end{pmatrix},$$

$$\Phi_{1,22} = \begin{pmatrix} \psi_{11} & 0 \\ 0 & \psi_{22} \end{pmatrix}, \quad \Phi_{2,22} = \begin{pmatrix} \psi_{11} \end{pmatrix}.$$

for $g = 1, 2$, where ψ_{11} is the common variance of ϵ_{i1} and ϵ_{i2}, ψ_{22} is the variance of ϵ_{i2}, and $\phi_{g,01}$ and $\phi_{g,00}$ are respectively the first and second-order moments of the unobservable variable x_g, $g = 1, 2$. We thus have the 12-dimensional parameter vector $\vartheta = (\tau', \phi')'$ with

$$\tau = (\alpha, \beta, \psi_{11}, \psi_{22})'$$

and

$$\phi = (\nu_1, \phi_{1,01}, \phi_{1,00}, \phi_{1,11}, \nu_2, \phi_{2,01}, \phi_{2,00}, \phi_{2,11}).$$

3.2 The Structural Model

Here, x_g is a random constituent of the model with mean μ_x, set equal across groups in the present example. We have the measurement equations of (35) and (36), and the structural equation common to both groups,

$$\begin{pmatrix} 1 \\ x_{ig} \\ Y_{ig} \end{pmatrix} = \begin{pmatrix} 0 & 0 & 0 \\ \mu_x & 0 & 0 \\ \alpha & 0 & \beta \end{pmatrix} \begin{pmatrix} 1 \\ x_{ig} \\ Y_{ig} \end{pmatrix} + \begin{pmatrix} 1 \\ x_{ig} - \mu_g \\ \zeta_{ig} \end{pmatrix}, \quad g = 1, 2. \tag{40}$$

Note that a new parameter, the mean μ_x of x, is in B. Also MLR is satisfied with the same Π_g as in (5) and (39) and with the appropriate substitution of the matrix B. Further,

$$\xi_i = (1, \zeta_{i1}, x_{i1} - \mu_x, \epsilon_{i1,1}, \epsilon_{i1,2})$$

and

$$\xi_{i2} = (1, \zeta_{i2}, x_{i2} - \mu_x, \epsilon_{i2,1})$$

We have the correspondence $\xi_{ig,0} = 1$, $\xi_{ig,1} = \zeta_{ig}$, $\xi_{ig,2} = x_{ig} - \mu$,for $g = 1, 2$, and $\xi_{i1,3} = (\epsilon_{i1,1}, \epsilon_{i1,2})'$, and $\xi_{i2,3} = \epsilon_{i2,1}$. The matrix Φ in MS is

$$\Phi_g = \operatorname{diag}(\nu_g, \Phi_{g,1}, \Phi_{g,2}, \Phi_{g,3}(\tau))$$

where $\Phi_{g,1} = (\phi_{g,11})$ corresponds to the variance of ζ_g, $\Phi_{g,2} = (\phi_{g,2})$ corresponds to the variance of x_g, $\Phi_{1,3} = \operatorname{diag}(\psi_{11}, \psi_{22})$ and $\Phi_{2,3} = \left(\ \psi_{11}\ \right)$; where ψ_{11} is the common variance of $\epsilon_{i1,1}$ and $\epsilon_{i1,2}$ and ψ_{22} is the variance of $\epsilon_{i2,1}$. Here $\vartheta = (\tau', \phi')'$ has 11 components,

$$\tau = (\alpha, \beta, \mu_x, \psi_{11}, \psi_{22})'$$

and

$$\phi = (\nu_1, \phi_{1,11}, \phi_{1,22}, \nu_2, \phi_{2,11}, \phi_{2,22}).$$

For both functional and structural models, ν_g is a "pseudo" parameter that is unrestricted across groups and that has a population value of 1. Note that in the case of the functional model it made no sense to restrict the means of x_g to be equal across groups.

The total number of distinct moments is 16 ($= \frac{4\times5}{2} + \frac{3\times4}{2}$). Hence, the number of degrees of freedom of the goodness-of-fit test statistic T^{**} is df= 4 ($= 16 - 12$) in the case of the functional model, and df= 5 ($= 16 - 11$) in the case of the structural model.

The Monte Carlo study generated 1000 sets of two-sample data according to the data generating process described below and the NA-WLS analysis of each two-sample data. A summary of the Monte Carlo results is presented in Tables 1 (the functional model) and 2 (the structural model).

Non-normal data was simulated for the x's and the ζ's as independent draws from the chi-square distribution with 1 df, conveniently scaled. The $\{\epsilon_{g,1i}\}_i$ and $\{\epsilon_{g,2i}\}_i$, $g = 1, 2$, were generated as iid draws from independent normal variables. Sample size was $n_1 = 800$ and $n_2 = 400$ in the structural model example, and $n_1 = 2800$ and $n_2 = 2200$ in the functional model example . The number of replications was set to 1000. In the functional model, a fixed set of values of x were used across the 1000 replications.

Given the described data generating process, conditions MLR and MS for Theorem 1 are clearly verified, for both the functional and structural models. We thus expect asymptotic correctness of the NA standard errors for the τ parameters (i.e., intercept and slope) and variances of $\epsilon_{g,1}$ and $\epsilon_{g,2}$. The goodness-of-fit test T^{**} is also expected to be asymptotically chi-square distributed. Note that this asymptotic correctness holds despite non-normality of ζ_g (disturbance term) and x_g ("true" values of the regressor), and the consequent non-normality of the observed variables. Correctness of the NA standard errors for estimates of the ϕ type of parameters (variances and covariances of non-normal constituents of the model) is not guaranteed. Such expectations of robustness are corroborated with the results shown by Tables 1 and 2.

Comparison of the entries in column $E(\hat{\vartheta})$ with the true values demonstrates the consistency of parameter estimates. Comparison of NA standard errors (column $E(se)$) with the empirical standard errors in column $sd(\hat{\vartheta})$ shows the consistency of standard errors for all parameters except for the variances of ζ_g in the functional model, and variances of ζ_g and x_g in the structural model. This consistency of standard errors for certain parameter estimates is corroborated by inspecting the deviations from 1 in column $\frac{E(se)}{sd(\hat{\vartheta})}$. The inspection of the columns giving empirical tails of a z-statistic shows also asymptotic normality and correctness of standard errors for the estimates of τ parameters. The empirical distribution of the goodness-of-fit test shows also a reasonably accurate fit to a chi-square distribution with the corresponding df. [2]

References

Amemiya, Y. 1985. "On the goodness-of-fit tests for linear statistical relationships." *Technical Report No. 10.* Econometric workshop, Stanford University.

Amemiya, Y. & Anderson, T.W. 1990. "Asymptotic chi-square tests for a large class of factor analysis models." *The Annals of Statistics, 3:* 1453-1463.

Anderson, T.W. 1989. "Linear latent variable models and covariance structures". *Journal of Econometrics, 41:* 91-119.

Anderson, T. W. 1987. "Multivariate linear relations." Pp. 9-36 in *Proceedings of the Second International Conference in Statistics,*edited by T. Pukkila & S. Puntanen. Tampere, Finland: University of Tampere.

Anderson, T.W. 1989. "Linear Latent Variable Models and Covariance Structures." *Journal of Econometrics, 41:* 91-119.

Anderson, T.W. & Amemiya, Y. 1988. "The asymptotic normal distribution of estimates in factor analysis under general conditions." *The Annals of Statistics, 16:*759-771.

Arminger, G. & Sobel, M.E., 1990. "Pseudo-maximum likelihood estimation of mean and covariance structures with missing data" *Journal of the American Statistical Association, 85:* 195- 203.

Bentler, P.M. 1983. "Simultaneous equation systems as moment structure models" *Journal of Econometrics, 22:* 13-42.

Bentler, P.M. 1989. *EQS Structural Equations Program Manual.* Los Angeles: BMDP Statistical Software, Inc.

Bollen, K.A., 1989. *Structural equations with latent variables.* New York: Wiley.

[2]The described WLS analysis can be carried out using standard software like LISREL or EQS. The "input" for a LISREL run on both type of models (functional and structural) is available from the author upon request.

Browne, M.W. 1984. "Asymptotically distribution-free methods for the analysis of covariance structures." *British Journal of Mathematical and Statistical Psychology, 37:* 62-83.

Browne, M.W. 1987. "Robustness in statistical inference in factor analysis and related models." *Biometrika, 74:* 375-384.

Browne, M.W. 1990. "Asymptotic Robustness of Normal Theory Methods for the Analysis of Latent Curves." Pp. 211-225 in *Statistical Analysis of Measurement Errors and Applications,* edited by P. J. Brown and W.A. Fuller. Rhode Island: American Mathematical Society.

Browne, M. W. & Shapiro, A. 1988. "Robustness of normal theory methods in the analysis of linear latent variable models." *British Journal of Mathematical and Statistical Psychology, 41:* 193-208.

Chamberlain, G. 1982. "Multivariate regression models for panel data." *Journal of Econometrics 18:* 5-46

Dham, P.F. & Fuller, W.A. (1986). *Generalized Least Squares Estimation of the Functional Multivariate Linear Errors-in-variables Model. Journal of Multivariate Analysis 19,* 132-141.

Hasabelnaby, N.A., Ware, J.H., and Fuller, W.A. (1989). "Indoor Air Pollution and Pulmonary Performance Investigating Errors in Exposure Assessment" (with comments), *Statistics in Medicine, 8,* 1109-1126.

Gleser, L. J. (1992). The importance of Assessing Measurement Reliability in Multivariate Regression *Journal of the American Statisticl Association 87,* 696-707.

Jöreskog, K. 1971. "Simultaneous factor analysis in several populations" *Psychometrika, 57:* 409-426.

Jöreskog, K. & Sörbom, D. 1989. *LISREL 7 A Guide to the Program and Applications.* (2nd ed.) Chicago: SPSS Inc

Kano, Y. (1993) "Asymptotic Properties of Statistical Inference Based on Fisher-Consistent Estimates in the Analysis of Covariance Structures" in *Statistical Modelling and Latent Variables,* edited by K. Haagen, D.J. Bartholomew and M. Deistler. Elsevier: Amsterdam.

Magnus J. & Neudecker, H. 1988. *Matrix differential calculus* . New York: Wiley.

Mooijaart, A. & Bentler, P. M. 1991. "Robustness of normal theory statistics in structural equation models." *Statistica Neerlandica, 45:* 159-171.

Muthén, B. 1987. *LISCOMP: Analysis of linear structural equations with a comprehensive measurement model* (User's Guide). Mooresville, IN: Scientific Software.

Muthén, B. Kaplan, D., & Hollis, M. (1987). On structural equation modeling with data that are not missing completely at random. *Psychometrika, 52,* 431-462.

Newey, W.K. 1988. "Asymptotic equivalence of closest moments and GMM estimators. " *Econometric Theory, 4*:336-340.

Satorra, A. 1989a. "Alternative test criteria in covariance structure analysis: a unified approach." *Psychometrika, 54:* 131-151.

Satorra, A. 1992b. "Asymptotic robust inferences in the analysis of mean and covariance structures." *Sociological Methodology 1992,* P.V. Marsden (edt.) pp. 249-278. Basil Blackwell: Oxford & Cambridge, MA

Satorra, A. 1993a. "Multi-Sample Analysis of Moment-Structures: Asymptotic Validity of Inferences Based on Second-Order Moments", in *Statistical Modelling and Latent*

Variables, edited by K. Haagen, D.J. Bartholomew and M. Deistler. Elsevier: Amsterdam.

Satorra, A. 1993b. "Asymptotic robust inferences in multiple-group analysis of augmented-moment structures", in pp. 211-229 *Multivariate Analysis: Future Directions 2*, edited by C.M. Cuadras and C.R. Rao. Elsevier: Amsterdam.

Satorra, A. & Bentler, P.M. 1990. "Model conditions for asymptotic robustness in the analysis of linear relations." *Computational Statistics & Data Analysis, 10:* 235-249.

Satorra, A. & Neudecker, H. 1994. On the asympotic optimality of alternative minimum-distance estimators in linear latent-variable models. *Econometric Theory, 10:* 867-883.

Shapiro, A. 1985. "Asymptotic equivalence of minimum discrepancy function estimators to G.L.S.estimators." *South African Statistical Journal, 19:*73-81.

Shapiro, A. 1986. "Asymptotic theory of overparameterized models." *Journal of the American Statistical Association, 81:* 142-149.

Table 1: Results of the simulation study. Structural model: $Y = \alpha + \beta x + \zeta$, $X_1 = x + \epsilon_1$, $X_2 = x + \epsilon_2$. Two subsamples of sizes $n_1 = 800$ and $n_2 = 400$ with the equation corresponding to X_2 missing in the second subsample. Distribution of x and ζ are independent chi-square of 1 df. Distribution of ϵ_1 and ϵ_2 is normal. Number of replications is 1000.

Estimates and standard errors

Parameters	$E(\hat{\vartheta})^a$	$sd(\hat{\vartheta})^b$	$E(se)^c$	$\frac{E(se)}{sd(\hat{\vartheta})}$	5% [d]	10%
$Var(e_1)= 1$	1.00	0.15	0.09	0.63	22.40	30.20
$Var(x_1)= 1$	0.99	0.14	0.06	0.42	43.90	52.00
$Var(u1) =0.3$	0.30	0.02	0.02	1.04	3.90	10.00
$Var(u2) =0.4$	0.40	0.03	0.03	0.96	6.90	11.90
$\beta =2$	2.00	0.05	0.05	0.98	5.10	9.30
$\alpha =1$	0.99	0.16	0.16	0.99	4.90	10.40
$\mu =3$	3.00	0.03	0.03	0.97	6.40	10.80
$Var(e_2)=1$	0.98	0.23	0.16	0.68	17.80	26.20
$Var(x_2)= 1$	0.99	0.19	0.08	0.45	38.10	45.90

Goodness-of-fit test, T^{**} (df=5)

Mean	Var	1%	5%	10%
5.13	10.49	1.10	4.90	9.80

[a] empirical mean across the 1000 replications

[b] standard deviation across 1000 replications

[c] empirical mean of the NA se

[d] 5% and 10% (nominal) two-sided tails for z-statistic $\frac{\hat{\vartheta}-\vartheta}{se(\hat{\vartheta})}$

Table 2: Results of the simulation study. Functional model: $Y = \alpha + \beta x + \zeta$, $X_1 = x + \epsilon_1$, $X_2 = x + \epsilon_2$. Two subsamples of sizes $n_1 = 2800$ and $n_2 = 2200$, with the equation corresponding to X_2 missing in the second subsample. Distribution of the disturbance term ϵ is chi-square of 1 df. The values of the x's are fixed across replications (and were generated as independent values from a uniform distribution). Distribution of ϵ_1 and ϵ_2 is normal. Number of replications is 500.

Estimates and standard errors

Parameters	$E(\hat{\vartheta})^a$	$sd(\hat{\vartheta})^b$	$E(se)^c$	$\frac{E(se)}{sd(\hat{\vartheta})}$	5% [d]	10%
$Var(e_1) = 1$	1.00	0.08	0.05	2.66	24.60	31.60
$Var(u_1) = 0.3$	0.30	0.01	0.01	1.01	6.00	11.20
$Var(u_2) = 0.4$	0.40	0.02	0.01	1.08	5.40	10.80
$\beta = 2$	2.00	0.01	0.01	1.03	6.40	10.60
$\alpha = 1$	1.00	0.02	0.02	1.01	5.80	11.60
$Var(e_2) = 1$	1.00	0.10	0.08	1.56	10.60	17.60

Goodness-of-fit test, T^{**} (df=4)

Mean	Var	5%	10%	20%
4.02	8.14	4.8	10.6	21.4

[a] empirical mean across the 1000 replications
[b] standard deviation across 1000 replications
[c] empirical mean of the NA se
[d] 5% and 10% (nominal) two-sided tails for z-statistic $\frac{\hat{\vartheta}-\vartheta}{se(\hat{\vartheta})}$

Non-Iterative Fitting of the Direct Product Model for Multitrait-Multimethod Matrices

M. W. Browne[1]
H. F. Strydom[2]

ABSTRACT The Composite Direct Product Model is a multiplicative model that conveniently decomposes a multitrait-multimethod correlation matrix into trait correlations, method correlations and communality indices. An iterative procedure is required to fit the model. Since the logarithms of absolute values of elements of the correlation matrix satisfy an additive model, the model may be fitted non-iteratively by generalized least squares. Antilogs of parameter estimates for the additive model then yield parameter estimates for the multiplicative model. The choice of weight matrix is considered and asymptotic properties of the estimators and test statistic are examined.

An example is used to compare the proposed estimates with conventional maximum likelihood estimates.

1. Introduction

Swain (1975) suggested a direct product model that is suitable for the analysis of multitrait-multimethod covariance matrices. The composite direct product (CDP) model (Browne, 1984b) is an extension of Swain's direct product model which is suitable for correlation matrices and allows for measurement error in the observations. Direct product models have desirable characteristics (Swain, 1975; Cudeck, 1988; Browne, 1989): they provide insight into the Campbell and Fiske (1959) conditions and have a multiplicative property suggested as appropriate for multitrait-multimethod correlation matrices by Campbell and O'Connell (1967). A number of applications of the CDP model have been reported (Bagozzi, 1993; Cudeck, 1988; Goffin & Jackson, 1992; Henly, Klebe, McBride & Cudeck, 1989; Lastovicka, Murry & Joachimsthaler, 1990).

Verhees and Wansbeek (1990, p. 238) expressed the ordinary least squares estimates for Swain's original direct product model *in closed form*, using the singular value decomposition of a rectangular matrix formed from elements of the covariance matrix. This approach, however, is not applicable to the CDP model. Browne (1984b) found that ordinary least squares, maximum likelihood and various generalized least squares parameter estimates for the CDP model all require an iterative computational procedure. An alternative parameterization of the CDP model which fits into the LISREL (Jöreskog, 1977) framework has been given by Wothke and Browne (1990), but an iterative computational procedure is still involved.

Non-iterative estimates for parameters in the CDP model will be developed here. We take the logs of the absolute values of correlation coefficients to transform the multiplicative direct product model to an additive model. Formulae for estimates can then be given in

---

[1]Department of Psychology and Department of Statistics, The Ohio State University, Columbus, Ohio
[2]Department of Statistics, Vista University, Pretoria, South Africa

closed form. Antilogs are taken of parameter estimates for the additive model to yield parameter estimates for the CDP model. A family of generalized least squares estimators is obtained and special cases arising from particular choices of the weight matrix are examined. Test statistics and measures of fit are provided. An example is then considered.

2. The Composite Direct Product Model

We consider situations where measurements are obtained under combinations of traits T and methods M. If there are t traits and m methods, $p = tm$ measurements are made on each of N subjects. Let $x(T_i M_r)$ denote the measurement of the i-th trait, T_i, using the r-th method, M_r. We assume that the p variables are ordered such that traits are nested within methods so that the vector, $\mathbf{x}$, of measurements is given by

$$\mathbf{x} = \{x(T_1M_1), x(T_2M_1), \ldots, x(T_tM_1), x(T_1M_2), x(T_2M_2), \ldots, x(T_tM_m)\}'$$

Consider a data model (Browne, 1989) where a $t \times 1$ vector, $\mathbf{z}_T$, representing trait effects and an $m \times 1$ vector, $\mathbf{z}_M$, representing method effects *interact multiplicatively* to yield the observation vector $\mathbf{x}$:

$$\mathbf{x} = \boldsymbol{\mu} + \mathbf{D}_\zeta^*(\mathbf{z}_M \otimes \mathbf{z}_T + \mathbf{u}), \tag{1}$$

where $\boldsymbol{\mu}$ is a $p \times 1$ mean vector with typical element $\mu(T_iM_r)$, $\mathbf{D}_\zeta^*$ is a $p \times p$ diagonal scaling matrix with typical diagonal element $\zeta(T_iM_r)$, $\otimes$ stands for the right direct product (Kronecker product) and $\mathbf{u}$ is a $p \times 1$ vector representing unique variates with typical element $u(T_iM_r)$. We assume that $\mathcal{E}\mathbf{z}_T = \mathbf{0}$, $\mathcal{E}\mathbf{z}_M = \mathbf{0}$, $\mathcal{E}\mathbf{u} = \mathbf{0}$, $\text{Cov}(\mathbf{z}_T, \mathbf{z}_M') = \mathbf{0}$, $\text{Cov}(\mathbf{z}_T, \mathbf{u}') = \mathbf{0}$, $\text{Cov}(\mathbf{z}_M, \mathbf{u}') = \mathbf{0}$, $\text{Cov}(\mathbf{z}_T, \mathbf{z}_T') = \mathbf{P}_T$, $\text{Cov}(\mathbf{z}_M, \mathbf{z}_M') = \mathbf{P}_M$ and $\text{Cov}(\mathbf{u}, \mathbf{u}') = \mathbf{D}_\psi$. The trait and method effects are standardized to have unit variances so that $\mathbf{P}_T$ and $\mathbf{P}_M$ are correlation matrices with typical elements $\rho(T_i, T_j)$ and $\rho(M_r, M_s)$ respectively. Unique variates are assumed to be uncorrelated so that $\mathbf{D}_\psi$ is a nonnegative definite diagonal matrix with typical diagonal element $\psi(T_iM_r)$.

The data model (1) generates the following structure for the covariance matrix, $\boldsymbol{\Sigma}$, of the observations (cf. Browne, 1989, expression (4.5))

$$\boldsymbol{\Sigma} = \mathbf{D}_\zeta^*(\mathbf{P}_M \otimes \mathbf{P}_T + \mathbf{D}_\psi)\mathbf{D}_\zeta^*. \tag{2}$$

Alternatively, the covariance structure in (2) is generated by a simplified three mode factor analysis data model (Browne, 1984b, Section 8).

We shall be concerned primarily with the correlation matrix

$$\mathbf{P} = \mathbf{D}_\sigma^{-1}\boldsymbol{\Sigma}\mathbf{D}_\sigma^{-1},$$

where

$$\mathbf{D}_\sigma^2 = \text{Diag}(\boldsymbol{\Sigma}) = \mathbf{D}_\zeta^{*2}(\mathbf{I} + \mathbf{D}_\psi),$$

which has the structure

$$\mathbf{P} = \text{Off}\{\mathbf{D}_\zeta(\mathbf{P}_M \otimes \mathbf{P}_T)\mathbf{D}_\zeta\} + \mathbf{I}, \tag{3}$$

where $\text{Off}\{\mathbf{W}\}$ represents the matrix formed from the off-diagonal elements of the square matrix $\mathbf{W}$ replacing all diagonal elements by zero. The diagonal matrix $\mathbf{D}_\zeta = \mathbf{D}_\sigma^{-1}\mathbf{D}_\zeta^*$ in (3) has the typical diagonal element

$$\zeta(T_iM_r) = \frac{1}{\sqrt{1+\psi(T_iM_r)}},$$

which is the correlation coefficient (cf. Browne, 1984b, formula (5.14)) between $x(T_iM_r)$ and its common part,

$$c(T_iM_r) = x(T_iM_r) - \zeta(T_iM_r) \times u(T_iM_r).$$

We shall refer to $\zeta(T_iM_r)$ as a "communality index". The squared communality index, $\zeta^2(T_iM_r)$, corresponds to the communality of factor analysis.

Consider a typical off-diagonal element of the manifest variable correlation matrix **P** in (3),

$$\rho(T_iM_r, T_jM_s) = \zeta(T_iM_r)\,\rho(T_i, T_j)\,\rho(M_r, M_s)\,\zeta(T_jM_s) \qquad j \neq i \text{ and/or } s \neq r, \tag{4}$$

where $\rho(T_i, T_j) = 1$ if $i = j$ and $\rho(M_r, M_s) = 1$ if $r = s$. Thus the model (3) expresses each manifest variable correlation coefficient as the product of four other correlation coefficients, which satisfy the inequalities

$$\begin{aligned} 0 \leq \zeta(T_iM_r) \leq 1 \qquad & i = 1, \ldots, t\,;\ r = 1, \ldots, m \\ -1 \leq \rho(T_i, T_j) \leq 1 \qquad & i = 1, \ldots, t\,;\ j = 1, \ldots, i-1 \\ -1 \leq \rho(M_r, M_s) \leq 1 \qquad & r = 1, \ldots, m\,;\ s = 1, \ldots, r-1 \end{aligned} \tag{5}$$

We shall now consider the transformation of the correlation coefficients $\rho(T_iM_r, T_jM_s)$ in (4) to yield an additive model.

3. The Log Absolute Correlation Structure

In order to carry out a transformation to an additive model we require an additional assumption, namely that no off-diagonal element of **P** is zero,

$$\rho(T_iM_r, T_jM_s) \neq 0 \qquad j \neq i \text{ and/or } s \neq r, \tag{6}$$

or, equivalently, that no parameter in the correlation structure (3), (4) is zero,

$$\begin{aligned} \zeta(T_iM_r) \neq 0 \qquad & i = 1, \ldots, t\,;\ r = 1, \ldots, m \\ \rho(T_i, T_j) \neq 0 \qquad & i = 1, \ldots, t\,;\ j = 1, \ldots, i-1 \\ \rho(M_r, M_s) \neq 0 \qquad & r = 1, \ldots, m\,;\ s = 1, \ldots, r-1. \end{aligned}$$

We may then take the logarithms of absolute values of correlation coefficients, $\rho(T_iM_r, T_jM_s)$, in (4) to obtain an additive model. Let

$$\eta(T_iM_r, T_jM_s) = \ln|\rho(T_iM_r, T_jM_s)|\,. \tag{7}$$

It follows from (4) that the log absolute correlation (LAC) model is:

$$\eta(T_iM_r, T_jM_s) = \lambda(T_iM_r) + \varphi(T_i, T_j) + \varphi(M_r, M_s) + \lambda(T_jM_s) \qquad j \neq i \text{ and/or } s \neq r, \tag{8}$$

where

$$\begin{aligned}
\lambda(T_i M_r) &= \ln \zeta(T_i M_r) \qquad i = 1, \dots, t;\ r = 1, \dots, m \\
\varphi(T_i, T_j) &= \ln \mid \rho(T_i, T_j) \mid \qquad i = 1, \dots, t\ ;\ j = 1, \dots, i-1 \\
\varphi(M_r, M_s) &= \ln \mid \rho(M_r, M_s) \mid \qquad r = 1, \dots, m\ ;\ s = 1, \dots, r-1.
\end{aligned} \tag{9}$$

and $\varphi(T_i, T_j) = 0$ if $i = j$, $\varphi(M_r, M_s) = 0$ if $r = s$. Use of (5) and (9) shows that the parameters in the LAC model satisfy the inequalities

$$\begin{aligned}
\lambda(T_i M_r) &\le 0 \qquad i = 1, \dots, t;\ r = 1, \dots, m \\
\varphi(T_i, T_j) &\le 0 \qquad i = 1, \dots, t\ ;\ j = 1, \dots, i-1 \\
\varphi(M_r, M_s) &\le 0 \qquad r = 1, \dots, m\ ;\ s = 1, \dots, r-1.
\end{aligned} \tag{10}$$

The LAC model (8) will be expressed in matrix notation as a linear model. Let $p^* = \frac{1}{2}p(p-1)$ stand for the number of distinct nondiagonal elements of $\mathbf{P}$, let $\boldsymbol{\eta}$ be a $p^* \times 1$ vector with typical element $\eta(T_i M_r, T_j M_s)$ and let $\boldsymbol{\theta}$ be a $q \times 1$ vector consisting of the parameters $\lambda(T_i M_r)$, $\varphi(T_i, T_j)$ and $\varphi(M_r, M_s)$.

In some situations parameters in the CDP model are constrained to be equal (cf. Bagozzi & Yi, 1992). This implies that the corresponding parameters (9) in the LAC model are equal. We may therefore allow for equality constraints by assigning the same element θ_k of $\boldsymbol{\theta}$ to several parameters. For example, if all method correlations forming the off-diagonal elements of $\mathbf{P}_M$ in (3) are required to be equal,

$$\rho(M_2, M_1) = \rho(M_3, M_1) = \rho(M_3, M_2) = \cdots = \rho(M_m, M_{m-1}),$$

we may let

$$\theta_1 = \varphi(M_2, M_1) = \varphi(M_3, M_1) = \varphi(M_3, M_2) = \cdots = \varphi(M_m, M_{m-1}).$$

Similarly, if all communality indices, $\zeta(T_i M_r)$, are required to be equal we take

$$\theta_2 = \lambda(T_1 M_1) = \lambda(T_2 M_1) = \cdots = \lambda(T_t M_m).$$

Consequently

$$q \le tm + \frac{1}{2}t(t-1) + \frac{1}{2}m(m-1).$$

Let $\mathbf{A}$ be a $p^* \times q$ design matrix with a row for each element of $\boldsymbol{\eta}$ and a column for each element of $\boldsymbol{\theta}$. Each element of $\mathbf{A}$ will represent the number of times the element of $\boldsymbol{\theta}$ corresponding to the column appears in the model (8) for the element of $\boldsymbol{\eta}$ corresponding to the row. Each row of $\mathbf{A}$ will have at most four elements equal to one and the remaining elements will be zero. If no equality constraints are imposed on parameters, the row of $\mathbf{A}$ corresponding to $\eta(T_i M_r, T_j M_s)$ will have ones in the columns corresponding to $\lambda(T_i M_r)$, $\lambda(T_j M_s)$, $\varphi(T_i, T_j)$, $i \ne j$, and $\varphi(M_r, M_s)$, $r \ne s$, and zeros elsewhere.

We may then write (8) in matrix notation as

$$\boldsymbol{\eta} = \boldsymbol{\eta}(\boldsymbol{\theta}) = \mathbf{A}\boldsymbol{\theta}. \tag{11}$$

Although the LAC model (8) holds whenever the CDP model (4) holds, the converse is not true. Since the transformation in (9) is not one to one unless all parameters are required to be positive, the LAC model does not imply the CDP model unless **P** is restricted to the set of correlation matrices with positive elements.

4. Fitting the Log Absolute Correlation Structure

Let **R** be a sample correlation matrix, with typical element $r(T_iM_r, T_jM_s)$, based on a sample of size $N = n+1$ drawn from a population with correlation matrix $\mathbf{P}_o$. We do not assume that $\mathbf{P}_o$ necessarily satisfies the model (3) exactly. Let

$$\mathrm{vec}_r(\mathbf{R}) = (r_{12}, r_{13}, r_{23}, \ldots, r_{p-1,p})'$$

stand for the $p^* \times 1$ vector formed from the *distinct nondiagonal* elements of a symmetric $p \times p$ matrix **R** and let $\mathbf{r} = \mathrm{vec}_r(\mathbf{R})$ and $\boldsymbol{\rho}_o = \mathrm{vec}_r(\mathbf{P}_o)$. The asymptotic distribution of $\boldsymbol{\delta}_r = n^{1/2}(\mathbf{r} - \boldsymbol{\rho}_o)$ is multivariate normal with a null mean vector and a covariance matrix which will be represented by

$$\boldsymbol{\Gamma}_r = \mathrm{Acov}(\boldsymbol{\delta}_r).$$

Specific expressions for the elements of $\boldsymbol{\Gamma}_r$ will be provided subsequently in Sections 4.2 and 4.3. We apply a similar transformation to that in (7) to the elements of **r**,

$$y(T_iM_r, T_jM_s) = \ln|r(T_iM_r, T_jM_s)|$$

and let **y** stand for the $p^* \times 1$ vector with typical element $y(T_iM_r, T_jM_s)$. If $\boldsymbol{\eta}_o$ is the $p^* \times 1$ vector obtained by applying (7) to the elements of $\boldsymbol{\rho}_o$, the asymptotic distribution of $\boldsymbol{\delta}_y = n^{1/2}(\mathbf{y} - \boldsymbol{\eta}_0)$ is multivariate normal with a null mean vector and a covariance matrix which will be represented by $\boldsymbol{\Gamma}_y$. Use of the Delta Theorem (e.g. Rao, 1973, Theorem 6a.2 (iii); Bishop, Fienberg & Holland, 1975, Theorem 14.6-2) shows that

$$\boldsymbol{\Gamma}_y = \mathbf{D}_\rho^{-1}\boldsymbol{\Gamma}_r\mathbf{D}_\rho^{-1} \tag{12}$$

where $\mathbf{D}_\rho$ is a $p^* \times p^*$ diagonal matrix with diagonal elements equal to the distinct nondiagonal elements of $\boldsymbol{\rho}_o$. Consequently expressions for $\boldsymbol{\Gamma}_y$ are easy to obtain in all situations where expressions for $\boldsymbol{\Gamma}_r$ are available.

We shall consider parameter estimates obtained by minimizing a generalized least squares discrepancy function of the type

$$F(\mathbf{y}, \boldsymbol{\eta}(\boldsymbol{\theta})) = (\mathbf{y} - \mathbf{A}\boldsymbol{\theta})'\mathbf{V}^{-1}(\mathbf{y} - \mathbf{A}\boldsymbol{\theta}), \tag{13}$$

where the $p^* \times p^*$ positive definite matrix **V** converges in probability to a fixed matrix $\bar{\mathbf{V}}$. The minimizer of $F(\mathbf{y}, \boldsymbol{\eta}(\boldsymbol{\theta}))$ is

$$\hat{\boldsymbol{\theta}} = \mathbf{H}^{-1}\mathbf{g} \tag{14}$$

where

$$\mathbf{g} = \mathbf{A}'\mathbf{V}^{-1}\mathbf{y} = -\frac{1}{2}\frac{\partial F}{\partial \boldsymbol{\theta}}\bigg|_{\boldsymbol{\theta}=0}, \tag{15}$$

$$\mathbf{H} = \mathbf{A}'\mathbf{V}^{-1}\mathbf{A} = \frac{1}{2}\frac{\partial^2 F}{\partial\theta\,\partial\theta'}, \qquad (16)$$

and the corresponding minimum is $\widehat{F} = F(\mathbf{y}, \eta(\widehat{\theta}))$. It is possible for an element of $\widehat{\theta}$ in (14) to exceed the upper bound of zero. In the LINLOG program (Browne & Strydom, 1991) a quadratic programming procedure is employed to minimize $F(\mathbf{y}, \eta(\theta))$ subject to the inequality constraints in (10).

We define the corresponding population vector θ_o as the minimizer of $F(\eta_o, \eta(\theta))$ so that

$$\theta_o = (\mathbf{A}'\bar{\mathbf{V}}^{-1}\mathbf{A})^{-1}\mathbf{A}'\bar{\mathbf{V}}^{-1}\eta_o.$$

If the corresponding minimum $F_o = F(\eta_o, \eta(\theta_o))$ is zero, η_o satisfies the linear model (11) exactly with $\theta = \theta_o$.

If **V** is a consistent estimator of Γ_y,

i.e.
$$\bar{\mathbf{V}} = \Gamma_y, \qquad (17)$$

and if n is large and F_o is small, the asymptotic distribution of $\delta_{\widehat{\theta}} = n^{1/2}(\widehat{\theta} - \theta_o)$ is approximately multivariate normal with a null mean vector and the covariance matrix (cf. Browne, 1984a, Corollary 2.1)

$$\mathrm{Acov}(\delta_{\widehat{\theta}}) = (\mathbf{A}'\Gamma_y^{-1}\mathbf{A})^{-1} \qquad (18)$$

so that the covariance matrix of $\widehat{\theta}$ may be estimated by

$$\widehat{\mathrm{Cov}}(\widehat{\theta}) = n^{-1}\mathbf{H}^{-1}. \qquad (19)$$

Then $\widehat{\theta}$ is asymptotically efficient within the class of minimum discrepancy estimators based on **y** (cf. Browne, 1984a, Proposition 3; Shapiro, 1985). The distribution of $n\widehat{F}$ is approximately noncentral chi-squared with $d = p^* - q$ degrees of freedom and noncentrality parameter nF_o (Browne, 1984a, Corollary 4.1).

Some caution should be exercised in regarding the test statistic $n\widehat{F}$ as providing a test of fit of the CDP model ((3), (4)) rather than of the LAC model (5). Whenever the CDP model (3) holds and $\bar{\mathbf{V}} = \Gamma_y$, the asymptotic distribution of $n\widehat{F}$ is central chi-squared. However, the asymptotic distribution of $n\widehat{F}$ will still be central chi-squared in situations when the CDP model does not hold, but the LAC model does. The power of the test, therefore, can be low (equal to the significance level) under certain alternatives to the CDP model. When it is known that no elements of **P** can be negative, the LAC model implies the CDP model and there are no difficulties in considering the test as a test of fit of the CDP model.

Since it is implausible that the LAC model holds exactly in a practical situation, it is preferable to make use of a measure of fit of the model. As a measure of fit of the LAC model it is convenient to make use of the root mean square error of approximation (RMSEA) suggested by Steiger and Lind (1980),

$$\varepsilon_a = \sqrt{\frac{F_o}{d}}.$$

An approximately unbiased point estimate is

$$\hat{\varepsilon}_a = \sqrt{\frac{n\hat{F} - d}{nd}} \tag{20}$$

and an interval estimate of ε_a may be obtained from a 90% confidence interval on the noncentrality parameter of a noncentral chi-squared distribution. A value of zero for the lower limit of the 90% confidence interval on the RMSEA implies that the test based on $n\hat{F}$ will not reject the null hypothesis that the LAC model fits perfectly (H_o: $\varepsilon_a = 0$) at the 95% level. Further details are given by Steiger (1989) and by Browne and Cudeck (1993). Again, the point estimate in (20) and the associated confidence interval are not always relevant to the CDP model.

We now consider some specific choices of **V**.

4.1 Ordinary Least Squares

We take $\mathbf{V} = \hat{\mathbf{V}} = \mathbf{I}$ so that $\hat{\theta}$ is obtained from (14) with

$$\mathbf{H} = \mathbf{A}'\mathbf{A} = \sum_{i=1}^{p^*} \mathbf{a}_i\mathbf{a}_i' \qquad \mathbf{g} = \mathbf{A}'\mathbf{y} = \sum_{i=1}^{p^*} \mathbf{a}_i y_i$$

where $\mathbf{a}_i'$ is the $1 \times q$ vector formed from the i-th row of **A** and y_i is the i-th element of **y**. It is not necessary to store the full $p^* \times q$ design matrix **A** and only one row need be evaluated at a time.

The main advantage of ordinary least squares is that computation of $\hat{\theta}$ is rapid and does not require much computer storage. The estimate, $\hat{\theta}$, is consistent but not asymptotically efficient. Since (17) does not apply, it is not, in general, true that the asymptotic covariance matrix of $\hat{\theta}$ is given by (18) and that nF has an asymptotic chi-squared distribution. Although an estimate of the asymptotic covariance matrix of $\hat{\theta}$ (Browne 1984a, Proposition 2) and a test statistic with an asymptotic chi-squared distribution (Browne, 1984a, Proposition 4) can be calculated, they are more complicated and involve additional computation. It would be easier to use one of the asymptotically efficient estimates suggested in Sections 4.2 and 4.3.

The ordinary least squares (OLS) estimate can be used to obtain starting values for an iterative procedure for obtaining estimates of parameters in the CDP model. It also is used in the first stage of the two stage generalized least squares procedure, GLSO, described in Section 4.3.

4.2 Asymptotically Distribution Free Methods

If the distribution of the observation vector variate, **x**, has finite fourth order moments, a typical element of Γ_r in (12) is (Isserlis, 1916; Hsu, 1949; Steiger & Hakstian, 1982)

$$\begin{aligned}[\Gamma_r]_{ij,k\ell} &= \mathrm{Acov}(\delta_{r_{ij}}, \delta_{r_{k\ell}}) \\ &= \frac{1}{4}\rho_{ij}\rho_{k\ell}(\rho_{iikk} + \rho_{jjkk} + \rho_{ii\ell\ell} + \rho_{jj\ell\ell}) - \frac{1}{2}\rho_{ij}(\rho_{iik\ell} + \rho_{jjk\ell}) - \frac{1}{2}\rho_{k\ell}(\rho_{ijkk} + \rho_{ij\ell\ell})\end{aligned} \tag{21}$$

where

$$\rho_{ijk\ell} = \frac{\mathcal{E}\{(x_i - \xi_i)(x_j - \xi_j)(x_k - \xi_k)(x_\ell - \xi_\ell)\}}{\sqrt{\sigma_{ii}\sigma_{jj}\sigma_{kk}\sigma_{\ell\ell}}}.$$

Let

$$s_{ii} = N^{-1}\sum_{t=1}^{N}(x_{it} - \bar{x}_i)^2$$

$$s_{ijk\ell} = N^{-1}\sum_{t=1}^{N}(x_{it} - \bar{x}_i)(x_{jt} - \bar{x}_j)(x_{kt} - \bar{x}_k)(x_{\ell t} - \bar{x}_\ell)$$

A consistent estimate, $\widehat{\gamma}_{r_{ij,k\ell}}$, of $\gamma_{r_{ij,k\ell}}$ may be obtained by substituting

$$r_{ijk\ell} = \frac{s_{ijk\ell}}{\sqrt{s_{ii}s_{jj}s_{kk}s_{\ell\ell}}} \quad \text{and} \quad r_{ij} = \frac{s_{ij}}{\sqrt{s_{ii}s_{jj}}}$$

for $\rho_{ijk\ell}$ and ρ_{ij} in (21). A simple alternative method for computing $\widehat{\mathbf{\Gamma}}_r$ has been described by Mooijaart (1985).

After $\widehat{\mathbf{\Gamma}}_r$ has been obtained, the weight matrix is obtained (cf. (12)) from

$$\mathbf{V} = \widehat{\mathbf{\Gamma}}_y = \mathbf{D}_r^{-1}\widehat{\mathbf{\Gamma}}_r\mathbf{D}_r^{-1},$$

the estimate $\widehat{\theta}$ is obtained from (14), $\widehat{F} = F(\mathrm{y}, \eta(\widehat{\theta}))$ is calculated using (13) and $\widehat{\mathrm{Cov}}(\widehat{\theta})$ is obtained from (19).

Under minimal assumptions concerning the distribution of observations, the ADF procedure yields a test statistic with an asymptotic chi-squared distribution and the standard errors of the asymptotic distribution of parameter estimators are known. There have been indications from random sampling experiments (Muthén & Kaplan, 1992) in related situations, however, that substantial sample sizes are needed for the asymptotic theory to yield adequate approximations, particularly if the model is large. This may be attributed to the imprecision of estimators of the fourth degree moments involved in the weight matrix, **V**. Also, since **V** is relatively large, heavy demands are made on computer storage.

4.3 Normality Assumptions

If the distribution of **x** is multivariate normal, a typical element of $\mathbf{\Gamma}_r$ is (Pearson & Filon, 1898)

$$[\mathbf{\Gamma}_r]_{ij,k\ell} = \frac{1}{2}\rho_{ij}\rho_{k\ell}(\rho_{ik}^2 + \rho_{i\ell}^2 + \rho_{jk}^2 + \rho_{j\ell}^2) + \rho_{ik}\rho_{j\ell} + \rho_{i\ell}\rho_{jk} - \rho_{ij}(\rho_{jk}\rho_{j\ell} + \rho_{ik}\rho_{i\ell}) - \rho_{k\ell}(\rho_{jk}\rho_{ik} + \rho_{j\ell}\rho_{i\ell}). \tag{22}$$

Unlike (21), expression (22) involves standardized second order moments alone. It would be possible to substitute consistent estimates for the correlations in (22) to yield elements of **V** which would be employed in the same way as in Section 4.2. Since no fourth order moments are involved, it is to be expected that asymptotic approximations would be useful at smaller sample sizes. The computational difficulties associated with the $p^* \times p^*$ weight matrix, **V**, would remain, however.

These computational difficulties may be avoided. Let $\mathbf{E}$ be a $p \times p$ symmetric matrix with null diagonal elements, and let $\mathbf{e} = \text{vec}_r(\mathbf{E})$. Using the relationship between $\boldsymbol{\Gamma}_r^{-1}$ and an information matrix under normality assumptions, Jennrich (1970) showed that

$$\mathbf{e}'\boldsymbol{\Gamma}_r^{-1}\mathbf{e} = \frac{1}{2}\text{tr}\left[\mathbf{E}\mathbf{P}^{-1}\right]^2 - \text{diag}'\left[\mathbf{E}\mathbf{P}^{-1}\right]\mathbf{T}^{-1}\text{diag}\left[\mathbf{E}\mathbf{P}^{-1}\right] \qquad (23)$$

where

$$\mathbf{T} = \mathbf{I} + \mathbf{P}{*}\mathbf{P}^{-1},$$

$\mathbf{P}$ represents a $p \times p$ population correlation matrix with typical element ρ_{ij}, $\text{diag}[\mathbf{W}]$ is a $p \times 1$ vector formed from the diagonal elements of a $p \times p$ matrix $\mathbf{W}$ and $\mathbf{Y}{*}\mathbf{Z}$ is the $p \times p$ Hadamard product of the $p \times p$ matrices $\mathbf{Y}$ and $\mathbf{Z}$, with typical element $[\mathbf{Y}{*}\mathbf{Z}]_{ij} = [\mathbf{Y}]_{ij} \times [\mathbf{Z}]_{ij}$. It follows from (12), (23) and the identity

$$\mathbf{D}_\rho\mathbf{e} = \text{vec}_r(\mathbf{E}{*}\mathbf{P})$$

that

$$\mathbf{e}'\boldsymbol{\Gamma}_y^{-1}\mathbf{e} = \frac{1}{2}\text{tr}\left[(\mathbf{E}{*}\mathbf{P})\mathbf{P}^{-1}\right]^2 - \text{diag}'\left[(\mathbf{E}{*}\mathbf{P})\mathbf{P}^{-1}\right]\mathbf{T}^{-1}\text{diag}\left[(\mathbf{E}{*}\mathbf{P})\mathbf{P}^{-1}\right] \qquad (24)$$

The result given in (24) expresses a quadratic form involving the large $p^* \times p^*$ matrix $\boldsymbol{\Gamma}_y$ in an equivalent form that involves no matrix of order greater than $p \times p$. It will now be employed to express the discrepancy function $F(\mathbf{y}, \boldsymbol{\eta}(\boldsymbol{\theta}))$ in (13) in a computationally efficient form.

Let $\widetilde{\mathbf{P}}$ be the consistent estimate of $\mathbf{P}_o$ that is used to form the elements of $\mathbf{V}$ from (22). We express the model, $\boldsymbol{\eta}(\boldsymbol{\theta})$, in (11) in matrix form, $\boldsymbol{\Upsilon}(\boldsymbol{\theta})$, with $\text{diag}[\boldsymbol{\Upsilon}(\boldsymbol{\theta})] = \mathbf{0}$ and

$$\text{vec}_r\{\boldsymbol{\Upsilon}(\boldsymbol{\theta})\} = \boldsymbol{\eta}(\boldsymbol{\vartheta}),$$

and substitute $\mathbf{E} = \{\mathbf{Y} - \boldsymbol{\Upsilon}(\boldsymbol{\theta})\}$ and $\mathbf{P} = \widetilde{\mathbf{P}}$ into (24) to obtain the following equivalent, but computationally more convenient, expression for the discrepancy function in (13):

$$F(\mathbf{y}, \boldsymbol{\eta}(\boldsymbol{\theta})) = \frac{1}{2}\text{tr}\left[\mathbf{B}_\theta^2\right] - \text{diag}'[\mathbf{B}_\theta]\widetilde{\mathbf{T}}^{-1}\text{diag}[\mathbf{B}_\theta], \qquad (25)$$

where

$$\mathbf{B}_\theta = \mathbf{B}(\boldsymbol{\theta}) = \left[\{\mathbf{Y} - \boldsymbol{\Upsilon}(\boldsymbol{\theta})\}{*}\widetilde{\mathbf{P}}\right]\widetilde{\mathbf{P}}^{-1}.$$

$$\widetilde{\mathbf{T}} = \mathbf{I} + \widetilde{\mathbf{P}}{*}\widetilde{\mathbf{P}}^{-1}$$

Equivalent expressions for the gradient $\mathbf{g}$ in (15) and Hessian $\mathbf{H}$ in (16) may be obtained from first and second derivatives of $F(\mathbf{y}, \boldsymbol{\eta}(\boldsymbol{\theta}))$. Let

$$\mathbf{B}_0 = \mathbf{B}(0) = \left(\mathbf{Y}{*}\widetilde{\mathbf{P}}\right)\widetilde{\mathbf{P}}^{-1},$$

$$\dot{\mathbf{B}}_k = -\frac{\partial \mathbf{B}}{\partial \theta_k} = \left(\mathbf{A}_k{*}\widetilde{\mathbf{P}}\right)\widetilde{\mathbf{P}}^{-1},$$

where

$$\mathbf{A}_k = \frac{\partial \Upsilon}{\partial \theta_k}$$

is a symmetric $p \times p$ matrix such that $\text{vec}_r(\mathbf{A}_k)$ is equal to the k-th column of the $p^* \times q$ design matrix $\mathbf{A}$ in (11). The diagonal elements of $\mathbf{A}_k$ are equal to 0.

Typical elements of $\mathbf{g}$ and $\mathbf{H}$ then are given by

$$g_k = -\frac{1}{2}\frac{\partial F}{\partial \theta_k}\bigg|_{\theta=0} = \frac{1}{2}\text{tr}[\mathbf{B}_0\dot{\mathbf{B}}_k] - \text{diag}'[\mathbf{B}_0]\widetilde{\mathbf{T}}^{-1}\text{diag}[\dot{\mathbf{B}}_k],$$

$$h_{k\ell} = \frac{1}{2}\frac{\partial^2 F}{\partial \theta_k \partial \theta_\ell} = \frac{1}{2}\text{tr}[\dot{\mathbf{B}}_k\dot{\mathbf{B}}_\ell] - \text{diag}'[\dot{\mathbf{B}}_k]\widetilde{\mathbf{T}}^{-1}\text{diag}[\dot{\mathbf{B}}_\ell].$$

Again, these expressions are convenient for computational purposes as no matrices of order greater than $p \times p$ are involved.

The estimate $\widehat{\theta}$ is obtained from (14), $\widehat{F} = F(\mathbf{y}, \eta(\widehat{\theta}))$ is calculated using (25) and $\widehat{\text{Cov}}(\widehat{\theta})$ is obtained from (19).

A possible choice for the consistent estimate, $\widetilde{\mathbf{P}}$, employed to obtain $\widehat{\theta}$, is to take $\widetilde{\mathbf{P}} = \mathbf{R}$. The resulting estimates will be referred to as GLSR estimates. This choice does not impose any structure on $\widetilde{\mathbf{P}}$. An alternative which has been found to give good results is a two stage procedure. In the first stage the methods of Sections 4.1 and 5.1 are used to obtain an ordinary least squares reconstructed correlation matrix, $\widehat{\mathbf{P}}_{\text{OLS}}$. In the second stage the methods of the present section are employed with $\widetilde{\mathbf{P}} = \widehat{\mathbf{P}}_{\text{OLS}}$, provided that $\widehat{\mathbf{P}}_{\text{OLS}}$ is positive definite. The resulting estimates will be referred to as GLSO estimates. In situations where $\widehat{\mathbf{P}}_{\text{OLS}}$ is not positive definite, GLSR estimates are used instead of GLSO estimates.

5. Transforming Back to the Multiplicative CDP Model

The elements of the vector $\widehat{\theta}$ in (14) are estimates, $\widehat{\lambda}(T_iM_r)$, $\widehat{\varphi}(T_i, T_j)$, $\widehat{\varphi}(M_r, M_s)$, of parameters in the additive LAC model and are not of direct interest. What we require, rather, are estimates, $\widehat{\zeta}(T_iM_r)$, $\widehat{\rho}(T_i, T_j)$, $\widehat{\rho}(M_r, M_s)$, of corresponding parameters in the CDP model.

5.1 Transformation of Parameters

Let $\widehat{\gamma}$ be a $q \times 1$ vector that corresponds to θ but consists of parameters, $\zeta(T_iM_r)$, $\rho(T_i, T_j)$, $\rho(M_r, M_s)$, in the CDP model. Absolute values of the elements of the estimate, $\widehat{\gamma}$, may be obtained from the elements of $\widehat{\theta}$ by the inverse transformation,

$$|\,\widehat{\gamma}_i\,| = \exp(\widehat{\theta}_i)$$

or specifically,

$$\begin{aligned} \widehat{\zeta}(T_iM_r) &= \exp\{\widehat{\lambda}(T_iM_r)\}, \\ |\,\widehat{\rho}(T_i, T_j)\,| &= \exp\{\widehat{\varphi}(T_i, T_j)\}, \\ |\,\widehat{\rho}(M_r, M_s)\,| &= \exp\{\widehat{\varphi}(M_r, M_s)\}. \end{aligned}$$

The signs of the trait correlations, $\rho(T_i, T_j)$, and method correlations, $\rho(M_r, M_s)$, are reflected by the signs of certain manifest variable correlations. Since (cf. (4))

$$\rho(T_i, T_j)\,\zeta(T_iM_r)\zeta(T_jM_r) = \rho(T_iM_r, T_jM_r), \qquad r = 1, \ldots, m,$$

and the communality indices $\zeta(T_iM_r)$, $\zeta(T_jM_r)$ cannot be negative, it follows that

$$\text{sign}\,\{\rho(T_i, T_j)\} = \text{sign}\,\{\rho(T_iM_r, T_jM_r)\}, \qquad r = 1, \ldots, m.$$

Therefore we may allocate signs to trait correlation estimates using the rule

$$\text{sign}\,\{\widehat{\rho}(T_i, T_j)\} = \text{sign}\left\{\sum_{r=1}^{m} r(T_iM_r, T_jM_r)\right\}, \tag{26}$$

although caution should be exercised in the interpretation of the result if the correlation coefficients summed on the right hand side of (26) are not all of the same sign. If the correlation coefficients involved in a sign clash do not differ significantly from zero, the sample size may not be large enough for one to be able to rely on asymptotic properties of the estimates and test statistic. If the correlation coefficients are large and the sign clash is unlikely to have occurred by chance, the direct product model may not be appropriate for the data.

Similarly, the rule

$$\text{sign}\Big\{\widehat{\rho}(M_r, M_s)\Big\} = \text{sign}\left\{\sum_{i=1}^{t} r(T_iM_r, T_iM_s)\right\} \tag{27}$$

may be used to allocate signs to method correlation estimates.

After $\widehat{\gamma}$ has been obtained, elements of the correlation matrix reconstructed according to the CDP model, $\widehat{\mathbf{P}} = \mathbf{P}(\widehat{\gamma})$, may be obtained from (4).

5.2 Asymptotic Distribution of Transformed Estimators

If $\bar{\mathbf{V}} = \mathbf{\Gamma}_y$, application of the delta theorem shows that the asymptotic distribution of $\delta_{\widehat{\gamma}} = n^{1/2}(\widehat{\gamma} - \gamma_o)$ is multivariate normal with a null mean vector and covariance matrix

$$\begin{aligned}\text{Acov}(\delta_{\widehat{\gamma}}) &= \mathbf{D}_\gamma \text{Acov}(\delta_{\widehat{\theta}})\mathbf{D}_\gamma \\ &= \mathbf{D}_\gamma (\mathbf{A}'\mathbf{\Gamma}_y^{-1}\mathbf{A})^{-1}\mathbf{D}_\gamma \end{aligned} \tag{28}$$

where $\mathbf{D}_\gamma$ is a $q \times q$ diagonal matrix with the elements of γ as diagonal elements. Consequently, if the methods of Section 4.2 are used, or the methods of Section 4.3 are used and the associated normality assumptions are appropriate, the estimator covariance matrix may be estimated using

$$\widehat{\text{Cov}}(\widehat{\gamma}) = n^{-1}\mathbf{D}_{\widehat{\gamma}}\mathbf{H}^{-1}\mathbf{D}_{\widehat{\gamma}}. \tag{29}$$

5.3 Asymptotic Efficiency of Transformed Estimators

If $\hat{\gamma}_e$ is asymptotically efficient within the class of minimum discrepancy estimators of γ_o based on **r**, the limiting distribution of $\delta_{\hat{\gamma}_e} = n^{1/2}(\hat{\gamma}_e - \gamma_o)$ is multivariate normal with a null mean vector and covariance matrix

$$\text{Acov}(\delta_{\hat{\gamma}_e}) = (\boldsymbol{\Delta}'_{\rho\gamma}\boldsymbol{\Gamma}_r^{-1}\boldsymbol{\Delta}_{\rho\gamma})^{-1}. \tag{30}$$

where

$$\boldsymbol{\Delta}_{\rho\gamma} = \frac{\partial\rho}{\partial\gamma'}$$

In the case of efficient estimation under normality assumptions for **x**, estimators for γ, the vector of standard deviations, σ, and the mean vector, μ, are obtained simultaneously. Both μ and σ are regarded as nuisance parameters, but are involved in the likelihood function. It can be shown, however, (cf. Browne, 1977, expression (3.28)) that the asymptotic covariance matrix of an efficient estimator for γ is still given by (30) with elements of $\boldsymbol{\Gamma}_r$ given by (22).

Use of the chain rule for matrix derivatives shows that

$$\begin{aligned}\boldsymbol{\Delta}_{\rho\gamma} &= \frac{\partial\rho}{\partial\eta'}\frac{\partial\eta}{\partial\theta'}\frac{\partial\theta}{\partial\gamma'} \\ &= \mathbf{D}_\rho\mathbf{A}\mathbf{D}_\gamma^{-1}\end{aligned} \tag{31}$$

Substitution of (31) into (30) and use of (12) yields

$$\begin{aligned}\text{Acov}(\delta_{\hat{\gamma}_e}) &= \mathbf{D}_\gamma\left(\mathbf{A}'\mathbf{D}_\rho\boldsymbol{\Gamma}_r^{-1}\mathbf{D}_\rho\mathbf{A}\right)^{-1}\mathbf{D}_\gamma \\ &= \mathbf{D}_\gamma(\mathbf{A}'\boldsymbol{\Gamma}_y^{-1}\mathbf{A})^{-1}\mathbf{D}_\gamma\end{aligned} \tag{32}$$

Comparison of (32) with (28) indicates that the transformation of the ADF estimator of Section 4.2 has the same asymptotic efficiency the ADF estimator based directly on **R**. Similarly the transformations of the GLSO and GLSR estimators of Section 4.3 will be asymptotically efficient and have the same asymptotic covariance matrix as the full information normal theory maximum likelihood estimator. The non-iterative estimators proposed here thus have the same asymptotic efficiency as corresponding estimators that require the iterative solution of an optimization problem.

6. An Example

Meier (1984) carried out an investigation of the construct validity of burnout using a multitrait-multimethod correlation matrix. Three traits, Burnout, Depression and Order, were measured using each of three measurement methods, Multi-point, True-False and Self-Ratings , on a sample of size 320. The sample correlation matrix, originally presented in Meier (1984, p. 215), is reproduced, with permission, in Table 1.

The pattern of occurrence of negative correlations in this matrix is consistent with a CDP model with negative trait correlations between Burnout and Order, ρ(B, O), and between Depression and Order, ρ(D, O), with all remaining trait and method correlations being positive. There are no sign inconsistencies in the correlation coefficients that are summed for the sign allocation rules in expressions (26) and (27).

Table 1: Construct Validity of Burnout. Sample Correlation Matrix

	B-MP	D-MP	O-MP	B-TF	D-TF	O-TF	B-SR	D-SR	O-SR
B-MP	1.00								
D-MP	0.57	1.00							
O-MP	− 0.18	− 0.13	1.00						
B-TF	0.61	0.65	− 0.09	1.00					
D-TF	0.57	0.67	− 0.13	0.69	1.00				
O-TF	− 0.17	− 0.17	0.74	− 0.14	− 0.23	1.00			
B-SR	0.65	0.53	− 0.14	0.63	0.59	− 0.12	1.00		
D-SR	0.55	0.62	− 0.12	0.54	0.63	− 0.13	0.60	1.00	
O-SR	− 0.14	− 0.23	0.70	− 0.12	− 0.20	0.73	− 0.10	− 0.13	1.00

Traits	B	Burnout	D	Depression	O	Order
Methods	MP	Multi-Point	TF	True-False	SR	Self-Ratings

Maximum Likelihood parameter estimates for the CDP model were obtained using the MUTMUM program (Browne, 1992) which makes use of an iterative Fisher Scoring procedure (Browne, 1984b, Section 6). Non-iterative OLS (Section 4.2), GLSO and GLSR (Section 4.3) estimates of parameters in the LAC model were obtained and transformed to estimates for the CDP model using the LINLOG program (Browne & Strydom, 1991). Results are reported in Table 2.

Parameter estimates yielded by ML and GLSO are close but those provided by GLSR differ more. The estimate of 1 for $\rho(\mathrm{MP},\mathrm{SR})$ yielded by OLS reflects an active equality constraint on the corresponding estimate of $\eta(\mathrm{MP},\mathrm{SR})$. Note that this constraint is not sufficient for the estimated method correlation correlation matrix to be nonnegative definite, since the estimate $\widehat{\rho}(\mathrm{MP},\mathrm{SR}) = 1$ does not result in equality of $\widehat{\rho}(\mathrm{MP},\mathrm{TF})$ and $\widehat{\rho}(\mathrm{SR},\mathrm{TF})$. Standard error estimates for ML, GLSO and GLSR, obtained from the diagonal elements of $\widehat{\mathrm{Cov}}(\widehat{\gamma})$ in (29), are also shown. They are very close.

Fit measures are shown in Table 3. Results yielded by ML and GLSO are very close but those from GLSR differ substantially and lead to a different conclusion. Random sampling experiments (Strydom, 1989) have suggested that the distribution of the GLSR test statistic converges more slowly to a chi-squared distribution as $n \to \infty$ than the ML and GLSO test statistics. Results from ML and GLSO alone will be interpreted here. The 90% confidence interval on ε_a is too wide for any firm conclusions concerning the fit of the model to be reached. The lower limit of the interval is below 0.05 which indicates close fit and the upper limit is above 0.08 which indicates mediocre fit. Since the lower limit of the interval is above zero the null hypothesis of perfect fit, H_o: $\varepsilon_a = 0$, is rejected at the 5% level. The null hypothesis of close fit, H_o: $\varepsilon_a \leq 0.05$, is not rejected since the lower limit of the interval is below 0.05.

Table 2: Construct Validity of Burnout. Parameter Estimates

	Estimates				Standard Errors		
Trait Corr.	ML	GLSO	GLSR	OLS	ML	GLSO	GLSR
$\rho(B,D)$	0.91	0.90	0.92	0.90	0.023	0.022	0.022
$\rho(B,O)$	− 0.19	− 0.22	− 0.34	− 0.19	0.058	0.065	0.072
$\rho(D,O)$	− 0.25	− 0.29	− 0.42	− 0.23	0.056	0.067	0.063
Meth. Corr.							
$\rho(MP,TF)$	0.94	0.93	0.95	0.86	0.028	0.029	0.028
$\rho(MP,SR)$	0.97	0.98	0.96	1.00	0.031	0.029	0.029
$\rho(TF,SR)$	0.89	0.90	0.91	0.93	0.030	0.032	0.029
Comm. Ind.							
ζ(B-MP)	0.78	0.79	0.81	0.89	0.030	0.019	0.029
ζ(D-MP)	0.80	0.80	0.86	0.86	0.029	0.021	0.027
ζ(O-MP)	0.83	0.84	0.85	0.81	0.029	0.033	0.029
ζ(B-TF)	0.83	0.85	0.88	0.83	0.024	0.026	0.024
ζ(D-TF)	0.88	0.89	0.89	0.93	0.023	0.020	0.022
ζ(O-TF)	0.95	0.95	0.93	0.99	0.030	0.031	0.028
ζ(B-SR)	0.82	0.82	0.85	0.76	0.028	0.035	0.027
ζ(D-SR)	0.80	0.79	0.82	0.71	0.029	0.038	0.029
ζ(O-SR)	0.87	0.85	0.88	0.87	0.032	0.029	0.032

Table 3: Construct Validity of Burnout. Fit Indices ($N = 320$)

	ML	GLSO	GLSR	OLS
$\widehat{F}$	0.143	0.153	0.070	0.627
RMSEA				
Point Estimate, $\widehat{\varepsilon}_a$.	0.061	0.064	0.014	
90% Conf. Interval	0.036 – 0.085	0.041 – 0.088	0.000 – 0.050	
Test Statistic	45.53	48.66	22.28	
Exceedance Probabilities				
H_o: $\varepsilon_a = 0$. (Perfect Fit)	0.001	0.001	0.383	
H_o: $\varepsilon_a \leq 0.05$. (Close Fit)	0.216	0.148	0.949	
Degrees of Freedom	21	21	21	21
No. of Parameters	24	15	15	15

The ML procedure operates on the covariance matrix and includes the diagonal elements of $\mathbf{D}_\sigma$ as nuisance parameters. Consequently the number of parameters for ML is greater than that for GLSO, GLSR and OLS which operate on the correlation matrix and involve no nuisance parameters. The degrees of freedom are the same for all methods considered.

6. Concluding Comments

The main advantage of the non-iterative method of Section 4.3 is that it only requires approximately as much computation as a single iteration of the Fisher scoring algorithm for maximum likelihood estimation described in Browne (1984b). If no inequality constraints, (10), are violated (or if the user is willing to tolerate correlation estimates that are greater than one in absolute value) the solution is given in closed form by (14). If some inequality constraints, (10), of the LAC model are violated, the program LINLOG (Browne & Strydom, 1991) uses the same quadratic programming procedure used in the program MUTMUM (Browne, 1992) to impose the corresponding inequality constraints (5) of the CDP model.

The occurrence of active inequality constraints, either with noniterative estimates or with maximum likelihood estimates, requires some care in the interpretation of estimates and the test statistic. A helpful discussion of this problem has been given by Dijkstra (1992).

A disadvantage of the non-iterative methods of Sections 4.2 and 4.3 is their sensitivity to near zero correlation coefficients. The non-iterative methods have attractive asymptotic properties but require the assumption that no elements of the population correlation matrix are zero. It is to be expected, therefore, that the sample size required for asymptotic theory to provide usable approximations to the distributions of estimators and the test statistic will depend strongly on whether or not there are near zero population correlation coefficients. Furthermore, the test statistic given here would have low power against any alternative hypothesis where the signs of some elements of a population correlation matrix satisfying the null hypothesis have been changed without yielding in an indefinite matrix.

It is worth bearing in mind that the CDP model considered here is a particular member of a family of CDP models considered in Browne (1984b). Other CDP models, where $\mathbf{D}_\zeta^*$ or $\mathbf{D}_\psi$ in (2) are required to have direct product structures, are not amenable to the present approach. This is based on a correlation matrix rather than a covariance matrix and avoids additive terms on the diagonal. It would, however, be easy to impose a direct product structure on $\mathbf{D}_\zeta$ in (3) by means of appropriate modifications to the design matrix, $\mathbf{A}$, in (11).

The extension of the present methods to multi-facet models of the form (cf. Browne, 1984b, equation (10.1))

$$\mathbf{P} = \mathrm{Off}\{\mathbf{D}_\zeta(\mathbf{P}_A \otimes \mathbf{P}_B \otimes \mathbf{P}_C \otimes \cdots)\mathbf{D}_\zeta\} + \mathbf{I}$$

would be straightforward and would involve modifications to the definition of the design matrix, $\mathbf{A}$, in (11) and the sign allocation procedure.

Acknowledgment. Much of this paper reports on material originally presented in a thesis written by H. F. Strydom, under supervision of M. W. Browne, in partial completion of the requirements for the M.Sc. degree in the Department of Statistics at the University of South Africa (Strydom, 1989).

REFERENCES

Bagozzi, R. P. (1993). Assessing construct validity in personality research: Applications to measures of self-esteem. *Journal of Research in Personality*, **27**, 49 – 87.

Bagozzi, R. P., & Yi, Y. (1992). Testing hypotheses about methods, traits, and communalities in the direct product model. *Applied Psychological Measurement*, **16**, 373–380.

Bishop, Y. M. M., Fienberg, S. E. & Holland, P. W. (1975). *Discrete multivariate analysis: Theory and Practice*. Cambridge: MIT Press.

Browne, M. W. (1977). The analysis of patterned correlation matrices by generalized least squares. *British Journal of Mathematical and Statistical Psychology*, **30**, 113 – 124.

Browne, M. W. (1984a). Asymptotically distribution-free methods in the analysis of covariance structures. *British Journal of Mathematical and Statistical Psychology*, **37**, 62-83.

Browne, M. W. (1984b). The decomposition of multitrait-multimethod matrices. *British Journal of Mathematical and Statistical Psychology*, **37**, 1–21.

Browne, M. W. (1989). Relationships between an additive model and a multiplicative model for multitrait-multimethod matrices. In R. Coppi & S. Bolasco (Eds.), *Multiway data analysis.* (pp. 507 – 520). Amsterdam: Elsevier Science Publishers BV.

Browne, M. W. (1992). *MUTMUM: User's Guide*. Department of Psychology, The Ohio State University.

Browne, M. W. & Cudeck, R. (1993). Alternative ways of assessing model fit. In: Bollen, K. A. & Long, J. S. (eds.) *Testing Structural Equation Models*. (pp. 136–162). Newbury Park: Sage.

Browne, M. W. & Strydom, H. F. (1991). *LINLOG: User's Guide*. Department of Psychology, The Ohio State University.

Campbell, D. T. & Fiske, D. W. (1959). Convergent and discriminant validation by the multitrait-multimethod matrix. *Psychological Bulletin*, **56**, 81–105.

Campbell, D. T. & O'Connell, E. S. (1967). Method factors in multitrait-multimethod matrices: Multiplicative rather than additive? *Multivariate Behavioral Research*, **2**, 409–426.

Cudeck, R. (1988). Multiplicative models and MTMM matrices. *Journal of Educational Statistics*, **13**, 131–147.

Dijkstra, T. K. (1992). On statistical inference with parameter estimates on the boundary of the parameter space. *British Journal of Mathematical and Statistical Psychology*, **45**, 289–309.

Goffin, R. D. & Jackson, D. N. (1992). Analysis of multitrait-multirater performance appraisal data: Composite direct product method versus confirmatory factor analysis. *Multivariate Behavioral Research*, **27**, 363–385.

Henly, S. J., Klebe, K. J., McBride, J. R. & Cudeck, R. (1989). Adaptive and conventional versions of the DAT: The first complete test battery comparison. *Applied Psychological Measurement*, **13**, 363–371.

Hsu, P. L. (1949). The limiting distribution of functions of sample means and application to testing hypotheses. *Proceedings of the First Berkeley Symposium on mathematical statistics and probability*, 359–402.

Isserlis, L. (1916). On certain probable errors and correlation coefficients of multiple frequency distributions with skew regression. *Biometrika*, **11**, 185–190.

Jennrich, R. I (1970). An asymptotic chi-square test for the equality of two correlation matrices. *Journal of the American Statistical Association*, **65**, 904–912.

Jöreskog, K. G. (1977). Structural equation models in the social sciences: Specification, estimation and testing. In P. R Krishnaiah (Ed.) *Applications of Statistics* (pp. 265–287). Amsterdam: North Holland.

Lastovicka, J. L., Murry, J. P. & Joachimsthaler, E. A. (1990). Evaluating the measurement validity of lifestyle typologies with qualitative measures and multiplicative factoring. *Journal of Marketing Research*, **2**, 11–23.

Meier, S. T. (1984). The construct validity of burnout. *Journal of Occupational Psychology*, **57**, 211–219.

Mooijaart, A. (1985). A note on computational efficiency in asymptotically distribution-free correlational models. *British Journal of Mathematical and Statistical Psychology*, **38**, 112–115.

Muthén, B. & Kaplan, D. (1992). A comparison of some methodologies for the factor analysis of non-normal Likert variables: A note on the size of the model. *British Journal of Mathematical and Statistical Psychology*, **45**, 19–30.

Pearson, K. & Filon, L. N. G. (1898). Mathematical contributions to the theory of evolution: IV. On the probable error of frequency constants and on the influence of random selection of variation and correlation. *Philosophical Transactions of the Royal Society of London, Series A*, **191**, 229–311.

Rao, C. R. (1973). *Linear Statistical Inference and its Applications, Second Edition.* New York: Wiley.

Shapiro, A. (1985). Asymptotic equivalence of minimum discrepancy estimators to GLS estimators, *South African Statistical Journal*, **19**, 73–81.

Steiger, J. H. (1989). *EzPATH: A supplementary module for SYSTAT and SYGRAPH.* Evanston, IL: SYSTAT Inc.

Steiger, J. H. & Hakstian, A. R. (1982). The asymptotic distribution of a correlation matrix: Theory and application. *British Journal of Mathematical and Statistical Psychology*, **35**, 208–215.

Steiger, J. H. & Lind, J. C. (1980). Statistically based tests for the number of common factors. Paper presented at the annual meeting of the Psychometric Society, Iowa City, IA.

Strydom, H. F. (1989). Non-iterative estimation of parameters in the composite direct product model. Unpublished MSc thesis. University of South Africa.

Swain, A. J. (1975). A class of factor analysis estimation procedures with common asymptotic properties. *Psychometrika*, **40**, 315–335.

Verhees, J. & Wansbeek, T. J. (1989). A multimode direct product model for covariance structure analysis. *British Journal of Mathematical and Statistical Psychology*, **43**, 231–240.

Wothke, W. & Browne, M. W. (1990). The direct product model for the MTMM matrix parameterized as a second order factor analysis model. *Psychometrika*, **55**, 255–262.

An EM Algorithm for ML Factor Analysis with Missing Data

Mortaza Jamshidian
BMDP Statistical Software*

August 28,1994
Revised: August 18, 1995

Abstract

EM algorithm is a popular algorithm for obtaining maximum likelihood estimates. Here we propose an EM algorithm for the factor analysis model. This algorithm extends a previously proposed EM algorithm to handle problems with missing data. It is simple to implement and is the most storage efficient among its competitors. We apply our algorithm to three examples and discuss the results. For problems with reasonable amount of missing data, it converges in reasonable time. For problems with large amount of missing data EM algorithm is usually slow. For such cases we successfully apply two EM acceleration methods to our examples. Finally, we discuss different methods of obtaining standard errors and in particular we recommend a method based on center difference approximation to the derivative.

1 Introduction

The Expectation-Maximization (EM) algorithm of Dempster, Laird, and Rubin (1977) is a popular algorithm for obtaining maximum likelihood estimates. It owes its popularity to its desirable global convergence properties and its simplicity in many applications. In the context of factor analysis an EM algorithm was first developed by Rubin and Thayer (1982). Their development, however, did not consider problems with missing data. Here we extend the Rubin and Thayer (1982) algorithm by giving an EM algorithm for the factor model which takes into account the missing data.

The estimation of the factor analysis model with missing data has recently received considerable attention. Jamshidian and Bentler (1994) give a short review. As they conclude from the literature, the normal theory maximum likelihood (ML) estimates are preferred over the mostly ad-hoc methods proposed. A number of algorithms have been proposed to compute ML estimates for the factor analysis model with missing data. Finkbeiner (1979) proposed a Fletcher-Powell algorithm which is based on the algorithm

*A major part of this research was done when the author was an assistant Professor at at Isfahan University of Technology.

of Gruvaeus and Jöreskog (1970). In this algorithm the inverse of the Fisher-information matrix is evaluated and used as the initial metric and subsequently the Fletcher-Powell update formulas determine the metric at each iteration. Lee (1986) proposed the Fisher-scoring algorithm, an iteratively reweighted Gauss-Newton algorithm, for a general covariance structure model with missing data and developed the formulas for the factor analysis model. Jamshidian and Bentler (1994) give an EM algorithm for a general mean and covariance structure model which in particular can be applied to the factor analysis model. Finally another class of algorithms has been proposed which uses the multiple group option of the existing complete data programs such as EQS (Bentler, 1992) and LISREL (Jöreskog and Sörbom, 1988). These algorithms treat every set of observations with the same missing data pattern as a group and then impose equality constraints on the parameters across groups (see e.g., Allison, 1987; Muthen, Kaplan, and Hollis, 1987; Arminger and Sobel, 1990).

The algorithms mentioned above, generally work well; however they do not enjoy the simplicity of the EM algorithm that we propose here. Both the Finkbeiner's (1979) and Lee (1986) algorithms require computation of the Fisher-information matrix. This computation, as compared to what is required in the EM algorithm developed below, is complicated (see e.g., Lee 1986). Finkbeiner's (1979) algorithm also requires a line search algorithm which is an additional complexity. A disadvantage of the Jamshidian and Bentler (1994) EM algorithm for the factor analysis model is that the M-step of their algorithm is iterative (for the definition of M-step see Section 2.2). Their algorithm, however, applies to a general mean and covariance structure model, a property that the algorithm that we develop here does not enjoy. Finally a practical limitation of the algorithms that group the data based on missing data patterns and use the existing software, is that they require that for each pattern of missing data enough cases be observed so that the sample variance-covariance matrix for each group (of missing data pattern) is positive definite.

In Section 2 we develop our EM algorithm for the factor analysis model with missing data. The algorithm that we propose does not have the complexities mentioned above and in addition it is more storage efficient than any of these algorithms. In Section 3 we apply our algorithm along with two of its accelerators to three examples and discuss the results. In Section 4 we discuss methods for computation of standard errors. In particular, we show how the approximation of the observed information matrix suggested by Jamshidian and Jennrich (1994c) can be implemented for our problem and we compare these approximations to the actual values. Finally in Section 5 we give a discussion and a summary.

2 The Model and the Algorithm

2.1 The factor analysis model

The basic model in factor analysis is

$$\mathbf{x}_i = \boldsymbol{\mu} + \Lambda \mathbf{f}_i + \mathbf{e}_i, \quad i = 1, \cdots, n, \tag{1}$$

where $\mathbf{x}_i$'s are p by 1 vectors of observed values with mean $\boldsymbol{\mu}$, Λ is a p by k ($k < p$) matrix of factor loadings, $\mathbf{f}_i$'s are vectors of latent common factors, and $\mathbf{e}_i$'s are vectors of unique scores. It is assumed that $\mathbf{f}_i$ and $\mathbf{e}_i$ are independently and normally distributed random vectors with mean zero and covariance matrices $\mathbf{\Phi}$ and $\mathbf{\Psi}$, respectively. From these assumptions $\mathbf{\Sigma}$, the variance-covariance matrix of $\mathbf{x}_i$ is given by

$$\mathbf{\Sigma} = \mathbf{\Lambda}\mathbf{\Phi}\mathbf{\Lambda}^T + \mathbf{\Psi}. \tag{2}$$

As is usual in the confirmatory factor analysis model, we allow arbitrary elements of $\mathbf{\Lambda}$ to be fixed to a constant. For the time being we assume that either all elements of $\mathbf{\Phi}$ are free or that $\mathbf{\Phi}$ is diagonal, and $\mathbf{\Psi}$ is assumed to be diagonal with the possibility that some of its diagonal elements are fixed to constant values. The case where arbitrary elements of $\mathbf{\Phi}$ are allowed to be restricted will be discussed in Section 2.3.

We further assume that some of the components of each $\mathbf{x}_i$ may not be observed and we let $\mathbf{y}_{o,i}$ and $\mathbf{y}_{m,i}$ respectively denote the observed and missing part of $\mathbf{x}_i$. We wish to estimate $\boldsymbol{\theta} = (\boldsymbol{\mu}, \mathbf{\Lambda}, \mathbf{\Phi}, \mathbf{\Psi})$, the free parameters in $\boldsymbol{\mu}$, $\mathbf{\Lambda}$, $\mathbf{\Phi}$, and $\mathbf{\Psi}$ from the density of the observed values $\mathbf{y}_{o,i}$. Let $\boldsymbol{\mu}_i(\boldsymbol{\theta})$ and $\mathbf{\Sigma}_i(\boldsymbol{\theta})$ be the mean and covariance matrix of $\mathbf{y}_{o,i}$, respectively. Then the log-likelihood based on $\mathbf{y}_{o,i}$ is

$$\mathcal{L}_y(\boldsymbol{\theta}) = (-N/2)\log(2\pi) - (1/2)\sum_{i=1}^{n}[\log|\mathbf{\Sigma}_i(\boldsymbol{\theta})| + \mathrm{trace}(\mathbf{\Sigma}_i^{-1}(\boldsymbol{\theta})C_i(\boldsymbol{\theta}))], \tag{3}$$

where $C_i(\boldsymbol{\theta}) = (\mathbf{y}_{o,i} - \boldsymbol{\mu}_i(\boldsymbol{\theta}))(\mathbf{y}_{o,i} - \boldsymbol{\mu}_i(\boldsymbol{\theta}))^T$ and $N = \sum_{i=1}^{n} p_i$, with p_i being the number of components of $\mathbf{y}_{o,i}$. We denote the value that maximizes (3) by $\widehat{\boldsymbol{\theta}}$, and refer to it as the Maximum Likelihood Estimate (MLE). Note that MLE's are valid as long as the distribution of the observed components, $y_{o,i}$, for each case is a marginal distribution of the complete data $\mathbf{x}_i$.

2.2 Computation of MLE using EM

As noted in the Introduction, the EM algorithm is a device for obtaining maximum likelihood estimates. Given a set of observed data Y with density $f_Y(Y|\boldsymbol{\theta})$ it obtains $\widehat{\boldsymbol{\theta}}$, the maximum likelihood estimate of $\boldsymbol{\theta}$, by introducing a function $\mathbf{h}$ and a second family of densities $f_X(X|\boldsymbol{\theta})$ related to the first by the requirement that if X has density $f_X(X|\boldsymbol{\theta})$, then $Y = \mathbf{h}(X)$ has density $f_Y(Y|\boldsymbol{\theta})$. The algorithm is determined by the choice of the second family $f_X(X|\boldsymbol{\theta})$. X is usually referred to as the *complete data* and Y is referred to as the *observed*, or *incomplete data*. Similarly f_X and f_Y are referred to as *complete* and *observed* log-likelihood, respectively.

Each iteration consists of an E-step and an M-step. Given $\boldsymbol{\theta}$, the E-step consists of computing

$$Q(\widetilde{\boldsymbol{\theta}}, \boldsymbol{\theta}) = E^*(\log f_X(X|\widetilde{\boldsymbol{\theta}})), \tag{4}$$

where $E^*(\cdot)$ denotes $E(\cdot|Y, \boldsymbol{\theta})$. The M-step consists of maximizing $Q(\widetilde{\boldsymbol{\theta}},\boldsymbol{\theta})$ with respect to $\widetilde{\boldsymbol{\theta}}$. Let $\widetilde{\boldsymbol{\theta}} = \boldsymbol{\theta}'$ denote the maximum point. Then, at each iteration $\boldsymbol{\theta}$ is replaced by $\boldsymbol{\theta}'$ and the E and the M steps are repeated generating a sequence of values of $\boldsymbol{\theta}$ that hopefully converges to a maximum likelihood estimate.

To obtain $\widehat{\boldsymbol{\theta}}$, we let $Y = (\mathbf{y}_{o,1}, \cdots, \mathbf{y}_{o,n})$ and $X = (\mathbf{z}_1, \cdots, \mathbf{z}_n)$, where $\mathbf{z}_i = \begin{pmatrix} \mathbf{x}_i \\ \mathbf{f} \end{pmatrix}$. Then the complete data log-likelihood up to an additive constant is given by

$$\mathcal{L}_X(\boldsymbol{\theta}) = (-n/2)(\log|\mathbf{\Phi}| + \log|\mathbf{\Psi}|) \tag{5}$$

$$-(1/2)\sum_{i=1}^{n}\text{trace}[\boldsymbol{\Psi}^{-1}(\mathbf{x}_i-\boldsymbol{\mu}-\boldsymbol{\Lambda}\mathbf{f}_i)(\mathbf{x}_i-\boldsymbol{\mu}-\boldsymbol{\Lambda}\mathbf{f}_i)^T+\boldsymbol{\Phi}^{-1}\mathbf{f}_i\mathbf{f}_i^T].$$

Given Y and $\boldsymbol{\theta}$, the E-step of the EM algorithm consists of obtaining $Q(\tilde{\boldsymbol{\theta}},\boldsymbol{\theta}) = E^*[\mathcal{L}_X(\tilde{\boldsymbol{\theta}})]$. Using (5) it can be shown that

$$\begin{aligned}Q(\tilde{\boldsymbol{\theta}},\boldsymbol{\theta}) = (-n/2)\{&\log|\tilde{\boldsymbol{\Phi}}|+\log|\tilde{\boldsymbol{\Psi}}|+\text{trace}[\tilde{\boldsymbol{\Phi}}^{-1}F^* \\ &+\tilde{\boldsymbol{\Psi}}^{-1}(S^*-2\bar{\mathbf{x}}^*\tilde{\boldsymbol{\mu}}^T-2V^*\tilde{\boldsymbol{\Lambda}}^T+\tilde{\boldsymbol{\mu}}\tilde{\boldsymbol{\mu}}^T+2\tilde{\boldsymbol{\Lambda}}\bar{\mathbf{f}}^*\tilde{\boldsymbol{\mu}}^T+\tilde{\boldsymbol{\Lambda}}F^*\tilde{\boldsymbol{\Lambda}}^T)]\},\end{aligned} \tag{6}$$

where $\tilde{\boldsymbol{\theta}} = (\tilde{\boldsymbol{\mu}},\tilde{\boldsymbol{\Lambda}},\tilde{\boldsymbol{\Phi}},\tilde{\boldsymbol{\Psi}})$, and

$$\bar{\mathbf{x}}^* = \frac{1}{n}\sum_{i=1}^{n}E^*(\mathbf{x}_i), \tag{7}$$

$$S^* = \frac{1}{n}\sum_{i=1}^{n}E^*(\mathbf{x}_i\mathbf{x}_i^T), \tag{8}$$

$$\bar{\mathbf{f}}^* = \frac{1}{n}\sum_{i=1}^{n}E^*(\mathbf{f}_i), \tag{9}$$

$$V^* = \frac{1}{n}\sum_{i=1}^{n}E^*(\mathbf{x}_i\mathbf{f}_i^T), \tag{10}$$

and

$$F^* = \frac{1}{n}\sum_{i=1}^{n}E^*(\mathbf{f}_i\mathbf{f}_i^T). \tag{11}$$

To complete the E-step we need to compute the expectations in (7)-(11). For simplicity of notation we drop the i indices on $\mathbf{x}_i$ and $\mathbf{f}_i$ and without loss of generality we assume that $\mathbf{x} = \begin{pmatrix}\mathbf{y}_o\\ \mathbf{y}_m\end{pmatrix}$, where $\mathbf{y}_o$ and $\mathbf{y}_m$ represent the observed and the missing part of $\mathbf{x}$. It can be shown that the joint density of $\begin{pmatrix}\mathbf{x}\\ \mathbf{f}\end{pmatrix}$ given $\boldsymbol{\theta}$ is normal with mean and variance-covariance matrix

$$\begin{pmatrix}\boldsymbol{\mu}_o\\ \boldsymbol{\mu}_m\\ \mathbf{0}\end{pmatrix}, \text{ and } \begin{pmatrix}\boldsymbol{\Sigma}_{oo} & \boldsymbol{\Lambda}_o\boldsymbol{\Phi}\boldsymbol{\Lambda}_m^T & \boldsymbol{\Lambda}_o\boldsymbol{\Phi}\\ \boldsymbol{\Lambda}_m\boldsymbol{\Phi}\boldsymbol{\Lambda}_o^T & \boldsymbol{\Sigma}_{mm} & \boldsymbol{\Lambda}_m\boldsymbol{\Phi}\\ \boldsymbol{\Phi}\boldsymbol{\Lambda}_o^T & \boldsymbol{\Phi}\boldsymbol{\Lambda}_m^T & \boldsymbol{\Phi}\end{pmatrix},$$

where $\boldsymbol{\mu}$ and $\boldsymbol{\Lambda}$ are partitioned according to the observed and missing data as $\boldsymbol{\mu} = \begin{pmatrix}\boldsymbol{\mu}_o\\ \boldsymbol{\mu}_m\end{pmatrix}$ and $\boldsymbol{\Lambda} = \begin{pmatrix}\boldsymbol{\Lambda}_o\\ \boldsymbol{\Lambda}_m\end{pmatrix}$, and $\boldsymbol{\Sigma}_{oo} = \boldsymbol{\Lambda}_o\boldsymbol{\Phi}\boldsymbol{\Lambda}_o^T+\boldsymbol{\Psi}_o$ and $\boldsymbol{\Sigma}_{mm} = \boldsymbol{\Lambda}_m\boldsymbol{\Phi}\boldsymbol{\Lambda}_m^T+\boldsymbol{\Psi}_m$. Now using the conditional distribution of $\begin{pmatrix}\mathbf{y}_m\\ \mathbf{f}\end{pmatrix}$ given $\mathbf{y}_o$ and $\boldsymbol{\theta}$ the expectations in (7) through (11) are

$$E^*(\mathbf{x}) = \begin{pmatrix}\mathbf{y}_o\\ \mathbf{y}_m^*\end{pmatrix}, \tag{12}$$

where $\mathbf{y}_m^* = \boldsymbol{\mu}_m+\boldsymbol{\Lambda}_m\boldsymbol{\Phi}\boldsymbol{\Lambda}_o^T\boldsymbol{\Sigma}_{oo}^{-1}(\mathbf{y}_o-\boldsymbol{\mu}_o)$,

$$E^*(\mathbf{x}\mathbf{x}^T) = \begin{pmatrix}\mathbf{y}_o\mathbf{y}_o^T & \mathbf{y}_o\mathbf{y}_m^{*T}\\ \mathbf{y}_m^*\mathbf{y}_o^T & E^*(\mathbf{y}_m\mathbf{y}_m^T)\end{pmatrix}, \tag{13}$$

where $E^*(\mathbf{y}_m\mathbf{y}_m^T) = \boldsymbol{\Sigma}_{mm} - \boldsymbol{\Lambda}_m\boldsymbol{\Phi}\boldsymbol{\Lambda}_o^T\boldsymbol{\Sigma}_{oo}^{-1}\boldsymbol{\Lambda}_o\boldsymbol{\Phi}\boldsymbol{\Lambda}_m^T + \mathbf{y}_m^*\mathbf{y}_m^{*T}$,

$$\mathbf{f}^* \equiv E^*(\mathbf{f}) = \boldsymbol{\Phi}\boldsymbol{\Lambda}_o\boldsymbol{\Sigma}_{oo}^{-1}(\mathbf{y}_o - \boldsymbol{\mu}_o), \tag{14}$$

$$E^*(\mathbf{f}\mathbf{f}^T) = \boldsymbol{\Phi} - \boldsymbol{\Phi}\boldsymbol{\Lambda}_o\boldsymbol{\Sigma}_{oo}^{-1}\boldsymbol{\Lambda}_o\boldsymbol{\Phi} + \mathbf{f}^*\mathbf{f}^{*T}, \tag{15}$$

and finally

$$E^*(\mathbf{x}\mathbf{f}^T) = \begin{pmatrix} \mathbf{y}_o\mathbf{f}^{*T} \\ \boldsymbol{\Lambda}_m\boldsymbol{\Phi} - \boldsymbol{\Lambda}_m\boldsymbol{\Phi}\boldsymbol{\Lambda}_o^T\boldsymbol{\Sigma}_{oo}^{-1}\boldsymbol{\Lambda}_o\boldsymbol{\Phi} + \mathbf{y}_m^*\mathbf{f}^{*T} \end{pmatrix}. \tag{16}$$

Thus equations (6)-(16) define the E-step completely.

As noted earlier, the M-step consists of maximizing $Q(\tilde{\boldsymbol{\theta}}, \boldsymbol{\theta})$ with respect to $\tilde{\boldsymbol{\theta}}$. Taking derivatives of (6) with respect to $\tilde{\boldsymbol{\theta}}$ we obtain

$$\frac{\partial Q(\tilde{\boldsymbol{\theta}}, \boldsymbol{\theta})}{\partial \tilde{\boldsymbol{\mu}}} = n\widetilde{\boldsymbol{\Psi}}^{-1}(\bar{\mathbf{x}}^* - \tilde{\boldsymbol{\Lambda}}\bar{\mathbf{f}}^* - \tilde{\boldsymbol{\mu}}) \tag{17}$$

$$\frac{\partial Q(\tilde{\boldsymbol{\theta}}, \boldsymbol{\theta})}{\partial \tilde{\boldsymbol{\Lambda}}} = n[\![\widetilde{\boldsymbol{\Psi}}^{-1}(V^* - \tilde{\boldsymbol{\mu}}\bar{\mathbf{f}}^{*T} - \tilde{\boldsymbol{\Lambda}}F^*)]\!]_{\boldsymbol{\Lambda}} \tag{18}$$

$$\frac{\partial Q(\tilde{\boldsymbol{\theta}}, \boldsymbol{\theta})}{\partial \tilde{\boldsymbol{\Phi}}} = (n/2)[\![\tilde{\boldsymbol{\Phi}}^{-1}(F^* - \tilde{\boldsymbol{\Phi}})\tilde{\boldsymbol{\Phi}}^{-1}]\!]_{\boldsymbol{\Phi}} \tag{19}$$

$$\frac{\partial Q(\tilde{\boldsymbol{\theta}}, \boldsymbol{\theta})}{\partial \widetilde{\boldsymbol{\Psi}}} = (n/2)[\![\widetilde{\boldsymbol{\Psi}}^{-1}(\tilde{G} - \widetilde{\boldsymbol{\Psi}})\widetilde{\boldsymbol{\Psi}}^{-1}]\!]_{\boldsymbol{\Psi}}, \tag{20}$$

where $\tilde{G} = S^* - 2\bar{\mathbf{x}}^*\tilde{\boldsymbol{\mu}}^T - 2V^*\tilde{\boldsymbol{\Lambda}}^T + \tilde{\boldsymbol{\mu}}\tilde{\boldsymbol{\mu}}^T + 2\tilde{\boldsymbol{\Lambda}}\bar{\mathbf{f}}^*\tilde{\boldsymbol{\mu}}^T + \tilde{\boldsymbol{\Lambda}}F^*\tilde{\boldsymbol{\Lambda}}^T$. Here $[\![A]\!]_{\boldsymbol{\Lambda}}$ is simply A with zeros inserted in positions corresponding to fixed parameters of $\boldsymbol{\Lambda}$, $[\![A]\!]_{\boldsymbol{\Phi}}$ is $2A - \operatorname{diag}(A)$ with zeros inserted in positions corresponding to fixed parameters in $\boldsymbol{\Phi}$ and $[\![\cdot]\!]_{\boldsymbol{\Psi}}$ is defined similarly. To obtain $\boldsymbol{\theta}'$, the maximizing value of $Q(\tilde{\boldsymbol{\theta}}, \boldsymbol{\theta})$ with respect to $\tilde{\boldsymbol{\theta}}$, we set equations (17) to (20) equal to zero. Then $\boldsymbol{\Lambda}'$ is obtained by solving the linear equation

$$[\![\boldsymbol{\Lambda}'B]\!]_{\boldsymbol{\Lambda}} = [\![V^* - \bar{\mathbf{x}}^*\bar{\mathbf{f}}^{*T}]\!]_{\boldsymbol{\Lambda}}, \tag{21}$$

where $B = F^* - \bar{\mathbf{f}}^*\bar{\mathbf{f}}^{*T}$, and

$$\boldsymbol{\mu}' = \bar{\mathbf{x}}^* - \boldsymbol{\Lambda}'\bar{\mathbf{f}}^*, \; \boldsymbol{\Phi}' = [\![F^*]\!]_{\boldsymbol{\Phi}}, \text{ and } \boldsymbol{\Psi}' = [\![G']\!]_{\boldsymbol{\Psi}}, \tag{22}$$

where $G' = S^* - 2\bar{\mathbf{x}}^*\boldsymbol{\mu}'^T - 2V^*\boldsymbol{\Lambda}'^T + \boldsymbol{\mu}'\boldsymbol{\mu}'^T + 2\boldsymbol{\Lambda}'\bar{\mathbf{f}}^*\boldsymbol{\mu}'^T + \boldsymbol{\Lambda}'F^*\boldsymbol{\Lambda}'^T$.

To summarize, given an initial value $\boldsymbol{\theta} = (\boldsymbol{\mu}, \boldsymbol{\Lambda}, \boldsymbol{\Phi}, \boldsymbol{\Psi})$, the EM algorithm proceeds as follows:

Step 1. Compute quantities (7) through (11).

Step 2. Solve (21) for $\boldsymbol{\Lambda}'$ and compute $\boldsymbol{\mu}', \boldsymbol{\Phi}', \boldsymbol{\Psi}'$ as in (22).

Step 3. Replace $\boldsymbol{\theta} = (\boldsymbol{\mu}, \boldsymbol{\Lambda}, \boldsymbol{\Phi}, \boldsymbol{\Psi})$ by $\boldsymbol{\theta}' = (\boldsymbol{\mu}', \boldsymbol{\Lambda}', \boldsymbol{\Phi}', \boldsymbol{\Psi}')$ and go to Step 1.

2.3 Arbitrary restrictions on Φ

Using (19), to obtain Φ' in the M-step of the EM algorithm, we need to solve the equation

$$[\![\, \Phi'^{-1}(F^* - \Phi')\Phi'^{-1} \,]\!]_{\Phi} = 0 \tag{23}$$

for Φ'. When Φ is free or diagonal, the solution Φ' of (23) is simple and is that given in (22). However, when there are arbitrary restrictions on Φ then equation (23) is not as simple to solve for Φ' and it generally requires an iterative procedure. The EM algorithm of Rubin and Thayer (1982) for the completely observed data problem also suffers from a similar problem. Jamshidian and Jennrich (1994a) gave a modification of the EM algorithm of Rubin and Thayer (1982) to handle the case of Φ with arbitrary restrictions. We recommend a similar modification of our algorithm here to handle this case. Instead of solving (23) for Φ' we recommend solving

$$[\![\, \Phi^{-1}(F^* - \Phi')\Phi^{-1} \,]\!]_{\Phi} = 0 \tag{24}$$

for Φ'. Equation (24) is linear in Φ' and is simple to solve. For our algorithm use of (24) in place of (23) can be justified by similar arguments made by Jamshidian (1988, pp. 66-64) and Jamshidian and Jennrich (1994a).

3 Examples

We illustrate performance of the EM algorithm of Section 2 on three examples. Since often the EM algorithm is accelerated, we also report on two quasi-Newton accelerators of EM, QN1 and QN2, proposed by Jamshidian and Jennrich (1994b). The QN1 algorithm uses the Broyden (1965) asymmetric update to obtain the zero of the EM step and QN2 views the EM step as a generalized gradient and uses the Broyden-Fletcher-Goldfarb-Shanno rank-two update to accelerate the EM algorithm. The QN1 algorithm is very simple to implement and it only requires the EM step. The QN2 algorithm is a bit more complicated in that it requires a line search algorithm in addition to the EM step (For more details see Jamshidian and Jennrich, 1994b).

For our examples, as Finkbeiner (1979) suggested, we replace the unobserved data for each variable with the mean of the observed values for that variable and use the maximum likelihood estimates of θ based on the completed data as initial values. Our convergence criteria is based on the value of *relative gradient* (Khalfan, Byrd, and Schnabel, 1993)

$$rg = \max_i \left[|[\mathbf{g}(\theta)]_i| \frac{\max\{|[\theta + \Delta\theta]_i|, 1\}}{\max\{|\mathcal{L}_y(\theta + \Delta\theta)|, 1\}} \right],$$

where $\mathbf{g}(\theta)$ is the gradient of $\mathcal{L}_y(\theta)$, $\Delta\theta = \theta' - \theta$, and $[\mathbf{v}]_i$ is the ith element of $\mathbf{v}$. We stop our algorithms when $rg < 10^{-4}$. This usually gives the value of the log-likelihood at the maximum point correct to about six significant digits and the parameter estimates accurate to about three to four significant digits.

Table 1 shows the results of applying EM and two of its accelerators, mentioned above, to three examples. For our first example we generated 300 data points from a multivariate normal distribution with mean zero and covariance matrix equal to that given on page 98 of Jöreskog and Sörbom (1988). This covariance matrix is based on the

Table 1: Number of iterations required by three algorithms on three examples.

Example	EM	QN1	QN2
1	59	16	12
2	289	64	34
3	492	89	38

nine psychological variables chosen by Jöreskog and Sörbom (1988) from the Holzinger and Swineford (1939) study. We created missing data by deleting 5% of the generated data points at random. This resulted in 31 missing data patterns with the number of missing variables for each pattern ranging from 1 to 3. There were 192 completely observed cases. We fitted the model described by Jöreskog and Sörbom (1988, page 104) to the resulting data set except that instead of fixing diagonal of $\mathbf{\Phi}$ to 1 we fixed $\mathbf{\Lambda}_{11}$, $\mathbf{\Lambda}_{42}$, and $\mathbf{\Lambda}_{73}$ to 1. Also, unlike Jöreskog and Sörbom (1988) who assume it to be zero, we estimated $\boldsymbol{\mu}$ as a free vector of parameters. There were a total of 31 parameters to be estimated. For this example the EM algorithm converged relatively fast requiring 59 iterations and QN1 and QN2 required 16 and 12 iterations respectively.

Our second example is similar to the confirmatory factor analysis example of Allison (1987). In his example he wanted to estimate the correlation between father's occupational status and father's educational attainment for black men in the U.S. There were a total of 2020 subjects that for 348 of them four variables were observed and for the remaining 1672 subjects only two of the four variables were observed. The sample means and covariance matrices for these two groups are given in Allison (1987), Table 1. Using these sample sizes, means, and covariances we generated a similar set of multivariate normal data and fitted the two-factor model given by Allison to our generated data. There were a total of 13 parameters to estimate. For this example the EM algorithm required 289 iterations, a relatively large number. This can be explained by the fact that the percentage of missing data here is relatively large, about 41%, and it is known that as the percentage of missing data increases the speed of convergence of EM algorithm decreases (See e.g., Little and Rubin, 1987). The QN1 algorithm converged in 64 iterations and the QN2 algorithm converged in 34 iterations, both converging in a reasonable time.

Our third example was used to examine the performance of our algorithm in a problem with a large number of parameters. For our model the factor loading matrix $\mathbf{\Lambda}$ was a 20 by 6 matrix with the structure $\begin{pmatrix} I \\ \bar{\Lambda} \end{pmatrix}$, where I is the identity matrix and is fixed and $\bar{\mathbf{\Lambda}}$ is a 14 by 6 matrix with free parameters. $\mathbf{\Phi}$ and $\mathbf{\Psi}$ were assumed to have free diagonal elements and $\boldsymbol{\mu}$ was a free vector. To generate our data we set elements in $\bar{\mathbf{\Lambda}}$ equal to uniform random numbers in the interval $(0,1)$, $\mathbf{\Phi} = I$, and diagonal elements of $\mathbf{\Psi}$ equal to uniform numbers in the interval $(0.1, 1.1)$. Then using (2) we obtained a covariance matrix and generated 200 multivariate normal data with mean zero and the covariance matrix obtained in this fashion. To have missing data we deleted two 25 by

10 matrices from the top left and bottom right of the data matrix. There were a total of 130 parameters to be estimated. For this example the EM algorithm required 492 iterations. The percentage of the missing data here was 12.5, not a very large number as compared to our second example. The large number of iterations required by the EM algorithm here is mainly due to the large number of parameters which is to be estimated. The QN1 and QN2 algorithms, however, converged in a reasonable number of iterations requiring 89 and 38 iterations respectively.

4 Computing Standard Errors

Here we discuss methods of obtaining the covariance matrix and in particular standard errors for the parameter estimates. Following Efron and Hinkley (1978) we propose using the inverse of he observed information matrix $H^{-1}(\widehat{\boldsymbol{\theta}})$, that is, minus the inverse of the second derivative of the log-likelihood function $\mathcal{L}_y(\boldsymbol{\theta})$ at $\widehat{\boldsymbol{\theta}}$, for this purpose. Often in factor analysis the inverse of the Fisher information matrix is used to obtain the standard errors. This is mainly because computation of the observed information matrix in this context is relatively complex and it is generally more complicated to compute than the Fisher Information matrix.

Especially in the context of missing data and the EM algorithm, there have been a few proposals for computing $H(\boldsymbol{\theta})$ efficiently. Louis (1982) gave a method for "finding the observed information matrix when using the EM algorithm." Implementation of his method for factor analysis is as complicated as computing $H(\boldsymbol{\theta})$ by directly differentiating $\mathcal{L}_y(\boldsymbol{\theta})$. Meng and Rubin (1991) also proposed a method for obtaining the observed information matrix. They called their method Supplement to the EM (SEM) algorithm. As the name suggests, it can be used with our EM algorithm to obtain the observed information matrix. A problem with this method is that it requires computing the second derivative of $Q(\widetilde{\boldsymbol{\theta}}, \boldsymbol{\theta})$ with respect to $\widetilde{\boldsymbol{\theta}}$. While this quantity is simpler to obtain than the observed information matrix itself, it is still a bit cumbersome to compute. A simpler method that uses numerical differentiation is proposed by Jamshidian and Jennrich (1994c). We explain this method briefly and recommend using it. As we will see shortly, this method gives good approximations to $H(\boldsymbol{\theta})$ and is simple to implement.

Jamshidian and Jennrich (1994c) proposed using the central difference approximation $\widetilde{H}(\boldsymbol{\theta})$, to approximate $H(\boldsymbol{\theta})$ as follows: Let

$$D_j(\boldsymbol{\theta}) = \frac{\mathbf{g}(\boldsymbol{\theta} + h_j\mathbf{e}_j) - \mathbf{g}(\boldsymbol{\theta} - h_j\mathbf{e}_j)}{2h_j}, \quad j = 1, \cdots, q, \tag{25}$$

where $\mathbf{g}(\boldsymbol{\theta})$ is the gradient of $\mathcal{L}_y(\boldsymbol{\theta})$ evaluated at $\boldsymbol{\theta}$, $\mathbf{e}_j$ is a the unit vector with all of its elements equal to zero except for its jth element which is equal to 1, h_j is a small number, and q is the number of parameters in $\boldsymbol{\theta}$. Let D be the q by q matrix with its jth column equal to $D_j(\boldsymbol{\theta})$ and set

$$\widetilde{H}(\boldsymbol{\theta}) = \frac{D + D^T}{2} \tag{26}$$

In order to use (25) and (26) $\mathbf{g}$ is the main quantity to compute. This, in our context, is simple. Fisher (1925) showed that

$$\mathbf{g}(\boldsymbol{\theta}) = E^*[\frac{\partial}{\partial \boldsymbol{\theta}}\mathcal{L}_x(\boldsymbol{\theta})]. \tag{27}$$

Exchanging the derivative with the expectation we have

$$\begin{aligned} \mathbf{g}(\boldsymbol{\theta}) &= \frac{\partial}{\partial \boldsymbol{\theta}'}\{E^*[\mathcal{L}_x(\boldsymbol{\theta}')]\}|_{\boldsymbol{\theta}'=\boldsymbol{\theta}}, \\ &= \frac{\partial}{\partial \boldsymbol{\theta}'}Q(\boldsymbol{\theta}',\boldsymbol{\theta})|_{\boldsymbol{\theta}'=\boldsymbol{\theta}}. \end{aligned} \tag{28}$$

Using (17)-(20) and (28) the elements of $\mathbf{g}(\boldsymbol{\theta})$ are the non-zero elements of

$$\frac{\partial \mathcal{L}_y}{\partial \boldsymbol{\mu}} = n\Psi^{-1}(\bar{\mathbf{x}}^* - \Lambda\bar{\mathbf{f}}^* - \boldsymbol{\mu}) \tag{29}$$

$$\frac{\partial \mathcal{L}_y}{\partial \Lambda} = n[\![\, \Psi^{-1}(V^* - \boldsymbol{\mu}\bar{\mathbf{f}}^{*T} - \Lambda F^*)\,]\!]_{\Lambda} \tag{30}$$

$$\frac{\partial \mathcal{L}_y}{\partial \Phi} = (n/2)[\![\, \Phi^{-1}(F^* - \Phi)\Phi^{-1}\,]\!]_{\Phi} \tag{31}$$

$$\frac{\partial \mathcal{L}_y}{\partial \Psi} = (n/2)[\![\, \Psi^{-1}(G - \Psi)\Psi^{-1}\,]\!]_{\Psi}, \tag{32}$$

where $G = (S^* - 2\bar{\mathbf{x}}^*\boldsymbol{\mu}^T - 2V^*\Lambda^T + \boldsymbol{\mu}\boldsymbol{\mu}^T + 2\Lambda\bar{\mathbf{f}}^*\boldsymbol{\mu}^T + \Lambda F^*\Lambda^T)$. Quantities (29)-(32) are simple to obtain since $\bar{\mathbf{x}}^*$, $\bar{\mathbf{f}}^*$, V^*, F^*, and S^* are readily available from the E-step of the EM algorithm.

To assess the accuracy of the approximation (26) we have coded both the actual second derivative formulas of $\mathcal{L}_y(\boldsymbol{\theta})$ and the approximations (26) and have used these codes on our examples of Section 3. We have used $h_j = max(\eta, \eta|\boldsymbol{\theta}_j|)$, as suggested by Dennis and Schnabel (1983, page 97), and set $\eta = 10^{-4}$. To compare the estimates $\widetilde{H}^{-1}(\hat{\boldsymbol{\theta}})$ to the actual value $H^{-1}(\hat{\boldsymbol{\theta}})$, as Jamshidian and Jennrich (1994c) recommended, we use the *Maximum Relative Error*

$$\text{MRE} = \max_{\ell} \frac{\ell^T(H^{-1}(\hat{\boldsymbol{\theta}}) - \widetilde{H}^{-1}(\hat{\boldsymbol{\theta}}))\ell}{\ell^T H^{-1}(\hat{\boldsymbol{\theta}})\ell}. \tag{33}$$

This is the upper bound in the relative error when $\ell^T H^{-1}(\hat{\boldsymbol{\theta}})\ell$ is approximated by $\ell^T \widetilde{H}^{-1}(\boldsymbol{\theta})\ell$ for arbitrary ℓ. The quantity $(1/n)\ell^T H^{-1}(\hat{\boldsymbol{\theta}})\ell$ is of interest because it is an estimate of the variance of an arbitrary linear combination $\ell^T\hat{\boldsymbol{\theta}}$ of the estimate $\hat{\boldsymbol{\theta}}$. For example, if $\ell = \mathbf{e}_j$, then the square root of $(1/n)\ell^T H^{-1}(\hat{\boldsymbol{\theta}})\ell$ would be the standard error for the jth element of $\hat{\boldsymbol{\theta}}$. The maximum relative error (33) for our three examples of Section 3 is as follows:

Example	1	2	3
MRE	8×10^{-8}	2×10^{-7}	5×10^{-8}

These maximum relative errors indicate that the approximation (26) is practically identical to the actual value of H obtained from the complicated routines that compute the second derivative of $\mathcal{L}_y(\boldsymbol{\theta})$ directly. This result is in agreement with the experience of Jamshidian and Jennrich (1994c) which promise good approximations for the observed information matrix using the center difference methods.

5 Summary and Discussion

We have proposed an EM algorithm for obtaining maximum likelihood estimates for the confirmatory factor analysis with missing data. This algorithm is simple to implement and it is the most storage efficient algorithm among its competitors. Another desirable property of the EM algorithm is that it often converges from almost any feasible starting value to a maximum point. When the amount of missing data is reasonable and the optimal parameter values are not in the boundary of the acceptable parameter space the algorithm often converges in a reasonable time. In cases where EM is slow it is still quite valuable for obtaining starting values, since, as is well known, its first few initial steps are quite effective. Also when EM is slow it can be accelerated. Usually starting from an arbitrary point a few EM steps are taken to obtain a good initial value and then an acceleration method is triggered. Fortunately simple and successful acceleration methods have been proposed in this context (see e.g., Jamshidian and Jennrich, 1993; Jamshidian and Jennrich, 1994b). In particular the QN1 algorithm used in Section 3 is quite simple to use and it is successful. With an additional complexity of having a line search algorithm the QN2 algorithm is even more successful. It converged quite fast for all of our examples of Section 3. Finally to obtain standard errors, the method of Jamshidian and Jennrich (1994c) is simple to implement and the required gradient values can easily be obtained using the Fisher (1925) result, as mentioned in Section 4.

References

[1] Allison, P. D. (1987). Estimation of linear models with incomplete data. In C. Clogg (Ed.), *Sociological Methodology,* (pp 71–103). San Francisco: Jossey-Bass.

[2] Arminger, G. & Sobel, M. E. (1990). Pseudo-maximum likelihood estimation of mean and covariance structure with missing data. *Journal of the American Statistical Association,* **85**, 195–203.

[3] Bentler, P. M. (1992). *EQS Structural Equations Program Manual.* Los Angeles: BMDP Statistical Software.

[4] Broyden, C. G. (1965). A Class of Methods for Solving Nonlinear Simultaneous Equations. *Math. Comp.*, **19**, 577-593.

[5] Dempster, A. P., Laird, N. M., & Rubin, D. B. (1977). Maximum Likelihood from Incomplete Data via the EM Algorithm (with Discussion). *Journal of the Royal Statistical Society,* **B39**, 1–38.

[6] Dennis, J. E., & Schnabel, R. B. (1983). *Numerical Methods for Unconstrained Optimization and Nonlinear Equations.* Englewood Cliffs, New Jersey: Prentice-Hall.

[7] Efron, B. & Hinkley, D. V. (1978). The observed versus the expected information. *Biometrika,* **65**, 457–487.

[8] Finkbeiner, C. (1979). Estimation for the multiple factor model when data are missing. *Psychometrika,* **44**, 409–420.

[9] Fisher, R. A. (1925). Theory of Statistical Estimation. *Proceedings of Cambridge Philosophical Society*, **22**, 700–725.

[10] Gruvaeus, G. T. & Jöreskog, K. G. (1970). *A computer program for minimizing a function of several variables,* (E.T.S. Res. Bull. RB770-14). PrinceFraton, N. J.: Educational Testing Service.

[11] Holzinger, K. J. & Swineford, F. (1939). A study in factor analysis: The stability of a bi-factor solution. *Supplementary Educational Monograph*, No. 48. Chicago: University of Chicago.

[12] Jamshidian, M. (1988). *Application of the Conjugate Gradient Methods in Statistical Computing*, University of California, Los Angeles, Ph.D. Thesis.

[13] Jamshidian, M. & Bentler, P. M. (1994). Using complete data routines for ML estimation of mean and covariance structures with missing data, submitted.

[14] Jamshidian, M. & Jennrich, R. I. (1994a). Conjugate Gradient Methods in Confirmatory Factor Analysis. *Computational Statistics and Data Analysis*, **17**, 247–263.

[15] Jamshidian, M. & Jennrich, R. I. (1994b). Quasi Newton acceleration of the EM algorithm, submitted.

[16] Jamshidian, M., & Jennrich, R. I. (1994c), Computing the observed information matrix using numerical differentiation, *manuscript under preparation.*

[17] Jamshidian, M., & Jennrich, R. I. (1993). Conjugate Gradient Acceleration of the EM Algorithm. *Journal of the American Statistical Association*, **88**, 221–228.

[18] Jöreskog, K. G. & Sörbom, D. (1988). *LISREL 7, A guide to the program and applications.* SPSS, Chicago.

[19] Khalfan, H. F., Byrd, R. H., & Schnabel, R. B. (1993). A Theoretical and Experimental Study of the Symmetric Rank-One Update. *SIAM Journal on Optimization*, **3**, 1–24.

[20] Lee, S. Y. (1986). Estimation for structural equation models with missing data. *Psychometrika*, **51**, 93–99.

[21] Little, R. J. A. & Rubin, D. B. (1987). *Statistical analysis with missing data.* New York: Wiley.

[22] Louis, T. A. (1982). Finding the Observed Information Matrix When Using the EM Algorithm. *Journal of the Royal Statistical Society,* **Ser. B**, **44**, 226-233.

[23] Meng, X. L. & Rubin, D. B. (1991). Using EM to Obtain Asymptotic Variance-Covariance Matrices: the SEM Algorithm. *Journal of the American Statistical Association,* **86**, 899–909.

[24] Muthen, B., Kaplan, D., & Hollis, M. (1987). On structural equation modeling with data that are not missing completely at random. *Psychometrika,* **52**, 431–462.

[25] Rubin, B. & Thayer, D. T. (1982). EM algorithm for ML factor analysis. *Psychometrika,* **47**, 69–76.

Optimal Conditionally Unbiased Equivariant Factor Score Estimators*

Peter M. Bentler and Ke-Hai Yuan
Department of Psychology and Center for Statistics
University of California, Los Angeles

Abstract

Some approaches to generating equivariant conditionally unbiased estimators of the factor scores in a confirmatory factor analytic model are defined. In order to select among these estimators, weighted (WLS) and generalized least squares (GLS) functions are optimized, optimal minimum variance estimators are obtained, and fifteen multivariate criterion functions are used to define optimality. Optimality is achieved by only the WLS and GLS estimators, which are found to be equivalent under standard conditions. It is shown that when the factor loading matrix consists of a simple cluster structure, and some conditions hold on the loadings and unique variances, the estimators simplify substantially and some simple estimators used in practice can be justified. The well-known Bartlett WLS estimator is not applicable when the unique variance matrix is singular, but the GLS estimator can be applied more generally. However, it also can break down when the observed variables are linearly dependent. A modified estimator remains valid.

1 Introduction

In factor analysis, interest is usually centered on the parameters of the factor model. However, the estimated values of the common factors, called factor scores, may also be required. These quantities are often used for diagnostic purposes as well as inputs to a subsequent analysis.

Suppose we have a factor analysis model

$$X = \mu + \Lambda\xi + \varepsilon, \tag{1.1}$$

where X is the observed p×1 random vector, μ is a p×1 vector of means, Λ is a p×k matrix of factor loadings, ξ is the unobserved k×1 vector of random effects called common factors, and ε is a p×1 vector of random errors called unique factors or variables. Under the usual assumptions of uncorrelatedness between factors and unique variables, the covariance structure of (1.1) is

$$\Sigma = \Lambda\Phi\Lambda' + \Psi, \tag{1.2}$$

where Φ is the covariance matrix of the factors and Ψ is the covariance matrix of the unique factors. Typically, Ψ is assumed to be diagonal, but along with Guttman (1955) we do not necessarily make this assumption. Also, we do not necessarily make the

* Research supported in part by USPHS grants DA00017 and DA01070, and by the New England Medical Center International Quality of Life Assessment (IQOLA) Project. The constructive comments of Yasuo Amemiya and Roderic P. McDonald are gratefully acknowledged.

standard assumption that Ψ is positive definite. It may be - and often is, in practice - singular. The matrices on the right of (1.2) are the parameters of the factor analysis model which, when restricted beyond the minimum needed for identification, give the confirmatory factor analysis model (e.g., Jöreskog, 1969).

Although there are some important problems associated with the "determinacy" of the factor scores (e.g., Guttman, 1955; Rozeboom, 1988; Vittadini, 1989), "estimation" of these scores remains a viable problem to most researchers. The problem of factor score estimation, and issues in the appropriate use of relevant statistical terminology, are essentially equivalent to the problem of estimation of random effects in linear models, see e. g., Robinson's (1991) review and the associated discussions. As noted by Robinson, "A convention has somehow developed that estimators of random effects are called predictors while estimators of fixed effects are called estimators." (p.15). Along with Robinson, we shall use the term "estimators" for random effects, recognizing that a factor score estimator $\hat{\xi}$ of ξ is not an estimator of unknown parameters in the usual sense because ξ is not a parameter vector. Rather, the "estimator" is a "predictor" of values for the unobserved random factor ξ_i for each of the i = 1, ..., n observations in a study. In keeping with common practice, in the notation (1.1) the observation subscript i is suppressed. In order to estimate ξ, the standard approach treats the parameter matrices of the factor model, given in (1.1) and (1.2), as known. In practice, these matrices must be replaced by their corresponding consistent estimators.

The estimation of factor scores for the orthogonal exploratory factor analysis model, with Φ=I and Ψ diagonal and positive definite, is a topic that has received considerable attention for half a century. Thurstone (1935), Bartlett (1937), Thomson (1936), and Anderson and Rubin (1956) are some classic references. Harris (1967) and McDonald and Burr (1967) describe the properties of the most popular methods. A recent empirical comparison among the classic and other methods, for a specialized situation, was made by Fava and Velicer (1992). These methods extend naturally to the model (1.2) with correlated factors; see, e.g., Lawley and Maxwell (1971) or Saris, de Pijper, and Mulder (1978).

The most popularly applied method is one of the oldest, the Thurstone-Thomson "regression" factor score estimator. This estimator can be obtained by minimizing the function

$$f_1(A) = \mathrm{tr}E(\bar{\xi}-\xi)(\bar{\xi}-\xi)', \tag{1.3}$$

where $\bar{\xi} = A'(X-\mu)$ for some p×k weight matrix A. The resulting estimator is

$$\bar{\xi} = \Phi\Lambda'\Sigma^{-1}(X-\mu). \tag{1.4}$$

Although optimization of (1.3) is a natural criterion, the estimator is conditionally biased. Conditionally unbiased estimators have the property that $E(\hat{\xi}|\xi)=\xi$, i.e., $E(A'\Lambda\xi + A'\varepsilon|\xi) = \xi$. With ε independent of ξ, this property requires that $A'\Lambda\xi = \xi$. Under the regularity condition that rank(Λ) = rank(A) = k, the unbiasedness requirement implies that the weight matrix A meets the condition (e.g., Lawley & Maxwell, 1971, eq. 8.14)

$$A'\Lambda = I. \tag{1.5}$$

In this paper we study only conditionally unbiased estimators.

There is not universal agreement that conditional unbiasedness is an important property of a factor score estimator, e.g., Bartholomew (1987, p. 68) considers this property "hardly relevant," but the more common attitude is that conditional unbiasedness is an appropriate criterion in certain situations such as "the comparison of different groups of units" (Saris et al., 1978, p. 88). For example, if a factor reflects general health, and one wants subjects having the same level of health to have the same estimated factor scores on average, the unbiasedness criterion is relevant.

The definitive conditionally unbiased estimator was introduced by Bartlett (1937). The standard way of obtaining his estimator is to minimize

$$f_2(A) = \varepsilon'\Psi^{-1}\varepsilon, \tag{1.6}$$

yielding the weighted least squares (WLS) estimator

$$\hat{\xi} = (\Lambda'\Psi^{-1}\Lambda)^{-1}\Lambda'\Psi^{-1}(X-\mu) \tag{1.7}$$

popularly known as the Bartlett estimator. Clearly, the weight matrix $A'_{WLS} = (\Lambda'\Psi^{-1}\Lambda)^{-1}\Lambda'\Psi^{-1}$ meets the unbiasedness condition (1.5). The estimator $\hat{\xi}$ also can be obtained by minimizing $f_1(A)$ under the constraint of unbiasedness $E(\hat{\xi}|\xi) = \xi$ (e.g., Lawley & Maxwell, 1971). It is the best linear conditionally unbiased estimator when Ψ is positive definite. An important limitation of this estimator, not emphasized in the literature, is that (1.7) is not optimal estimator when Ψ is singular. In practice, furthermore, one or more unique variances often are estimated at 0, making Bartlett's estimator impossible to apply. In this paper, we shall study some procedures for generating unbiased estimators to determine an optimal estimator that remains well-defined with zero unique variances.

It is apparent from (1.1) that, in standard models (i.e., models in which loading restrictions are only fixed zeros or identification constraints), the scale of the factor scores is arbitrary. To retain this feature in our factor score estimator, we shall restrict our study to "equivariant" estimators. We shall define an equivariant factor score estimator $\hat{\xi}$ as one that is rescaled in the same way the factor scores are rescaled, that is, as $\xi \to D\xi \Rightarrow \hat{\xi} \to D\hat{\xi}$ for any arbitrary positive definite diagonal matrix D. This is an important condition since under the factor model $X = \mu + \Lambda\xi + \varepsilon$, it is also true that $X = \mu + \Lambda D^{-1}D\xi + \varepsilon = \mu + \Lambda^*\xi^* + \varepsilon$, with $\Lambda^* = \Lambda D^{-1}$ and $\xi^* = D\xi$.

Our study of equivariant conditionally unbiased estimators will lead us to an estimator that is more general than the Bartlett estimator. However, it is equivalent to the Bartlett estimator in "regular" cases, i.e., those in which Λ, Φ, and Ψ are positive definite. In developing our estimator, we also obtain a number of new properties of the Bartlett estimator.

2 A Generalized Least Squares Estimator

An alternative to the Bartlett estimator can be motivated in many ways. Suppose that in the factor model (1.1) the factor scores ξ are fixed rather than random. Then variation in ε is reflected in X, and another natural criterion for estimation of factor scores is to minimize the generalized least squares (GLS) function

$$f_3(A) = \varepsilon' \Sigma^{-1} \varepsilon. \tag{2.1}$$

Taking derivatives of (2.1) with respect to the unknown factor scores ξ, and solving, yields the GLS estimator

$$\tilde{\xi} = (\Lambda'\Sigma^{-1}\Lambda)^{-1}\Lambda'\Sigma^{-1}(X-\mu). \tag{2.2}$$

With $A'_{GLS} = (\Lambda'\Sigma^{-1}\Lambda)^{-1}\Lambda'\Sigma^{-1}$, our GLS estimator clearly meets condition (1.5) for a conditionally unbiased estimator. In contrast to the Bartlett estimator, (2.2) is still operational when Ψ is singular. As pointed out in personal communication by Y. Amemiya, a parameterization of Σ^{-1} under the model also can yield an alternative way to write (2.2). Furthermore, rescaling the factors by a diagonal matrix D similarly rescales (2.2). Thus, simple substitution, along with the fact that rescaling the factors by D will rescale the covariance matrix of these factors to $D\Phi D$, yields the following.

Lemma 1. In standard models, the factor score estimators (1.4), (1.7), and (2.2) are equivariant.

Note that (1.4) and (2.2) remain equivariant when Ψ is singular, while (1.7) is not defined.

In practice, the GLS estimator $\tilde{\xi}$ is available in two forms. The first, say $\tilde{\xi}_1$, is based on a consistent parameter estimator for Λ and Σ under the hypothesized model, yielding $\tilde{\xi}_1 = (\tilde{\Lambda}'\tilde{\Sigma}^{-1}\tilde{\Lambda})^{-1}\tilde{\Lambda}'\tilde{\Sigma}^{-1}(X-\mu)$. The second, say $\tilde{\xi}_2$, is based on a consistent parameter estimator for Λ under the model, but an unstructured estimator for Σ. The natural such estimator is the sample covariance matrix S, yielding $\tilde{\xi}_2 = (\tilde{\Lambda}'S^{-1}\tilde{\Lambda})^{-1}\tilde{\Lambda}'S^{-1}(X-\mu)$. If the model is correct, and the sample is large enough, $\tilde{\xi}_1$ and $\tilde{\xi}_2$ will be the same, but in smaller samples and/or under some misspecification they may behave differently. Note also that the two estimators have a different tolerance to linear dependence among the measured variables: $\tilde{\xi}_1$ allows dependence among the sample data variates, but not the model variates, while $\tilde{\xi}_2$ allows dependence among the model variates, but not the data variates.

Actually, our GLS estimator is not new. When $|\Psi| \neq 0$, it is just the Bartlett WLS estimator. We will prove that (2.2) and (1.7), i.e., $\tilde{\xi}$ and $\hat{\xi}$, are identical in regular cases.

Lemma 2. $\tilde{\xi}$ and $\hat{\xi}$ are equal when Ψ is nonsingular.

Proof: From (1.2), it follows that with Ψ nonsingular,

$$\Sigma^{-1} = \Psi^{-1} - \Psi^{-1}\Lambda(\Lambda'\Psi^{-1}\Lambda + \Phi^{-1})^{-1}\Lambda'\Psi^{-1}. \tag{2.3}$$

Hence,

$$\Lambda'\Sigma^{-1}\Lambda = \Lambda'\Psi^{-1}\Lambda - \Lambda'\Psi^{-1}\Lambda(\Lambda'\Psi^{-1}\Lambda + \Phi^{-1})^{-1}\Lambda'\Psi^{-1}\Lambda. \tag{2.4}$$

Denoting $\Delta = \Lambda'\Psi^{-1}\Lambda$, we get

$$(\Lambda'\Sigma^{-1}\Lambda)^{-1} = \Delta^{-1} + \Phi. \tag{2.5}$$

Using (2.3) again we get

$$\Lambda'\Sigma^{-1} = \Lambda'\Psi^{-1} - \Delta(\Delta+\Phi^{-1})^{-1}\Lambda'\Psi^{-1}$$
$$= \Phi^{-1}(\Delta+\Phi^{-1})^{-1}\Lambda'\Psi^{-1}$$
$$= (\Delta^{-1}+\Phi)^{-1}\Delta^{-1}\Lambda'\Psi^{-1} \tag{2.6}$$

From (2.5) and (2.6), we have $\tilde{\xi} = (\Lambda'\Sigma^{-1}\Lambda)^{-1}\Lambda'\Sigma^{-1}(X-\mu) = \Delta^{-1}\Lambda'\Psi^{-1}(X-\mu) = \hat{\xi}$.

Thus our estimator (2.2) is just another way to compute the Bartlett scores in regular cases. For the exploratory orthogonal factor model, formula (2.2) was given by Heermann (1963), Schönemann and Steiger (1976), and Hakstian, Zidek, and McDonald (1977). Heermann (1963, p. 171) developed it as a "univocal" estimator of orthogonal factors, that is, one that yields estimates that correlated only with their corresponding true factor scores. Schönemann and Steiger gave (2.2) as a definitional statement involving components, not as a factor score estimator having optimal properties. Hakstian et al. noted that, except for standardization, A_{GLS} is equivalent to Heermann's W_n. They also showed that A_{GLS} is the solution to the optimization problem min tr(A'SA) subject to constraint (1.5), and proved that $A_{WLS} = A_{GLS}$. Lemma 2 is a minor extension to the confirmatory factor model.

Hakstian, Zidek, and McDonald's (1977) derivation of (2.2) was the first to demonstrate an optimal property for $\tilde{\xi}$, but they did not show any optimality for $\tilde{\xi}$ when it does not equal the Bartlett estimator. Our derivation of (2.2) via (2.1) is, to our knowledge, new, and below, we shall use more general approaches to find optimal factor score estimators. Our special interest lies in evaluating their performance when regularity conditions break down, i.e., when the Bartlett estimator is not optimal.

3 Minimum Variance Estimators

Now we try a new approach to generating a class of unbiased factor score estimators and inquire how to select an optimal member of this class. Let V be an unknown p×p symmetric matrix, not necessarily positive definite, such that $\Lambda'V\Lambda$ is full rank. In view of (1.5), the factor score weight matrix

$$A' = (\Lambda'V\Lambda)^{-1}\Lambda'V, \tag{3.1}$$

can be used to yield conditionally unbiased factor score estimators. Premultiplying (1.1) by (3.1), it follows that any member of this class has the property that the factor score estimator has the additive decomposition

$$\hat{\xi} = (\Lambda'V\Lambda)^{-1}\Lambda'V(X-\mu) = \xi + \delta, \tag{3.2}$$

where $\delta = (\Lambda'V\Lambda)^{-1}\Lambda'V\varepsilon$. Different choices of V lead to different conditionally unbiased estimators $\hat{\xi} = (\Lambda'V\Lambda)^{-1}\Lambda'V(X-\mu)$. For any such estimator, the covariance matrices of $\hat{\xi}$ and δ, given by $\Sigma_{\hat{\xi}}$ and Σ_δ, respectively, are

$$\Sigma_{\hat{\xi}} = (\Lambda'V\Lambda)^{-1}\Lambda'V\Sigma V\Lambda(\Lambda'V\Lambda)^{-1} \tag{3.3}$$

and

$$\Sigma_\delta = (\Lambda' V\Lambda)^{-1}\Lambda' V\Psi V\Lambda(\Lambda' V\Lambda)^{-1}, \tag{3.4}$$

with

$$\Sigma_{\hat{\xi}} = \Phi + \Sigma_\delta. \tag{3.5}$$

Since (3.3)-(3.5) hold for any V, we inquire whether there are optimal choices for V. An optimal V is one in which the residual covariance matrix (3.4) is minimal, in the sense that any other V would yield a covariance matrix that is larger, where the matrix difference is measured by a positive semidefinite difference. In view of the similarity of (3.2), (2.2), and (1.7), the results will not be surprising.

First we address the question of whether an optimal V to minimize (3.4) also will minimize (3.3). In view of the relation (3.5), we immediately get the following.

Lemma 3. Let V_{opt} be a weight matrix yielding covariance matrices (3.3)-(3.4) given by $\Sigma_{\hat{\xi}_{opt}}$ and $\Sigma_{\delta_{opt}}$. For any other choice of V yielding $\Sigma_{\hat{\xi}}$ and Σ_δ, the matrix difference $\Sigma_{\hat{\xi}}$ - $\Sigma_{\hat{\xi}_{opt}}$ is positive semidefinite if and only if Σ_δ - $\Sigma_{\delta_{opt}}$ is positive semidefinite.

Proof: In view of (3.5), for V_{opt}, $\Sigma_{\hat{\xi}_{opt}} = \Phi + \Sigma_{\delta_{opt}}$. Subtracting this from (3.5) for an arbitrary V yields $\Sigma_{\hat{\xi}}$ - $\Sigma_{\hat{\xi}_{opt}}$ = Σ_δ - $\Sigma_{\delta_{opt}}$. Obviously, if the left-hand side is positive semidefinite, so is the right-hand side; and vice versa.

Lemma 3 clarifies that the problem of finding a minimal residual covariance matrix $\Sigma_{\delta\,opt}$ is equivalent to the problem of finding a minimal estimated factor score covariance matrix $\Sigma_{\hat{\xi}}$. Based on the WLS and GLS results described above, two obvious choices for V_{opt} are $V = \Sigma^{-1}$ and $V = \Psi^{-1}$. The latter choice requires that the residual matrix Ψ in the factor model (1.1)-(1.2) be positive definite. Let us see what we can do without this assumption. We develop the general case first, which has not previously been discussed in the literature.

3.1 Ψ May Be Singular

Lemma 4. If $V = \Sigma^{-1}$ in (3.1)-(3.4), the GLS score estimator (2.2) is optimal in the sense that it yields the minimum $\Sigma_{\hat{\xi}}$, $\Sigma_{\hat{\xi}\,opt} = (\Lambda'\Sigma^{-1}\Lambda)^{-1}$.

Proof: Substituting $V = \Sigma^{-1}$ into (3.3) yields $\Sigma_{\hat{\xi}\,opt} = (\Lambda'\Sigma^{-1}\Lambda)^{-1}$. This is optimal since, for $\Sigma_{\hat{\xi}}$ based on any other choice of V, the matrix difference $\Sigma_{\hat{\xi}} - \Sigma_{\hat{\xi}\,opt}$ is given by $B'\Sigma B$, where $B' = [(\Lambda' V\Lambda)^{-1}\Lambda' V - \Sigma_{\hat{\xi}_{opt}}\Lambda'\Sigma^{-1}]$. This difference is positive semidefinite.

In other words, the choice $V = \Sigma^{-1}$ yields a minimal covariance matrix for the estimated factor scores. Lemma 4 and the optimal matrix $\Sigma_{\hat{\xi}\,opt} = (\Lambda'\Sigma^{-1}\Lambda)^{-1}$ are implicit in Hakstian et. al. (1977). By Lemma 3, Lemma 4 also must yield the minimal covariance matrix (3.4) for the residual scores. Combining Lemmas 3 and 4, we get the following result.

Theorem 1. Providing it is defined, the GLS estimator (2.2) yields lower-bound estimator and residual covariance matrices (3.3) and (3.4) given by $\Sigma_{\hat{\xi}\,\mathrm{opt}} = (\Lambda'\Sigma^{-1}\Lambda)^{-1}$ and $\Sigma_{\delta\mathrm{opt}} = (\Lambda'\Sigma^{-1}\Lambda)^{-1}\Lambda'\Sigma^{-1}\Psi\Sigma^{-1}\Lambda(\Lambda'\Sigma^{-1}\Lambda)^{-1}$, related by $\Sigma_{\hat{\xi}\,\mathrm{opt}} = \Phi + \Sigma_{\delta\,\mathrm{opt}}$.

If Ψ is severely rank deficient, the rank of $\Lambda'\Sigma^{-1}\Psi\Sigma^{-1}\Lambda$ could be less than k. In that case, the covariance matrix $\Sigma_{\delta\,\mathrm{opt}}$ may be singular. This does not disturb the optimality properties of the GLS estimator. In regular cases, of course Ψ will be full rank and we get further results.

3.2 Ψ is Positive Definite

First we consider the effect of using the weight matrix $V = \Psi^{-1}$.

Lemma 5. If $V = \Psi^{-1}$ in (3.1)-(3.4), the WLS score estimator (1.7) is optimal in the sense that it yields the minimum Σ_δ, $\Sigma_{\delta_{\mathrm{opt}}} = (\Lambda'\Psi^{-1}\Lambda)^{-1}$.

Proof: Substituting $V = \Psi^{-1}$ into (3.4) yields $\Sigma_{\delta_{\mathrm{opt}}} = (\Lambda'\Psi^{-1}\Lambda)^{-1}$. This is optimal since, for Σ_δ based on any other choice of V, the matrix difference $\Sigma_\delta - \Sigma_{\delta_{\mathrm{opt}}}$ is given by $C'\Sigma C$, where $C' = [(\Lambda'V\Lambda)^{-1}\Lambda'V - \Sigma_{\delta\mathrm{opt}}\Lambda'\Psi^{-1}]$. This difference is positive semidefinite.

We now seem to be in a quandary. Lemmas 4 and 5 based on the two choices $V = \Psi^{-1}$ and $V = \Sigma^{-1}$ make it appear that two different estimators yield two different types of optimality. This is a misleading impression since, as shown in Lemma 2, these estimators are identical in regular cases, and as shown in Lemma 3, the optimizing solutions must coincide. We combine Lemmas 2-5, with equation (2.5), to obtain the following special case of Theorem 1.

Corollary 1. Providing they are defined, the WLS and GLS estimators (1.7) and (2.2) are equivalent, and yield lower-bound estimator and residual covariance matrices (3.3) and (3.4) given by $\Sigma_{\hat{\xi}\,\mathrm{opt}} = (\Lambda'\Sigma^{-1}\Lambda)^{-1}$ and $\Sigma_{\delta\,\mathrm{opt}} = (\Lambda'\Psi^{-1}\Lambda)^{-1}$, related by $\Sigma_{\hat{\xi}\,\mathrm{opt}} = \Phi + \Sigma_{\delta\,\mathrm{opt}}$.

The corollary shows that the relation (2.5) conceptually represents the relation (3.5) under the choice of optimal V. Next we take a different approach based on classical multivariate statistical criteria to see whether any other optimal conditionally unbiased estimators can be obtained.

4 An Optimization Approach to Factor Score Estimation

4.1 Estimator Definition

It follows from (1.1) that if T is any p×k matrix that makes $T'\Lambda$ nonsingular, we can premultiply (1.1) by $(T'\Lambda)^{-1}T'$, yielding

$$\hat{\xi} = (T'\Lambda)^{-1}T'(X-\mu) = \xi + (T'\Lambda)^{-1}T'\varepsilon = \xi + \delta. \tag{4.1}$$

Evidently, $\hat{\xi} = (T'\Lambda)^{-1}T'(X-\mu)$ defines a class of factor score estimators, with the particular estimator type given by the particular choice of T. These estimators are unbiased since the weight matrix $A' = (T'\Lambda)^{-1}T'$ meets condition (1.5) for unbiasedness. One choice of T is $T = V\Lambda$ for some matrix V. This yields the weight matrix (3.1) and score estimator (3.2) that specializes to optimal Bartlett and GLS estimators. However, we want to explore T without the restriction that it necessarily has decomposition $T = V\Lambda$. Bartholomew (1987, p. 67) previously gave the decomposition (4.1), and showed that the factor score estimator that makes the variances of δ_i as small as possible is the Bartlett estimator. We extend his analysis and this result.

Our analysis will be based on the covariance matrices generated by decomposition (4.1). For any T, the covariance matrix of the estimator has the decomposition

$$\Sigma_{\hat{\xi}} = \Phi + \Sigma_\delta, \tag{4.2}$$

where $\Sigma_{\hat{\xi}} = (T'\Lambda)^{-1}T'\Sigma T(\Lambda'T)^{-1}$, and $\Sigma_\delta = (T'\Lambda)^{-1}T'\Psi T(\Lambda'T)^{-1}$. As before, $\Sigma_{\hat{\xi}}$ is the covariance matrix of the factor score estimator, and Σ_δ is the covariance matrix of the residual $(\hat{\xi}-\xi)$. Clearly, the specific choice of T will determine the properties of these covariance matrices, since, in (4.2), we take Φ to be fixed. If Σ_δ is in some sense small, $\Sigma_{\hat{\xi}}$ will be close to Φ. We shall propose various ways of measuring this closeness, and determine the estimators that optimize these measures.

A problem with using (4.2) to define how small Σ_δ and/or $\Sigma_{\hat{\xi}}$ are, or how close these matrices are to Φ, is that (4.2) depends on the particular scale chosen for the factor scores. If the factor scores are rescaled by D, the covariance matrix of these factors, Φ, becomes $D\Phi D$. As a consequence the decomposition (4.2) becomes

$$D\Sigma_{\hat{\xi}}D = D\Phi D + D\Sigma_\delta D. \tag{4.3}$$

Since the scale of the factor scores is arbitrary, and a good factor score estimator should have a smaller Σ_δ, the function chosen to optimize that will select a particular T from the set of all possible T matrices should be one that does not depend on the choice of D, i.e., it should be scale invariant, and indifferent to choice of (4.2) or (4.3). A basis for this idea comes from the following lemma.

Lemma 5. Suppose in model (1.1) the factor scores are rescaled so that $X = \mu + \Lambda D^{-1}D\xi + \varepsilon = \Lambda^*\xi^* + \varepsilon$, with $\Lambda^* = \Lambda D^{-1}$ and $\xi^* = D\xi$. If T defined in (4.1) remains invariant to such a rescaling, the factor score estimator is equivariant.

Proof: Evaluating $\hat{\xi} = (T'\Lambda)^{-1}T'(X-\mu)$ with the rescaled Λ^* and original T, we obtain $\hat{\xi}^* = D(T'\Lambda)^{-1}T'(X-\mu) = D\hat{\xi}$.

A consequence of Lemma 5 is that if we can find an optimal T that is invariant to rescaling of the factor scores, we will have obtained an equivariant factor score estimator. We approach this by optimizing some function $h(T) = F(\Sigma_{\hat{\xi}}, \Phi)$ to estimate T, where

F(A,B) is some measure of "distance" from A to B. Such a procedure will produce the same estimated T if F(.,.) satisfies:

$$F(D\Sigma_{\hat{\xi}}D, D\Phi D) = F(\Sigma_{\hat{\xi}}, \Phi). \tag{4.4}$$

4.2 Optimization Criteria

In view of (4.4), we want to choose a factor score estimator by optimizing some function that is scale invariant with respect to the decompositions given in (4.2) or (4.3). Although there are a lot of possible choices, the decompositions can be compared to the decompositions in multivariate analysis of variance (e.g., Anderson, 1984, Chapter 8), and the associated functions can be applied here. In particular, we may choose the function h as, $h_1(T) = \dfrac{|\Sigma_{\hat{\xi}}|}{|\Phi|}$, $h_2(T) = tr(\Sigma_{\hat{\xi}}^{-1}\Phi)$, or $h_3(T) = tr(\Phi^{-1}\Sigma_{\hat{\xi}})$. However, since $\Sigma_{\hat{\xi}} = \Phi + \Sigma_\delta$, the following functions also seem to be natural candidates for h: $h_4(T) = \dfrac{|\Sigma_\delta|}{|\Sigma_{\hat{\xi}}|}$, $h_5(T) = \dfrac{|\Sigma_\delta|}{|\Phi|}$, $h_6(T) = tr(\Phi^{-1}\Sigma_\delta)$, $h_7(T) = tr(\Sigma_\delta^{-1}\Phi)$, $h_8(T) = tr(\Sigma_{\hat{\xi}}^{-1}\Sigma_\delta)$, $h_9(T) = tr(\Sigma_\delta^{-1}\Sigma_{\hat{\xi}})$, $h_{10}(T) = \lambda_{min}(\Sigma_{\hat{\xi}}^{-1}\Phi)$, $h_{11}(T) = \lambda_{max}(\Phi^{-1}\Sigma_{\hat{\xi}})$, $h_{12}(T) = \lambda_{max}(\Phi^{-1}\Sigma_\delta)$, $h_{13}(T) = \lambda_{min}(\Sigma_\delta^{-1}\Phi)$, $h_{14}(T) = \lambda_{max}(\Sigma_{\hat{\xi}}^{-1}\Sigma_\delta)$, $h_{15}(T) = \lambda_{min}(\Sigma_\delta^{-1}\Sigma_{\hat{\xi}})$, where λ_{min} and λ_{max} are the smallest and the largest eigenvalues, respectively, of the corresponding matrix. These are not the only functions that could be considered, but they provide a good starting point for developing factor score estimators.

4.3 Estimators by Optimization

In order to estimate T by minimizing the "distance" between $\Sigma_{\hat{\xi}}$ and Φ we need to maximize for some of the above functions and minimize for the other functions. It is obvious that some of the functions are related. We will give the relationships among these functions by considering the relative eigenvalues of these matrices, and for each equivalence class we will give the optimal T. In this section we assume that Λ, Φ, and Ψ are rank k, k, and p, respectively.

Denote the eigenvalue decomposition of $\Phi^{-\frac{1}{2}}\Sigma_{\hat{\xi}}\Phi^{-\frac{1}{2}}$ as

$$\Phi^{-\frac{1}{2}}\Sigma_{\hat{\xi}}\Phi^{-\frac{1}{2}} = L\Gamma L', \tag{4.5}$$

where $L = (l_1, \cdots, l_k)$, and $\Gamma = diag(\gamma_1, \cdots, \gamma_k)$ with $\gamma_1 \ge \gamma_2 \ge ... \ge \gamma_k \ge 1$. It follows that all the functions in section 4.2 can be written in terms of $\gamma_1, \gamma_2, ..., \gamma_k$. In particular, we have:

$$h_1=\prod_{i=1}^{k}\gamma_i,\ h_2=\sum_{i=1}^{k}\frac{1}{\gamma_i},\ h_3=\sum_{i=1}^{k}\gamma_i,\ h_4=\prod_{i=1}^{k}(1-\frac{1}{\gamma_i}),\ h_5=\prod_{i=1}^{k}(\gamma_i-1),\ h_6=\sum_{i=1}^{k}(\gamma_i-1),$$

$$h_7=\sum_{i=1}^{k}\frac{1}{(\gamma_i-1)},\quad h_8=\sum_{i=1}^{k}(1-\frac{1}{\gamma_i}),\quad h_9=\sum_{i=1}^{k}\frac{\gamma_i}{\gamma_i-1},\quad h_{10}=\frac{1}{\gamma_1},\quad h_{11}=\gamma_1,\quad h_{12}=\gamma_1-1,$$

$$h_{13}=\frac{1}{\gamma_1-1},\ h_{14}=1-\frac{1}{\gamma_1},\ h_{15}=\frac{\gamma_1}{\gamma_1-1}.$$

It is apparent that h_{10} to h_{15} are equivalent. In addition, $h_8 = k - h_2$, $h_6 = h_3 - k$, $h_7 = h_9 - k$. Even though h_1, h_4 and h_5 have a relation, it is not obvious that the three functions have the same solution. We conclude that by providing details on h_1, h_3, h_4, h_5, h_7, h_8 and h_{10}, we can get solutions to all the above 15 functions.

Before providing the technical details, we make a general observation. Note that besides the functions above, any function that defines some "distance" between $\Sigma_{\hat{\xi}}$ and Φ based on the eigenvalues γ_i ($i = 1, ..., k$) are scale invariant. For example, $h=\sum_{i=1}^{k}\gamma_i^2$ is a such function. From the decomposition (4.2), $\Sigma_{\hat{\xi}}=\Phi+\Sigma_\delta$, since $\Sigma_\delta\geq 0$, it is clear that all the eigenvalues are bounded by 1.0, that is, $\gamma_i \geq 1.0$, i=1, ..., k. Any optimal T should make the eigenvalues as close to 1 as possible, and the choice of optimization function is basically a choice of how to define this closeness.

Now we begin to work on the functions h_1, h_3, h_4, h_5, h_7, h_8 and h_{10} in sequence and refer to these procedures as method 1 to method 10 respectively.

Method 1: minimizing $h_1(T)=\frac{|\Sigma_{\hat{\xi}}|}{|\Phi|}$ to estimate T.

As Φ is fixed, minimizing $h_1(T)$ is to minimize $|\Sigma_{\hat{\xi}}|=\frac{|T'\Sigma T|}{|T'\Lambda|^2}$. Let $g_1(T)=\frac{|T'\Lambda|^2}{|T'\Sigma T|}$, then minimizing $h_1(T)$ is equivalent to maximizing $g_1(T)$. Let B be any arbitrary full rank p×k matrix. Making the transformation $T=\Sigma^{-\frac{1}{2}}B$, we can rewrite $g_1(T)$ as $g_1(B)=\frac{\left|B'\Sigma^{-\frac{1}{2}}\Lambda\right|^2}{|B'B|}$. By the Cauchy-Schwarz inequality $\left|B'\Sigma^{-\frac{1}{2}}\Lambda\right|^2\leq|B'B||\Lambda'\Sigma^{-1}\Lambda|$, we have $g_1(B)\leq|\Lambda'\Sigma^{-1}\Lambda|$, and equality holds when $B=\Sigma^{-\frac{1}{2}}\Lambda$. So when $T=\Sigma^{-1}\Lambda$, the function $h_1(T)$ attains its minimum, and the corresponding factor score estimator is the GLS estimator $\hat{\xi}_1=(\Lambda'\Sigma^{-1}\Lambda)^{-1}\Lambda'\Sigma^{-1}(X-\mu)$. By Lemma 2, this is equivalent to the WLS Bartlett estimator.

Method 2: maximizing $h_2(T)=\mathrm{tr}(\Sigma_{\hat{\xi}}^{-1}\Phi)$ to estimate T.

Maximizing $h_2(T)$ is to maximize $h_2(T)=\mathrm{tr}[(T'\Sigma T)^{-1}T'\Lambda\Phi\Lambda'T]$. As before, we make the transformation $T=\Sigma^{-\frac{1}{2}}B$, where B is an arbitrary full rank p×k matrix. Then $h_2(T)$ can be written as $h_2(B)=\mathrm{tr}[(B'B)^{-1}B'\Sigma^{-\frac{1}{2}}\Lambda\Phi\Lambda'\Sigma^{-\frac{1}{2}}B]$. Now let $B=\Gamma Q$, where Γ

is a matrix with orthogonal columns and Q is a nonsingular square matrix. Then we can rewrite $h_2(B)$ as $h_2(\Gamma)=tr(\Gamma'\Sigma^{-\frac{1}{2}}\Lambda\Phi\Lambda'\Sigma^{-\frac{1}{2}}\Gamma)$. Let $\Sigma^{-\frac{1}{2}}\Lambda\Phi\Lambda'\Sigma^{-\frac{1}{2}}=LDL'$ be the eigenvector-eigenvalue decomposition of $\Sigma^{-\frac{1}{2}}\Lambda\Phi\Lambda'\Sigma^{-\frac{1}{2}}$ with eigenvectors $L=(l_1,\cdots,l_p)$ and eigenvalues $D=diag(d_1,\cdots,d_p)$, where $d_1\geq d_2\geq\cdots\geq d_p$. Consequently, we have $\max_{\Gamma'\Gamma=I_k} h_2(\Gamma)=\sum_{i=1}^{k}d_i$, and the maximization is attained by choosing Γ as $\Gamma=(l_1,\cdots,l_k)$. This means that Γ satisfies

$$\Sigma^{-\frac{1}{2}}\Lambda\Phi\Lambda'\Sigma^{-\frac{1}{2}}\Gamma=\Gamma D_k, \tag{4.6}$$

where $D_k=diag(d_1,\cdots,d_k)$. From (4.6) we have $\Sigma^{-1}\Lambda=\Sigma^{-\frac{1}{2}}\Gamma D_k(\Phi\Lambda'\Sigma^{-\frac{1}{2}}\Gamma)^{-1}$. So by choosing $Q=D_k(\Phi\Lambda'\Sigma^{-\frac{1}{2}}\Gamma)^{-1}$, we get the estimator $T=\Sigma^{-1}\Lambda$. The corresponding factor score estimator is $\hat{\xi}_2=(\Lambda'\Sigma^{-1}\Lambda)^{-1}\Lambda'\Sigma^{-1}(X-\mu)$. This is identical to $\hat{\xi}_1$, the GLS estimator, and, by Lemma 2, to the WLS estimator.
Method 3: minimizing $h_3(T)=tr(\Phi^{-1}\Sigma_{\hat{\xi}})$ to estimate T.

We can write $h_3(T)$ as $h_3(T)=tr[(T'\Lambda\Phi\Lambda'T)^{-1}T'\Sigma T]$. As before, let $T=\Sigma^{-\frac{1}{2}}B$. Then $h_3(T)$ can be written as $h_3(B)=tr[(B'\Sigma^{-\frac{1}{2}}\Lambda\Phi\Lambda'\Sigma^{-\frac{1}{2}}B)^{-1}(B'B)]$. Furthermore, let $B=\Gamma Q$, where Γ is a matrix with orthogonal columns and Q is a nonsingular square matrix. It follows that $h_3(T)$ becomes $h_3(\Gamma)=tr(\Gamma'\Sigma^{-\frac{1}{2}}\Lambda\Phi\Lambda'\Sigma^{-\frac{1}{2}}\Gamma)^{-1}$. By comparing $h_3(\Gamma)$ with $h_2(\Gamma)$, and using the same strategy as there, we obtain $T=\Sigma^{-1}\Lambda$ minimizing $h_3(T)$. Thus the factor score estimator is $\hat{\xi}_3=(\Lambda'\Sigma^{-1}\Lambda)^{-1}\Lambda'\Sigma^{-1}(X-\mu)$, the GLS estimator. By Lemma 2, it is also the Bartlett estimator.
Method 4: minimizing $h_4(T)=\dfrac{|\Sigma_\delta|}{|\Sigma_{\hat{\xi}}|}$ to estimate T.

Since $h_4(T)=\dfrac{|T'\Psi T|}{|T'\Sigma T|}$, making the familiar transformation $T=\Sigma^{-\frac{1}{2}}B$, $h_4(T)$ can be written as $h_4(B)=\dfrac{|B'\Sigma^{-\frac{1}{2}}\Psi\Sigma^{-\frac{1}{2}}B|}{|B'B|}$. As in method 3, we let $B=\Gamma Q$, where Γ is a matrix with orthogonal columns and Q is a nonsingular square matrix. Then we can rewrite $h_4(B)$ as $h_4(\Gamma)=|\Gamma'\Sigma^{-\frac{1}{2}}\Psi\Sigma^{-\frac{1}{2}}\Gamma|$. Denoting the eigenvalue decomposition of $\Sigma^{-\frac{1}{2}}\Psi\Sigma^{-\frac{1}{2}}$ as

$$\Sigma^{-\frac{1}{2}}\Psi\Sigma^{-\frac{1}{2}}=LDL', \tag{4.7}$$

where $L=(l_1,\cdots,l_p)$, $D=\text{diag}(d_1,\cdots,d_p)$ with $d_1\le d_2\le\cdots\le d_p$, we have $\min_{\Gamma'\Gamma=I_k} h_4(\Gamma)=\prod_{i=1}^{k} d_i$. The minimization is attained by choosing Γ as $\Gamma=(l_1,\cdots,l_k)$. So Γ satisfies

$$\Sigma^{-\frac{1}{2}}\Psi\Sigma^{-\frac{1}{2}}\Gamma=\Gamma D_k, \tag{4.8}$$

where $D_k=\text{diag}(d_1,\cdots,d_k)$. From (4.8) we have

$$\Sigma^{-1}\Psi\Sigma^{-\frac{1}{2}}\Gamma=\Sigma^{-\frac{1}{2}}\Gamma D_k. \tag{4.9}$$

From (1.2), assuming a full rank matrix Ψ, we have

$$\Sigma^{-1}=\Psi^{-1}-\Psi^{-1}\Lambda(\Lambda'\Psi^{-1}\Lambda+\Phi^{-1})^{-1}\Lambda'\Psi^{-1}. \tag{4.10}$$

Putting (4.10) into (4.9), we get

$$[I-\Psi^{-1}\Lambda(\Lambda'\Psi^{-1}\Lambda+\Phi^{-1})^{-1}\Lambda']\Sigma^{-\frac{1}{2}}\Gamma=\Sigma^{-\frac{1}{2}}\Gamma D_k.$$

This can be rewritten as

$$\Psi^{-1}\Lambda(\Lambda'\Psi^{-1}\Lambda+\Phi^{-1})^{-1}\Lambda'\Sigma^{-\frac{1}{2}}\Gamma=\Sigma^{-\frac{1}{2}}\Gamma(I-D_k),$$

from which we have

$$\Sigma^{-\frac{1}{2}}\Gamma(I-D_k)(\Lambda'\Sigma^{-\frac{1}{2}}\Gamma)^{-1}(\Lambda'\Psi^{-1}\Lambda+\Phi^{-1})=\Psi^{-1}\Lambda.$$

It follows that if we take $Q=(I-D_k)(\Lambda'\Sigma^{-\frac{1}{2}}\Gamma)^{-1}(\Lambda'\Psi^{-1}\Lambda+\Phi^{-1})$, we can choose $T=\Psi^{-1}\Lambda$, and the corresponding optimal factor score estimator is $\hat{\xi}_4=(\Lambda'\Psi^{-1}\Lambda)^{-1}\Lambda'\Psi^{-1}(X-\mu)$. This is Bartlett's WLS factor score estimator, but by Lemma 2 it is equal to the GLS estimator.

Method 5: minimizing $h_5(T)=\dfrac{|\Sigma_\delta|}{|\Phi|}$ to estimate T.

Since Φ is a fixed matrix, minimizing $h_5(T)$ is the same as minimizing $|\Sigma_\delta|=\dfrac{|T'\Psi T|}{|T'\Lambda|^2}$.

Let $g_5(T)=\dfrac{|T'\Lambda|^2}{|T'\Psi T|}$, then minimizing $h_5(T)$ is equivalent to maximizing $g_5(T)$. Using the same steps as in method 1, we can get $\hat{\xi}_5=(\Lambda'\Psi^{-1}\Lambda)^{-1}\Lambda'\Psi^{-1}(X-\mu)$, which is the same estimator as was obtained in method 4.

Method 7: maximizing $h_7(T)=\text{tr}(\Sigma_\delta^{-1}\Phi)$ to estimate T.

Maximizing $h_7(T)$ is to maximize $h_7(T)=\text{tr}[(T'\Psi T)^{-1}T'\Lambda\Phi\Lambda'T]$. By comparing $h_7(T)$ to $h_3(T)$, and following similar steps as above, we can get $\hat{\xi}_7=(\Lambda'\Psi^{-1}\Lambda)^{-1}\Lambda'\Psi^{-1}(X-\mu)$. This is again Bartlett's estimator, and hence the GLS estimator.

Method 10: maximizing $h_{10}(T) = \lambda_{min}(\Sigma_{\xi}^{-1}\Phi)$ to estimate T.

As $\lambda_{min}(\Sigma_{\xi}^{-1}\Phi) = \lambda_{min}[(T'\Sigma T)^{-1}T'\Lambda\Phi\Lambda'T]$, using the same procedure as in method 3, we have $\lambda_{min}(\Sigma_{\xi}^{-1}\Phi) = \lambda_{min}[\Gamma'\Sigma^{-\frac{1}{2}}\Lambda\Phi\Lambda'\Sigma^{-\frac{1}{2}}\Gamma]$, where Γ is a p×k matrix such that $\Gamma'\Gamma=I_k$. Let $\Sigma^{-\frac{1}{2}}\Lambda\Phi\Lambda'\Sigma^{-\frac{1}{2}} = LDL'$ be the eigenvalue decomposition of $\Sigma^{-\frac{1}{2}}\Lambda\Phi\Lambda'\Sigma^{-\frac{1}{2}}$ with $L = (l_1,\cdots,l_p)$, $D = \text{diag}(d_1,\cdots,d_k,0,\cdots,0)$ and $d_1 \geq d_2 \geq\cdots\geq d_k$. It follows that $\lambda_{min}[\Gamma'\Sigma^{-\frac{1}{2}}\Lambda\Phi\Lambda'\Sigma^{-\frac{1}{2}}\Gamma] \leq d_k$, and the maximization is attained by choosing Γ as $\Gamma = (l_1,\cdots,l_k)$. So Γ satisfies

$$\Sigma^{-\frac{1}{2}}\Lambda\Phi\Lambda'\Sigma^{-\frac{1}{2}}\Gamma = \Gamma D_k, \tag{4.11}$$

where $D_k = \text{diag}(d_1,\cdots,d_k)$. This has the same structure as in method 3, so we have $T=\Sigma^{-1}\Lambda$, and the corresponding factor score estimator is $\hat{\xi}_{10} = (\Lambda'\Sigma^{-1}\Lambda)^{-1}\Lambda'\Sigma^{-1}(X-\mu)$. This is again the GLS estimator, but by Lemma 2 it is also the Bartlett estimator.

4.4 Conclusions

To our knowledge it has not been previously shown that the GLS and Bartlett WLS estimators, $\tilde{\xi} = (\Lambda'\Sigma^{-1}\Lambda)^{-1}\Lambda'\Sigma^{-1}(X-\mu)$ and $\hat{\xi} = (\Lambda'\Psi^{-1}\Lambda)^{-1}\Lambda'\Psi^{-1}(X-\mu)$, optimize the 15 functions defined above. They are the only optimal unbiased equivariant estimators that we have been able to find, and, by Lemma 2, they are equal.

An important practical feature of the Bartlett estimator (1.7) is that only a small matrix and a diagonal matrix needs to be inverted. A critical disadvantage is that when Ψ is singular, Bartlett's estimator cannot be used. In fact, Lemma 2 then will not hold, and the two estimators will not necessarily be equivalent. However, as long as Σ is positive definite, our GLS estimator $\tilde{\xi}$ will be defined. Thus it should be used in practical computations when Ψ is singular, though it could also be used more generally at the expense of some additional computations.

5 Simple Factor Score Estimators

In real applications, the technically justified factor score estimation formulas are often not used and simpler estimators are substituted instead (see, e.g., Comrey & Lee, 1992, sect. 10.3; Gorsuch, 1983, ch. 12). In this section we study some special cases of the optimal estimators that may provide a justification for not using the general formulae. In particular, we concentrate on models in which the factor loading matrix has no variables with complexity greater than one. Many applications of confirmatory factor models use such simple cluster structures, in which each measured variable is influenced by a single latent common factor. We additionally investigate the form that our optimal estimators take when the nonzero elements of $\Lambda = (\lambda_{ij})$ are of similar magnitude, and the elements in

$\Psi = \text{diag}(\psi_1,\cdots,\psi_p)$ are approximately equal. We shall see that simpler estimators also can be optimal. For simplicity, we use the Bartlett version of our optimal formulae.

When the factor loading matrix has a simple cluster structure, each variable is influenced by only one factor and an error term. Each row of $\Lambda = (\lambda_{ij})$ has only one nonzero element. More precisely, denote K_j, j=1, ..., k as the subsets of {1, 2, ..., p} such that $\lambda_{ij} \neq 0$ for $i \in K_j$. As in each row of Λ, there is only one element not equal zero, so $K_l \cap K_m = \phi$, the empty set, and $\bigcup_{j=1}^{k} K_j = \{1, 2, ..., p\}$. If Ψ is diagonal, then the Bartlett estimator becomes

$$\hat{\xi} = (\Lambda'\Psi^{-1}\Lambda)^{-1}\Lambda'\Psi^{-1}(X-\mu) = (\hat{\xi}_1^{(B)},\cdots,\hat{\xi}_k^{(B)})'$$

with (5.1a)

$$\hat{\xi}_j^{(B)} = (\sum_{i\in K_j} \lambda_{ij}^2 \psi_i^{-1})^{-1} \sum_{i\in K_j} \lambda_{ij}\psi_i^{-1}(X_i-\mu_i), j=1, ..., k.$$

Because K_j, j=1, ..., k are disjoint, $\hat{\xi}_j^{(B)}$ is only a linear combination of X_i, $i \in K_j$, while different $\hat{\xi}_j^{(B)}$ will depend on different X_i. This means that if X_i, $i \in K_j$ are indicators of a common factor ξ_j, the estimate $\hat{\xi}_j^{(B)}$ only depends on those X_i that have ξ_j as the common factor. As a result, in this situation the common practice of creating a factor score estimator by considering for a given factor only those variables that load on the factor, is completely justified. However, the optimal equivariant unbiased estimator will use weights computed by (5.1a).

Additional insight into (5.1a) can be obtained by assuming, for simplicity, that all of the variables that load on a given factor are in sequence in the vector X_j. Then Λ will contain blocks Λ_j that are nonzero (j=1, ..., k), and the remaining entries will be zero. If Ψ is diagonal, it can be similarly partitioned into diagonal blocks Ψ_j. Thus, the general factor model (1.1) can be specialized for each set of variables as $X_j = \mu_j + \Lambda_j\xi_j + \varepsilon_j$. It follows that, in matrix terms, (5.1a) can be written as

$$\hat{\xi}_j^{(B)} = (\Lambda_j'\Psi_j^{-1}\Lambda_j)^{-1}\Lambda_j'\Psi_j^{-1}(X_j-\mu_j). \tag{5.1b}$$

Now suppose further that the unique variances of those variables defining a given factor are all equal, that is, $\Psi_j = c_j I$ for some constant c_j. Then (5.1b) specializes still further to

$$\hat{\xi}_j^{(B)} = (\Lambda_j'\Lambda_j)^{-1}\Lambda_j'(X_j-\mu_j). \tag{5.2}$$

The part in the left parentheses is just a scaling value which may not matter in some applications. Then the optimal factor score estimator is just a weighted sum of the measured variables in a cluster, the weight for each variable being its factor loading. Although this type of scoring is often dismissed by experts (e.g., Fava & Velicer, 1992), it has a long history; it is, for example, method 4 in Harris (1967), and method 3 in Comrey and Lee (1992). Note further in (5.2) that if the nonzero loadings are equal in absolute value, then $\Lambda_j = b_j e_j$, where e_j is a vector with elements ±1, and the optimal factor score estimator is, except for a constant, given by the sign-weighted composite of variables defining the factor. A variant of such a weighting, with small loadings replaced by zeros, is often used in practice (e.g., Comrey & Lee, 1992; Gorsuch, 1983). It seems plausible that if the assumptions associated with (5.1)-(5.2) are approximately rather than exactly satisfied, these simple estimators may still behave quite well. Let us examine this claim in some detail.

We shall examine the sign-weighted estimator more precisely. In particular, let $\hat{\xi}_s = (\hat{\xi}_1^{(s)},\cdots,\hat{\xi}_k^{(s)})'$ with $\hat{\xi}_j^{(s)} = a_j \sum_{i \in K_j} \text{sign}(\lambda_{ij})(X_i - \mu_i)$ for some a_j, j=1, ..., k. Each factor score estimator $\hat{\xi}_j^{(s)}$ uses a simple summation of only those X_i that are indicators of ξ_j as a common factor, to estimate ξ_j. Under the unbiasedness constraint $E(\hat{\xi}_s|\xi) = \xi$, we can instantly get a_j. In matrix form we can express $\hat{\xi}_s$ as

$$\hat{\xi}_s = A\Pi'(X-\mu), \tag{5.3}$$

where $\Pi = (\pi_{ij})$ is a p×k matrix such that $\pi_{ij} = \text{sign}(\lambda_{ij})$, and $A = \text{diag}(a_1,\cdots,a_k)$ is a matrix to be decided. Since $E(\hat{\xi}_s|\xi) = \xi$ we have

$$A = (\Pi'\Lambda)^{-1} = \text{diag}[(\sum_{i \in K_1}|\lambda_{i1}|)^{-1},\cdots,(\sum_{i \in K_k}|\lambda_{ik}|)^{-1}].$$

So we have $\hat{\xi}_s = (\Pi'\Lambda)^{-1}\Pi'(X-\mu)$ with

$$\hat{\xi}_j^s = \frac{1}{\sum_{i \in K_j} |\lambda_{ij}|} \sum_{i \in K_j} \text{sign}(\lambda_{ij})(X_i - \mu_i), \text{ j=1, ..., k.}$$

As the Bartlett score estimate is the best estimate under $E(\hat{\xi}_s|\xi) = \xi$ and $|\Psi| \neq 0$, we know that var($\hat{\xi}_s$)≥var($\hat{\xi}$). We now show that when $\frac{|\lambda_{ij}|}{\psi_i}$, $i \in K_j$ are approximately equal, the simple estimator $\hat{\xi}_s$ will perform reasonably well.

From (5.3) we have

$$\text{var}(\hat{\xi}_s) = \Phi + (\Pi'\Lambda)^{-1}\Pi'\Psi\Pi(\Lambda'\Pi)^{-1}$$

$$= \Phi + \text{diag}(\frac{\sum_{i \in K_1}\psi_i}{(\sum_{i \in K_1}|\lambda_{i1}|)^2},\cdots,\frac{\sum_{i \in K_k}\psi_i}{(\sum_{i \in K_k}|\lambda_{ik}|)^2}). \tag{5.4}$$

It is known that

$$\text{var}(\hat{\xi}) = \Phi + (\Lambda'\Psi^{-1}\Lambda)^{-1}$$

$$= \Phi + \text{diag}[(\sum_{i \in K_1}\lambda_{i1}^2\psi_i^{-1})^{-1},\cdots,(\sum_{i \in K_k}\lambda_{ik}^2\psi_i^{-1})^{-1}]. \tag{5.5}$$

In (5.4) and (5.5) only $|\lambda_{ij}|$ and λ_{ij}^2 are involved, so we can take λ_{ij} as positive. By the Cauchy-Schwarz inequality, we have

$$(\sum \lambda_{ij})^2 = (\sum \lambda_{ij}\psi_i^{-\frac{1}{2}}\psi_i^{\frac{1}{2}})^2 \leq (\sum \lambda_{ij}^2\psi_i^{-1})(\sum \psi_i). \tag{5.6}$$

From (5.6) we have

$$\frac{\sum_{i \in K_j} \psi_i}{(\sum_{i \in K_j} |\lambda_{ij}|)^2} \geq (\sum_{i \in K_j} \lambda_{ij}^2 \psi_i^{-1})^{-1}, \tag{5.7}$$

and the equality holds when $\lambda_{ij}\psi_i^{-\frac{1}{2}} = c_j\psi_i^{\frac{1}{2}}$, $i \in K_j$ for some c_j. When $\frac{|\lambda_{ij}|}{\psi_i}$, $i \in K_j$ are approximately constant, (5.7) should be approximately an equality. In this case, (5.4) and (5.5) are almost equal, and the simple score estimator $\hat{\xi}_s$ is also approximately optimal.

6 Estimation with Singular Ψ

In this section we will consider the problem of how to estimate ξ when Ψ is singular but Σ is nonsingular. When Ψ is singular, the formulas in (2.3) to (2.6) are no longer valid, and Lemma 2 does not apply. We shall study what happens to $\tilde{\xi}$ and $\hat{\xi}$ if we use the Moore-Penrose inverses Ψ^+ of Ψ and $(\Lambda'\Psi^+\Lambda)^+$ of $\Lambda'\Psi^+\Lambda$ instead of their regular inverses. Additionally, it is important to evaluate whether the estimator of the data variate values under the model, i.e., the predicted values of X obtained from the common factors, say $\tilde{X} = \mu + \Lambda\tilde{\xi}$, remains identical under such a choice.

First we consider the case of Ψ singular, with $\Lambda'\Psi^+\Lambda$ full rank. Lemma 2 on the equivalence of WLS and GLS estimators no longer holds. We look at the least squares functions, minimum variance, and eigenvalue functions in turn. Bartlett's WLS function (1.6) is no longer defined, An obvious modification to (1.6) is to consider minimizing

$$f_4(A) = \varepsilon'\Psi^+\varepsilon, \tag{6.1}$$

yielding what we may call the modified WLS estimator

$$\hat{\xi}^+ = (\Lambda'\Psi^+\Lambda)^{-1}\Lambda'\Psi^+(X-\mu). \tag{6.2}$$

Since $A' = (\Lambda'\Psi^+\Lambda)^{-1}\Lambda'\Psi^+$, the unbiasedness condition is still met, and computations can verify equivariance. However, note that when Ψ is diagonal, Ψ^+ will contain one or more diagonal zeros and the corresponding observed variables in X will receive zero weight in estimating the factor scores. It does not seem to be a desirable feature to give zero weight to a variable that is completely in the common factor space.

Since Σ will be positive definite even if Ψ is singular, we can also consider minimizing the GLS function $f_3(A) = \varepsilon'\Sigma^{-1}\varepsilon$ as before. The resulting estimator is just (1.7) as before, namely, $\tilde{\xi} = (\Lambda'\Sigma^{-1}\Lambda)^{-1}\Lambda'\Sigma^{-1}(X-\mu)$. Thus this estimator remains the same under singularity of Ψ. But it is not generally true that $\tilde{\xi} = \hat{\xi}^+$.

Turning next to the estimators obtained by choices of V in (3.1)-(3.2), clearly V can be singular in the weight matrix $A' = (\Lambda' V \Lambda)^{-1}\Lambda' V$, permitting unbiased estimators to be defined. Lemma 3 is maintained, so that the GLS estimator yields the minimum covariance matrix of the factor score estimator $\Sigma_{\hat{\xi}_{opt}} = (\Lambda' \Sigma^{-1}\Lambda)^{-1}$. However, the modified WLS estimator requires a change to Lemma 4. Denoting R(A) as the column space of A, then we have the following lemma.

Lemma 4'. If $V = \Psi^+$ in (3.1)-(3.4) and $R(\Lambda) \subset R(\Psi)$, the modified WLS score estimator (6.2) is optimal in the sense that it yields the minimum Σ_δ, $\Sigma_{\delta_{opt}} = (\Lambda' \Psi^+ \Lambda)^{-1}$.

When the column space of Λ does not belong to $R(\Psi)$, $V = \Psi^+$ is usually not the best choice for V. It would seem natural to use Lemmas 3 and 4' to yield a modified version of Theorem 1, but it does not seem to be true that $\Sigma_{\hat{\xi}_{opt}} = \Phi + \Sigma_{\delta_{opt}}$ under this new definition.

With regard to the 15 optimization functions defined on the decomposition $\hat{\xi} = (T'\Lambda)^{-1}T'(X-\mu) = \xi + \delta$ given by (4.1), we find that the estimators remain valid when Ψ is singular. In this decomposition, we need only assume that $T'\Lambda$ and $T'\Psi T$ are nonsingular, even though Ψ may be singular. This is appropriate if Ψ is only minimally rank deficient [rank(Ψ)≥k], which is surely the most likely case since there are typically only a few improper or nonexistent unique variables. Also, both $\Sigma_{\hat{\xi}} = (T'\Lambda)^{-1}T'\Sigma T(\Lambda' T)^{-1}$ and $\Sigma_\delta = (T'\Lambda)^{-1}T'\Psi T(\Lambda' T)^{-1}$ continue to be positive definite, and all 15 optimization functions discussed in section 4 continue to be defined.

Since in method 1, method 2, method 3 and method 10 we do not need to assume that Ψ is nonsingular, as long as Λ is of full column rank, we certainly can define $\hat{\xi} = (T'\Lambda)^{-1}T'(X-\mu)$ for some T. So the score estimator $\tilde{\xi} = (\Lambda'\Sigma^{-1}\Lambda)^{-1}\Lambda'\Sigma^{-1}(X-\mu)$ still keeps the optimal properties of unbiasedness and equivariance. As h_1, h_2, h_3, h_{10} do not depend on whether Ψ is singular or not, they still give the GLS score estimator when optimized. Functions h_6, h_8 are equivalent to h_3 and h_2 respectively, and are also properly defined when Ψ is singular and will again give the GLS score estimator. When Ψ is singular, h_7, h_9, h_{13}, h_{15}, may not be defined properly. While h_5 and h_4 are 0 for some T, this certainly attains the minimum but may not be a good estimator. Functions h_{11}, h_{12} and h_{14} are also properly defined and are equivalent to h_{10}. In the process of minimizing h_4, we did not use the full rank property before we attain (4.8). This function is also properly defined if rank(Ψ)=p-1, and will give the GLS score estimator if we choose $T = \Sigma^{-\frac{1}{2}}\Gamma$. When the rank of Ψ is less than p-1, Γ in (4.8) is not uniquely defined, and method 4 will break down. For functions h_5, h_7, h_9, h_{13}, h_{15}, even if the rank of Ψ is p-1, one may obtain improper solutions. Let us take h_5, for example. Suppose $\Psi = \mathrm{diag}(\psi_1, \cdots, \psi_p)$, $T' = (t_{(1)}, \cdots, t_{(p)})$, $\Lambda' = (\lambda_{(1)}, \cdots, \lambda_{(p)})$, with $\psi_{i_0} = 0$, and $t_{(i)}$ and $\lambda_{(i)}$ are $k\times 1$ vectors. Then $T'\Psi T = \sum_{i\neq i_0} t_{(i)}\psi_i t_{(i)}'$ and $T'\Lambda = \sum_{i\neq i_0} t_{(i)}\lambda_{(i)}' + t_{(i_0)}\lambda_{(i_0)}'$. Take $t_{(i_0)} = c\lambda_{(i_0)}$ and denote $W = \sum_{i\neq i_0} t_{(i)}\lambda_{(i)}'$, we then have $|\Sigma_\delta| = \dfrac{|T'\Psi T|}{|W|(1 + c\lambda_{(i_0)}' W^{-1}\lambda_{(i_0)})}$. As

$|T'\Psi T|$ does not depend on c, let c approach infinity and fix other $t_{(i)}$ at any point, then the function h_5 attains its minimum at 0. Certainly, this is not a good estimate of T. So our conclusion is: we should use h_1, h_2, h_3, h_6, h_8 h_{10}, h_{11}, h_{12} or h_{14} as objective functions to estimate the factor scores when Ψ is singular, and the resulting estimator is $\tilde{\xi} = (\Lambda'\Sigma^{-1}\Lambda)^{-1}\Lambda'\Sigma^{-1}(X-\mu)$.

Next let us consider what happens when $\Lambda'\Psi^{+}\Lambda$ also is singular. First, consider the minimization problem (6.1). Setting the first partial derivatives to zero yields the equation $(\Lambda'\Psi^{+}\Lambda)\xi = \Lambda'\Psi^{+}(X-\mu)$. This cannot be solved to precisely yield something like $(\Lambda'\Psi^{+}\Lambda)^{+}\Lambda'\Psi^{+}(X-\mu)$, analogous to the modified Bartlett estimator (6.2) $\hat{\xi}^{+} = (\Lambda'\Psi^{+}\Lambda)^{-1}\Lambda'\Psi^{+}(X-\mu)$. Rather, we obtain

$$(\Lambda'\Psi^{+}\Lambda)^{+}(\Lambda'\Psi^{+}\Lambda)\hat{\xi}^{++} = (\Lambda'\Psi^{+}\Lambda)^{+}\Lambda'\Psi^{+}(X-\mu), \tag{6.3}$$

and the left-hand side is a projective transformation of the desired $\hat{\xi}^{++}$ which cannot be eliminated. Now suppose we ignore this inconvenience, and decide to define $A' = (\Lambda'\Psi^{+}\Lambda)^{+}\Lambda'\Psi^{+}$ as our desired weight matrix. For this A', the unbiasedness condition (1.5) no longer holds. Thus, although we can use this A' to compute a factor score estimator, it no longer has a key property. The weaker condition $A'\Lambda\xi = \xi$ for unbiasedness also is not met.

On the other hand, singularity of $\Lambda'\Psi^{+}\Lambda$ has no impact on our GLS function and estimator (2.1)-(2.2), and $\tilde{\xi} = (\Lambda'\Sigma^{-1}\Lambda)^{-1}\Lambda'\Sigma^{-1}(X-\mu)$ is defined as before. There is no reason to suppose that $\Lambda'\Sigma^{-1}\Lambda$ is singular just because $\Lambda'\Psi^{+}\Lambda$ is singular, since we maintain the assumption that Λ has full column rank (if it does not, the basic model (1.1)-(1.2) must be reparameterized first).

To illustrate what happens when we use (6.2) and the right-hand side of (6.3) as a factor score estimators, we will evaluate the following artificial examples.

Example 1. Let p=3, k=2, $\Psi = \begin{pmatrix} 1 & 0 & 0 \\ 0 & 1 & 0 \\ 0 & 0 & 0 \end{pmatrix}$, $\Lambda = \begin{pmatrix} 1 & 0 \\ 1 & 0 \\ 0 & 1 \end{pmatrix}$, $\Phi = \begin{pmatrix} 1 & 0 \\ 0 & 1 \end{pmatrix}$, and μ be null.

From (1.2) we have $\Sigma = \begin{pmatrix} 2 & 1 & 0 \\ 1 & 2 & 0 \\ 0 & 0 & 1 \end{pmatrix}$ and is nonsingular. As $\Psi^{+} = \Psi$ we have the following estimators

$$\tilde{\xi} = (\Lambda'\Sigma^{-1}\Lambda)^{-1}\Lambda'\Sigma^{-1}X = \begin{pmatrix} \frac{1}{2}(x_1 + x_2) \\ x_3 \end{pmatrix}, \quad (\Lambda'\Psi^{+}\Lambda)^{+}\Lambda'\Psi^{+}X = \begin{pmatrix} \frac{1}{2}(x_1 + x_2) \\ 0 \end{pmatrix}.$$

It is apparent that the factor score estimates are not identical. Similarly, the respective predicted values of the observed variables are

$$\tilde{X} = \Lambda\tilde{\xi} = \begin{pmatrix} \frac{1}{2}(x_1 + x_2) \\ \frac{1}{2}(x_1 + x_2) \\ x_3 \end{pmatrix} \text{ and } \hat{X} = \Lambda(\Lambda'\Psi^+\Lambda)^+\Lambda'\Psi^+X = \begin{pmatrix} \frac{1}{2}(x_1 + x_2) \\ \frac{1}{2}(x_1 + x_2) \\ 0 \end{pmatrix},$$

which also are not equal. In both factor scores and the predicted data, there is a degenerate variate when using the right-hand-side of (6.3); the predicted variate for x_3 is 0, which is definitely not a good score estimator. On the other hand $\tilde{\xi}$ uses x_3 to predict x_3, which of course is a good predicted observed variable value.

Example 2. Let p=5, k=2, $\Psi = \text{diag}(0,1,1,1,1)$, $\Lambda' = \begin{pmatrix} 1 & 1 & 1 & 0 & 0 \\ 0 & 0 & 0 & 1 & 1 \end{pmatrix}$, $\Phi = \begin{pmatrix} 1 & 0 \\ 0 & 1 \end{pmatrix}$, and μ be null. From $\Lambda'\Psi^+\Lambda = \begin{pmatrix} 2 & 0 \\ 0 & 2 \end{pmatrix}$ and is nonsingular, we have

$$\hat{\xi}^+ = (\Lambda'\Psi^+\Lambda)^+\Lambda'\Psi^+X = \begin{pmatrix} \frac{1}{2}(x_2 + x_3) \\ \frac{1}{2}(x_4 + x_5) \end{pmatrix}.$$

As the first component of ξ equals x_1 almost surely, a good score estimator should have x_1 as the first component. The GLS estimator $\tilde{\xi}$ has this property with $\tilde{\xi} = \begin{pmatrix} x_1 \\ \frac{1}{2}(x_4 + x_5) \end{pmatrix}$.

7 Estimation with Singular Σ

In the previous section we discussed score estimation when Ψ is singular, and we recommended using the GLS estimator $\tilde{\xi} = (\Lambda'\Sigma^{-1}\Lambda)^{-1}\Lambda'\Sigma^{-1}(X - \mu)$. It is clear that this expression requires the inverse of Σ to exist and we may ask what to do when Σ is singular. As stated above, we could use the inverse of S, providing the data variables are not linearly dependent. However, this practical suggestion is an incomplete evaluation of the problem.

Obviously, the GLS function (2.1) is not defined, and we may consider replacing it by its obvious modified GLS analogue

$$f_5(A) = \varepsilon'\Sigma^+\varepsilon, \tag{7.1}$$

where Σ^+ is the Moore-Penrose inverse of Σ. This function has a well-defined minimum and the optimal solution

$$\tilde{\xi}^+ = (\Lambda'\Sigma^+\Lambda)^{-1}\Lambda'\Sigma^+(X - \mu), \tag{7.2}$$

providing that $\Lambda'\Sigma^+\Lambda$ is invertible. This should not be a problem just because Ψ is singular. With regards to the minimum variance properties of the estimator (7.2), a modified form of Lemma 3 still holds.

Lemma 3'. If $V = \Sigma^+$ and $R(\Lambda) \subset R(\Sigma)$ in (3.1)-(3.4), the modified GLS score estimator (7.2) is optimal in the sense that it yields the minimum $\Sigma_{\hat{\xi}}$, $\Sigma_{\hat{\xi}\,opt} = (\Lambda'\Sigma^+\Lambda)^{-1}$.

Now we turn to the most serious case, when either F or $\Lambda'\Lambda$, or both, may be singular, in addition to Ψ being singular. When Λ is not full rank, the optimizer (7.2) is not defined, and the weight matrix $A' = (\Lambda'\Sigma^+\Lambda)^+\Lambda'\Sigma^+$ that one might consider using no longer meets the unbiasedness condition (1.5). Lemma 3' cannot be modified to apply. Similarly, the 15 optimization functions in section 4 break down, since the basic estimator $A' = (T'\Lambda)^{-1}T'$, when modified to the form $A' = (T'\Lambda)^+T'$, no longer meets the unbiasedness condition (1.5). So all the functions in section 4 are not defined properly. Nevertheless, we still may be able to get a good estimator under some conditions. For this, we need to borrow a theorem from the theory of linear models for estimating fixed effects.

Theorem 2. If $X \in R(\Lambda:\Psi)$ and $R(\Lambda) \subset R(\Sigma)$, then given ξ, the best linear estimator of $\Lambda\xi$ is $\Lambda\tilde{\xi}_g$, where $\tilde{\xi}_g = (\Lambda'\Sigma^+\Lambda)^+\Lambda'\Sigma^+(X-\mu)$.

Proof. The proof is the same as that of Theorem 15 of Magnus and Neudecker (1988, p. 282), and when $|\Phi| \neq 0$, the condition $R(\Lambda) \subset R(\Sigma)$ in the theorem is automatically satisfied.

As $E(\Lambda\tilde{\xi}_g|\xi) = \Lambda\xi$ and $Var(\Lambda\tilde{\xi}_g|\xi)$ does not depend on ξ, from the formula $var(Z) = E\,var(Z|Y) + var[E(Z|Y)]$, the estimator $\Lambda\tilde{\xi}_g$ also gives the best unbiased estimate of $\Lambda\xi$ unconditionally. Because $(DAD)^+ \neq D^+A^+D^+$ in general even for a diagonal matrix D (Albert, 1972, Theorem 4.11), the estimator $\Lambda\tilde{\xi}_g$ does not possess the equivariance property as in the nonsingular case. However, if $\Lambda'\Sigma^+\Lambda$ is diagonal, it is still equivariant as in the following example.

Example 3. Let p=3, k=2, $\Psi = \begin{pmatrix} 1 & 0 & 0 \\ 0 & 1 & 0 \\ 0 & 0 & 0 \end{pmatrix}$, $\Lambda = \begin{pmatrix} 1 & 0 \\ 0 & 1 \\ 0 & 0 \end{pmatrix}$, $\Phi = \begin{pmatrix} 1 & 0 \\ 0 & 1 \end{pmatrix}$, and μ be null.

From (1.2) we have $\Sigma = \begin{pmatrix} 2 & 0 & 0 \\ 0 & 2 & 0 \\ 0 & 0 & 0 \end{pmatrix}$ and is singular. So $\tilde{\xi}$ and $\hat{\xi}$ are not properly defined. But we still can compute $\tilde{\xi}_g$. As $\Psi^+ = \Psi$ and $\Sigma^+ = diag(\frac{1}{2}, \frac{1}{2}, 0)$, we have

$\tilde{\xi}_g = (\Lambda'\Sigma^+\Lambda)^+\Lambda'\Sigma^+X = \begin{pmatrix} x_1 \\ x_2 \end{pmatrix}$ and $\Lambda\tilde{\xi}_g = \begin{pmatrix} x_1 \\ x_2 \\ 0 \end{pmatrix}$. In this example, the estimated values predict the true X exactly. In the original model we must have $x_3 = 0$, almost surely.

8 Discussion

In this paper we developed several different approaches towards defining unbiased equivariant factor score estimators: optimizing a GLS function, solving a minimum variance problem, and optimizing a variety of eigenvalue problems. Rather remarkably, in the regular situation where all parameter matrices of the factor model are full rank, all solutions point to Bartlett's (1937) WLS estimator as being optimal. We did not find a new optimal unbiased estimator. We did find, however, that an alternative computational form could be developed for Bartlett's estimator. This is our GLS estimator, which was previously mentioned by Schönemann and Steiger (1979) but not proposed by them as a factor score estimator.

The unbiasedness condition $A'\Lambda=I$ would have permitted other factor score estimators to appear as the solution to our optimization problems. For example, the estimator $\hat{\xi}=(\Lambda'\Lambda)^{-1}\Lambda'(X-\mu)$ with $A'=(\Lambda'\Lambda)^{-1}\Lambda'$ is unbiased. This is method 3 in Harris (1967), the so-called ideal variable method of Holzinger and Harman (1941) or the least squares method of Horst (1965). Although this estimator solves an optimization problem (see method 1 in McDonald & Burr, 1967), it does not optimize any of the functions considered in this paper. This is not to say that additional optimization functions might not also be informative, but this topic is beyond the scope of the current paper.

In spite of the optimality of the WLS estimator, it seems to be rarely used in practice. One important reason must be that, in typical applications of confirmatory factor analysis, the unique variance matrix is often found to be estimated as improper with negative variance estimates or zero variance estimates. We did not study the effect of negative variance estimates, since they clearly do not conform to the basic factor analytic assumptions. Zero variance estimates, however, make the WLS estimator impossible to apply, and the equivalence of the WLS and GLS estimators breaks down. While we were able to define a modified Bartlett estimator, it seems to have the peculiar feature of giving no weight to a variable that might be heavily influenced by a particular factor. On the other hand, the GLS estimator remains well defined with zero unique variances, and should be used in practice.

We studied some consequences that accrue when the factor loading matrix is not full rank. In practice this condition is easily avoided, for example by reparameterization, so that the associated results are probably not important for practice and need not be further discussed.

The GLS estimator breaks down when the model covariance matrix is singular. This can happen in pathological applications, but it is generally precluded by standard methods such as maximum likelihood estimation. While use of a Moore-Penrose inverse based GLS function does not lead to an optimal factor score estimator, we were able to show that the model-reproduced measured variables remain well defined with certain optimal properties.

We also investigated some conditions under which the Bartlett estimator would simplify to yield the type of practical estimators sometimes recommended in practice. These have to do with simple cluster structures, equality of unique variances, and equality of factor loadings. It seems that, when variables have a simple cluster structure, and the indicators of a factor have approximately equal unique variances, the common practice of using factor loadings as weights to generate factor scores has some justification. If the factor loadings are also approximately equal, a sign-weighted composite also may hardly lose in precision of estimation.

9 References

Albert, A. (1972). *Regression and the Moore-Penrose Pseudoinverse*. Academic Press, New York.

Anderson, T. W. (1984). *An Introduction to Multivariate Statistical Analysis*. Wiley, New York.

Anderson, T. W., & Rubin, H. (1956). Statistical inference in factor analysis. In J. Neyman (Ed.), *Proceedings of the Third Berkeley Symposium on Mathematical Statistics and Probability* (pp. 111-150). University of California Press, Berkeley.

Bartholomew, D. J. (1987). *Latent Variable Models and Factor Analysis*. Oxford, New York.

Bartlett, M. S. (1937). The statistical conception of mental factors. *British Journal of Psychology* **28** 97-104.

Comrey, A. L., & Lee, H. B. (1992). *A First Course in Factor Analysis*. Erlbaum, Hillsdale, NJ.

Fava, J. L., & Velicer, W. F. (1992). An empirical comparison of factor, image, component, and scale scores. *Multivariate Behavioral Research* **27** 301-322.

Gorsuch, R. L. (1983). *Factor Analysis*. Erlbaum, Hillsdale, NJ.

Guttman, L. (1955). The determinacy of factor score matrices with implications for five other basic problems of common-factor theory. *British Journal of Statistical Psychology* **8** 65-81.

Hakstian, A. R., Zidek, J. V., & McDonald, R. P. (1977). Best univocal estimates of orthogonal common factors. *Psychometrika* **42** 627-630.

Harris, C. W. (1967). On factors and factor scores. *Psychometrika* **32** 363-379.

Heermann, E. F. (1963). Univocal or orthogonal estimators of orthogonal factors. *Psychometrika* **28** 161-172.

Holzinger, K. J., & Harman, H. H. (1941). *Factor Analysis - A Synthesis of Factorial Methods*. University of Chicago Press, Chicago.

Horst, P. (1965). *Factor Analysis of Data Matrices*. Holt, Rinehart & Winston, New York.

Jöreskog, K. G. (1969). A general approach to confirmatory maximum likelihood factor analysis. *Psychometrika* **34** 183-202.

Lawley, D. N., & Maxwell, A. E. (1971). *Factor Analysis as a Statistical Method*. Butterworth, London.

Magnus, J. R., & H. Neudecker (1988). *Matrix Differential Calculus with Applications in Statistics and Econometrics*. Wiley, New York.

McDonald, R. P., & Burr, E. J. (1967). A comparison of four methods of constructing factor scores. *Psychometrika* **32** 381-401.

Robinson, G. K. (1991). That BLUP is a good thing: The estimation of random effects. *Statistical Science* **6** 15-51.

Rozeboom, W. W. (1988). Factor indeterminacy: The saga continuous. *British Journal of Mathematical and Statistical Psychology* **41** 209-226.

Saris, W. E., de Pijper, M., & Mulder, J. (1978). Optimal procedures for estimation of factor scores. *Sociological Methods & Research* **7** 85-106.

Schönemann, P. H., & Steiger, J. H. (1979). Regression component analysis. *British Journal of Mathematical and Statistical Psychology* *29* 175-189.

Thurstone, L. L. (1935). *The Vectors of Mind*. University of Chicago Press, Chicago.

Thomson, G. (1936). Some points of mathematical technique in the factorial analysis of ability. *Journal of Educational Psychology* **27** 37-54.

Vittadini, G. (1989). Indeterminacy problems in the LISREL model. *Multivariate Behavioral Research* **24** 397-414.

Williams, J. S. (1978). A definition of the common factor-analysis model and the elimination of problems of factor score indeterminacy. *Psychometrika* **43** 293-306.

Lecture Notes in Statistics

For information about Volumes 1 to 43 please contact Springer-Verlag

Vol. 44: D.L. McLeish, Christopher G. Small, The Theory and Applications of Statistical Inference Functions. vi, 124 pages, 1987.

Vol. 45: J.K. Ghosh (Editor), Statistical Information and Likelihood. 384 pages, 1988.

Vol. 46: H.-G. Müller, Nonparametric Regression Analysis of Longitudinal Data. vi, 199 pages, 1988.

Vol. 47: A.J. Getson, F.C. Hsuan, {2}-Inverses and Their Statistical Application. viii, 110 pages, 1988.

Vol. 48: G.L. Bretthorst, Bayesian Spectrum Analysis and Parameter Estimation. xii, 209 pages, 1988.

Vol. 49: S.L. Lauritzen, Extremal Families and Systems of Sufficient Statistics. xv, 268 pages, 1988.

Vol. 50: O.E. Barndorff-Nielsen, Parametric Statistical Models and Likelihood. vii, 276 pages, 1988.

Vol. 51: J. Hüsler, R.-D. Reiss (Editors). Extreme Value Theory, Proceedings, 1987. x, 279 pages, 1989.

Vol. 52: P.K. Goel, T. Ramalingam, The Matching Methodology: Some Statistical Properties. viii, 152 pages, 1989.

Vol. 53: B.C. Arnold, N. Balakrishnan, Relations, Bounds and Approximations for Order Statistics. ix, 173 pages, 1989.

Vol. 54: K.R. Shah, B.K. Sinha, Theory of Optimal Designs. viii, 171 pages, 1989.

Vol. 55: L. McDonald, B. Manly, J. Lockwood, J. Logan (Editors), Estimation and Analysis of Insect Populations. Proceedings, 1988. xiv, 492 pages, 1989.

Vol. 56: J.K. Lindsey, The Analysis of Categorical Data Using GLIM. v, 168 pages, 1989.

Vol. 57: A. Decarli, B.J. Francis, R. Gilchrist, G.U.H. Seeber (Editors), Statistical Modelling. Proceedings, 1989. ix, 343 pages, 1989.

Vol. 58: O.E. Barndorff-Nielsen, P. Bl¾sild, P.S. Eriksen, Decomposition and Invariance of Measures, and Statistical Transformation Models. v, 147 pages, 1989.

Vol. 59: S. Gupta, R. Mukerjee, A Calculus for Factorial Arrangements. vi, 126 pages, 1989.

Vol. 60: L. Gyorfi, W. Härdle, P. Sarda, Ph. Vieu, Nonparametric Curve Estimation from Time Series. viii, 153 pages, 1989.

Vol. 61: J. Breckling, The Analysis of Directional Time Series: Applications to Wind Speed and Direction. viii, 238 pages, 1989.

Vol. 62: J.C. Akkerboom, Testing Problems with Linear or Angular Inequality Constraints. xii, 291 pages, 1990.

Vol. 63: J. Pfanzagl, Estimation in Semiparametric Models: Some Recent Developments. iii, 112 pages, 1990.

Vol. 64: S. Gabler, Minimax Solutions in Sampling from Finite Populations. v, 132 pages, 1990.

Vol. 65: A. Janssen, D.M. Mason, Non-Standard Rank Tests. vi, 252 pages, 1990.

Vol 66: T. Wright, Exact Confidence Bounds when Sampling from Small Finite Universes. xvi, 431 pages, 1991.

Vol. 67: M.A. Tanner, Tools for Statistical Inference: Observed Data and Data Augmentation Methods. vi, 110 pages, 1991.

Vol. 68: M. Taniguchi, Higher Order Asymptotic Theory for Time Series Analysis. viii, 160 pages, 1991.

Vol. 69: N.J.D. Nagelkerke, Maximum Likelihood Estimation of Functional Relationships. V, 110 pages, 1992.

Vol. 70: K. Iida, Studies on the Optimal Search Plan. viii, 130 pages, 1992.

Vol. 71: E.M.R.A. Engel, A Road to Randomness in Physical Systems. ix, 155 pages, 1992.

Vol. 72: J.K. Lindsey, The Analysis of Stochastic Processes using GLIM. vi, 294 pages, 1992.

Vol. 73: B.C. Arnold, E. Castillo, J.-M. Sarabia, Conditionally Specified Distributions. xiii, 151 pages, 1992.

Vol. 74: P. Barone, A. Frigessi, M. Piccioni, Stochastic Models, Statistical Methods, and Algorithms in Image Analysis. vi, 258 pages, 1992.

Vol. 75: P.K. Goel, N.S. Iyengar (Eds.), Bayesian Analysis in Statistics and Econometrics. xi, 410 pages, 1992.

Vol. 76: L. Bondesson, Generalized Gamma Convolutions and Related Classes of Distributions and Densities. viii, 173 pages, 1992.

Vol. 77: E. Mammen, When Does Bootstrap Work? Asymptotic Results and Simulations. vi, 196 pages, 1992.

Vol. 78: L. Fahrmeir, B. Francis, R. Gilchrist, G. Tutz (Eds.), Advances in GLIM and Statistical Modelling: Proceedings of the GLIM92 Conference and the 7th International Workshop on Statistical Modelling, Munich, 13-17 July 1992. ix, 225 pages, 1992.

Vol. 79: N. Schmitz, Optimal Sequentially Planned Decision Procedures. xii, 209 pages, 1992.

Vol. 80: M. Fligner, J. Verducci (Eds.), Probability Models and Statistical Analyses for Ranking Data. xxii, 306 pages, 1992.

Vol. 81: P. Spirtes, C. Glymour, R. Scheines, Causation, Prediction, and Search. xxiii, 526 pages, 1993.

Vol. 82: A. Korostelev and A. Tsybakov, Minimax Theory of Image Reconstruction. xii, 268 pages, 1993.

Vol. 83: C. Gatsonis, J. Hodges, R. Kass, N. Singpurwalla (Editors), Case Studies in Bayesian Statistics. xii, 437 pages, 1993.

Vol. 84: S. Yamada, Pivotal Measures in Statistical Experiments and Sufficiency. vii, 129 pages, 1994.

Vol. 85: P. Doukhan, Mixing: Properties and Examples. xi, 142 pages, 1994.

Vol. 86: W. Vach, Logistic Regression with Missing Values in the Covariates. xi, 139 pages, 1994.

Vol. 87: J. Müller, Lectures on Random Voronoi Tessellations.vii, 134 pages, 1994.

Vol. 88: J. E. Kolassa, Series Approximation Methods in Statistics.viii, 150 pages, 1994.

Vol. 89: P. Cheeseman, R.W. Oldford (Editors), Selecting Models From Data: AI and Statistics IV. xii, 487 pages, 1994.

Vol. 90: A. Csenki, Dependability for Systems with a Partitioned State Space: Markov and Semi-Markov Theory and Computational Implementation. x, 241 pages, 1994.

Vol. 91: J.D. Malley, Statistical Applications of Jordan Algebras. viii, 101 pages, 1994.

Vol. 92: M. Eerola, Probabilistic Causality in Longitudinal Studies. vii, 133 pages, 1994.

Vol. 93: Bernard Van Cutsem (Editor), Classification and Dissimilarity Analysis. xiv, 238 pages, 1994.

Vol. 94: Jane F. Gentleman and G.A. Whitmore (Editors), Case Studies in Data Analysis. viii, 262 pages, 1994.

Vol. 95: Shelemyahu Zacks, Stochastic Visibility in Random Fields. x, 175 pages, 1994.
Vol. 96: Ibrahim Rahimov, Random Sums and Branching Stochastic Processes. viii, 195 pages, 1995.

Vol. 97: R. Szekli, Stochastic Ordering and Dependence in Applied Probability. viii, 194 pages, 1995.
Vol. 98: Philippe Barbe and Patrice Bertail, The Weighted Bootstrap. viii, 230 pages, 1995.

Vol. 99: C.C. Heyde (Editor), Branching Processes: Proceedings of the First World Congress. viii, 185 pages, 1995.

Vol. 100: Wlodzimierz Bryc, The Normal Distribution: Characterizations with Applications. viii, 139 pages, 1995.

Vol. 101: H.H. Andersen, M.Højbjerre, D. Sørensen, P.S.Eriksen, Linear and Graphical Models: for the Multivariate Complex Normal Distribution. x, 184 pages, 1995.

Vol. 102: A.M. Mathai, Serge B. Provost, Takesi Hayakawa, Bilinear Forms and Zonal Polynomials. x, 378 pages, 1995.

Vol. 103: Anestis Antoniadis and Georges Oppenheim (Editors), Wavelets and Statistics. vi, 411 pages, 1995.

Vol. 104: Gilg U.H. Seeber, Brian J. Francis, Reinhold Hatzinger, Gabriele Steckel-Berger (Editors), Statistical Modelling: 10th International Workshop, Innsbruck, July 10-14th 1995. x, 327 pages, 1995.

Vol. 105: Constantine Gatsonis, James S. Hodges, Robert E. Kass, Nozer D. Singpurwalla(Editors), Case Studies in Bayesian Statistics, Volume II. x, 354 pages, 1995.

Vol. 106: Harald Niederreiter, Peter Jau-Shyong Shiue (Editors), Monte Carlo and Quasi-Monte Carlo Methods in Scientific Computing. xiv, 372 pages, 1995.

Vol. 107: Masafumi Akahira, Kei Takeuchi, Non-Regular Statistical Estimation. vii, 183 pages, 1995.

Vol. 108: Wesley L. Schaible (Editor), Indirect Estimators in U.S. Federal Programs. iix, 195 pages, 1995.

Vol. 109: Helmut Rieder (Editor), Robust Statistics, Data Analysis, and Computer Intensive Methods. xiv, 427 pages, 1996.

Vol. 110: D. Bosq, Nonparametric Statistics for Stochastic Processes. xii, 169 pages, 1996.

Vol. 111: Leon Willenborg, Ton de Waal, Statistical Disclosure Control in Practice. xiv, 152 pages, 1996.

Vol. 112: Doug Fischer, Hans-J. Lenz (Editors), Learning from Data. xii, 450 pages, 1996.

Vol. 113: Rainer Schwabe, Optimum Designs for Multi-Factor Models. viii, 124 pages, 1996.

Vol. 114: C.C. Heyde, Yu. V. Prohorov, R. Pyke, and S. T. Rachev (Editors), Athens Conference on Applied Probability and Time Series Analysis Volume I: Applied Probability In Honor of J.M. Gani. viii, 424 pages, 1996.

Vol. 115: P.M. Robinson, M. Rosenblatt (Editors), Athens Conference on Applied Probability and Time Series Analysis Volume II: Time Series Analysis In Memory of E.J. Hannan. viii, 448 pages, 1996.

Vol. 116: Genshiro Kitagawa and Will Gersch, Smoothness Priors Analysis of Time Series. x, 261 pages, 1996.

Vol. 117: Paul Glasserman, Karl Sigman, David D. Yao (Editors), Stochastic Networks. xii, 298, 1996.

Vol. 118: Radford M. Neal, Bayesian Learning for Neural Networks. xv, 183, 1996.

Vol. 119: Masanao Aoki, Arthur M. Havenner, Applications of Computer Aided Time Series Modeling, ix, 329 pages, 1997.

Vol. 120: Maia Berkane, Latent Variable Modeling and Applications to Causality, vi, 288 pages, 1997.